Lecture Notes in Artificial Intelligence 6284

Edited by R. Goebel, J. Siekmann, and W. Wahlster

Subseries of Lecture Notes in Computer Science

Kumiyo Nakakoji Yohei Murakami
Elin McCready (Eds.)

New Frontiers
in Artificial Intelligence

JSAI-isAI 2009 Workshops
LENLS, JURISIN, KCSD, LLLL
Tokyo, Japan, November 19-20, 2009
Revised Selected Papers

 Springer

Series Editors

Randy Goebel, University of Alberta, Edmonton, Canada
Jörg Siekmann, University of Saarland, Saarbrücken, Germany
Wolfgang Wahlster, DFKI and University of Saarland, Saarbrücken, Germany

Volume Editors

Kumiyo Nakakoji
SRA Inc., Key Technology Laboratory
2-32-8 Minami-Ikebukuro, Toyoshima-Ku
Tokyo 171-8513, Japan
E-mail: kumiyo@sra.co.jp

Yohei Murakami
National Institute of Information and Communications Technology
3-5 Hikaridai, Seika-cho, Soraku-gun
Kyoto 619-0289, Japan
E-mail: yohei@nict.go.jp

Elin McCready
Aoyama Gakuin University, Department of English
4-4-25 Shibuya, Shibuya-ku
Tokyo 150-8366, Japan
E-mail: mccready@cl.aoyama.ac.jp

Library of Congress Control Number: 2010931529

CR Subject Classification (1998): I.2, H.3, H.4, F.1, H.2.8, I.5

LNCS Sublibrary: SL 7 – Artificial Intelligence

ISSN   0302-9743
ISBN-10   3-642-14887-5 Springer Berlin Heidelberg New York
ISBN-13   978-3-642-14887-3 Springer Berlin Heidelberg New York

springer.com

© Springer-Verlag Berlin Heidelberg 2010
Printed in Germany

Typesetting: Camera-ready by author, data conversion by Scientific Publishing Services, Chennai, India
Printed on acid-free paper   06/3180

Preface

JSAI (The Japanese Society for Artificial Intelligence) is a premier academic society that focuses on artificial intelligence in Japan and was established in 1986. JSAI publishes journals of the JSAI and bimonthly transactions, and hosts 19 special interest groups. The JSAI annual conference attracts several hundred attendees each year.

JSAI-isAI (JSAI International Symposia on Artificial Intelligence) 2009 was the First International Symposium, which hosted three co-located international workshops and one satellite workshop that had been selected by the JSAI-isAI 2009 Organizing Committee. This is in succession to the international workshops co-located with the JSAI annual conferences since 2001. JSAI-isAI 2009 was successfully held during November 19–20 in Tokyo, Japan; 158 people from 16 countries participated in JSAI-isAI 2009.

This volume of "New Frontiers in Artificial Intelligence: JSAI-isAI 2009 Workshops" is the proceedings of JSAI-isAI 2009. The organizers of the four workshops, LENLS, JURISIN, KCSD, and LLLL, hosted by JSAI-isAI 2009, selected 24 papers out of 61 presentations. This has resulted in the excellent selection of papers that are representative of some of the topics of AI research both in Japan and in other parts of the world.

LENLS (Logic and Engineering of Natural Language Semantics) is an annual international workshop on formal semantics and pragmatics. LENLS hosted by JSAI-isAI 2009 was the sixth event in the series. LENLS focuses on the formal and theoretical aspects of natural language, which demonstrates one of the strengths of Japanese AI studies. JURISIN (Juris-Informatics) was the third event, focusing on law and AI, and broader areas of legal issues from the perspective of information science. One of the interesting aspects of JURISIN is its inter-disciplinary nature, involving people from a wide variety of backgrounds, such as computer science, social science, philosophy, and law. KCSD (Knowledge Collaboration in Software Development) had already been held twice and this was the third event in the series. The workshop focuses on the collaborative nature of software developers and explores the application of AI to support such collaboration. The recent trend of open-source software development encourages researchers to look into open-source software developmental data and apply data-mining techniques and develop collaborative tools to support software development as knowledge activities. LLLL (Learning with Logics and Logics for Learning) was the sixth event in the series. The workshop focuses both on the application of logic to represent data and rules for machine learning and knowledge discovery, and on the development of procedural semantics to algebraic and local inference based on machine learning procedures. The most interesting aspect of LLLL is its bidirectional approach as its name stands. It provides a unique forum for researchers to explore the synergy between the two approaches.

It is our great pleasure that we are able to share some part of the outcome
of these fascinating workshops through this volume. We hope the readers of this
book find a way to grasp the state-of-the art research outcomes of JSAI-isAI
2009, and may be motivated to participate in future JSAI-isAI events.

May 2010   Kumiyo Nakakoji
                                                    Yohei Murakami
                                                    Elin McCready

Organization and Editorial Board

The paper selection of each co-located international workshop was made by the Program Committee of the workshop. Upon the decisions of the paper selections, each chapter was edited by the co-located international workshop committees. The entire contents and structure of the book were managed and edited by the chief editors.

Volume Editors

Primary Volume Editor: Kumiyo Nakakoji (SRA Inc., Japan)
Yohei Murakami (National Institute of Information and Communications
    Technology, Japan)
Elin McCready (Aoyama Gakuin University, Japan)

Chapter Editors (Program Chairs)

JURISIN-2009 Chapter:	Seiichiro Sakurai (Meiji Gakuin University, Japan)
	Ken Satoh (National Institute of Informatics, Japan)
KCSD-2009 Chapter:	Masao Ohira (Nara Institute of Science and Technology, Japan)
	Yunwen Ye (SRA Inc., Japan)
LENLS-6 Chapter:	Daisuke Bekki (Ochanomizu University, Japan)
LLLL-2009 Chapter:	Akihiro Yamamoto (Kyoto University, Japan)

Table of Contents

Part III: Logic and Engineering of Natural Language Semantics

Part IV: Learning with Logics and Logics for Learning

Part I
Juris-Informatics

Third International Workshop on Juris-Informatics

Seiichiro Sakurai

Meiji Gakuin University

The Third International Workshop on Juris-Informatics (JURISIN 2009) was held on Nov. 19-20, 2009 at Campus Innovation Center Tokyo in Tokyo, Japan, with a support of The Japanese Society for Artificial Intelligence in association with First JSAI International Symoposia on AI (JSAI-isAI 2009). This workshop was organized to study legal issues from the perspective of informatics. Law is one of the oldest practical applications of computer science. Though lots of legal reasoning systems have been developed thus far, they were not supported by the lawyers, or they didn't have a positive impact on jurisprudence. One of the reasons is that legal reasoning mechanisms currently implemented are too simple from the lawyer's viewpoint. Another reason is that legal reasoning has been studied mainly from the viewpoint of logical aspects, but it has not been studied so much from the viewpoint of natural language processing. If we can bring lawyers and informatics people and natural language processing people together, we can expect great advances in both informatics and jurisprudence by implementing legal reasoning systems clear to what lawyers expect.

The main purpose of the JURISIN workshop is to discuss both the fundamental and practical issues in juris-informatics among people from diverse backgrounds such as law, social science, information and intelligent technology, logic and philosophy, including the conventional "AI and law" area. The program committee (PC) was organized with the help of leading researchers in AI and Law as follows; Kevin Ashley (Univ. Pittsburgh, USA), Trevor Bench-Capon (Univ. Liverpool, UK), Phan Minh Dung (AIT, Thailand), Aditya Ghose (Univ. Wollongong, Australia), Guido Governatori (Univ. of Queensland, Australia), Tokuyasu Kakuta (Univ. Nagoya, Japan), Makoto Nakamura (JAIST, Japan), Katsumi Nitta (Tokyo Tech, Japan), Henry Prakken(Univ. Utrecht & Groningen, The Netherlands), Giovanni Sartor (European Univ. Institute, Italy), Ken Satoh (NII and Sokendai, Japan), Hajime Sawamura (Niigata Univ., Japan), Akira Shimazu (JAIST, Japan), Satoshi Tojo (JAIST, Japan), Katsuhiko Toyama (Univ. Nagoya, Japan), Takahira Yamaguchi (Keio Univ., Japan), John Zeleznikow (Victoria Univ., Australia).

Though the announcement period was short, fifteen papers were submitted. Each paper was reviewed by three PC members. While one paper was cancelled, eleven papers were accepted in total. The collection of papers covers various topics such as legal reasoning, argumentation theory, legal ontology, computer-aided law education, use of informatics and AI in law, and so on. The workshop was a provoking and stimulating opportunity for new research areas.

K. Nakakoji, Y. Murakami, and E. McCready (Eds.): JSAI-isAI, LNAI 6284, pp. 3–4, 2010.

After the workshop, seven papers were submitted for the post proceedings. They were reviewed by PC members again and six papers were selected. Followings are their synopses.

Andrade et al. analyze the usefulness of Best Alternative to Negotiated Agreement (BATNA) and Worst Alternative to Negotiated Agreement (WATNA) in online dispute resolution. The software agents consider the space between BATNA and WATNA as a useful element to be taken into account when making or accepting a proposal. Based on such software agents, they present a system for the resolution of labour disputes.

Boonchom and Soonthornphisaj develop an algorithm using ant colony to automatically extends a seed ontology that was initially created by law experts. The initial ontology will be extended to the weighted ontology by using Supreme Court sentences as a corpus. They investigate the experimental results and found that the weighted ontology provides significant improvement on the retrieval process.

Maeno et al. propose a text processing method which aids mediation trainees in reflecting on how they reached an agreement from their dialogue. The text processing method is an improved variant of the Data Crystallization algorithm, which visualizes the inter-topic associations which foreshadow the intentional or unintentional subsequent development of topics far apart in time.

Robino et al. present an efficient implementation of temporal defeasible logic, and they argue that it can be used to efficiently capture the legal concepts of persistence, retroactivity and periodicity. In particular, they illustrate how the system works with a real life example of a regulation.

Stenetorp and Li propose to draw a link from the quantitative notion of coherence to legal reasoning. They evaluate the stories of the plaintiff and the defendant in a legal case as rival theories by measuring how well they cohere when accounting for the evidence. They argue that their notion of coherence gives rise to an important measure to the quality of a case, and allow rival cases to be compared in a quantitative manner.

Stieghahn and Engel show the necessity of incorporating the requirements of sets of legislation into access control. After describing the legislation model, various types of contextual information, and their relationship, they introduce a new policy-combining algorithm that respects the different precedence of laws of different controlling authorities. Then they demonstrate how laws may be transformed into policies using the eXtensible Access Control Markup Language.

Finally, we wish to express our gratitude to all those who submitted papers, PC members, discussant and attentive audience.

Using BATNAs and WATNAs in Online Dispute Resolution

Francisco Andrade[1], Paulo Novais[2], Davide Carneiro[2],
John Zeleznikow[3], and José Neves[2]

[1] Escola de Direito, Universidade do Minho, Braga, Portugal
fandrade@direito.uminho.pt
[2] DI-CCTC, Universidade do Minho, Braga, Portugal
{pjon,dcarneiro,jneves}@di.uminho.pt
[3] School of Management and Information Systems, Victoria University, Melbourne, Australia
john.zeleznikow@vu.edu.au

Abstract. When contracting through software agents, disputes will inevitably arise. Thus there is an urgent need to find alternatives to litigation for resolving conflicts. Methods of Online Dispute Resolution (ODR) need to be considered to resolve such disputes. Having agents understanding what the dispute is about, managing all interaction between the parties and even formulating proposed solutions is an important innovation. Hence it is of the utmost relevance that the agents may be able to recognise and evaluate the facts, the position of the parties and understand all the relevant data. In many circumstances, risk management and avoidance will be a crucial point to be considered. In this sense we analyze the usefulness of a parallel concept to BATNA – Best Alternative to Negotiated Agreement, that of a WATNA – Worst Alternative to Negotiated Agreement, allowing the software agents to consider the space between BATNA and WATNA as a useful element to be taken into account when making or accepting a proposal. These software agents embodied with intelligent techniques are integrated in an architecture designed to provide support to the ODR in a system we have developed for the resolution of labour disputes - UMCourt. In this context software agents are used to compute and provide the parties with the best and worst alternative to a negotiated agreement.

Keywords: On-Line Dispute Resolution, Negotiation, BATNA, WATNA.

1 Introduction

When moving to a global information society, new needs have appeared in the field of dispute resolution, since disputes can now take place between virtually any two entities in the world. With the integration of new communication technologies into our daily lives, traditional Alternative Dispute Resolution (ADR) mechanisms including mediation, conciliation, negotiation or modified arbitration and jury proceedings ([10] and [30]) have slowly started to adapt, giving birth to what is now known as Online Dispute Resolution (ODR).

K. Nakakoji, Y. Murakami, and E. McCready (Eds.): JSAI-isAI, LNAI 6284, pp. 5–18, 2010.
© Springer-Verlag Berlin Heidelberg 2010

ODR allows for the moving of already traditional alternative dispute resolution methods "from a physical to virtual place" [3]. This provides the parties with an easier course than litigation, for dealing simply and efficiently with disputes, saving both "temporal and monetary costs" [12]. This new model for dispute resolution aims at being an online alternative to litigation and traditional ADR. It can expand the possibilities of common ADR systems as, with the introduction of entities with enhanced abilities, increases the generation of solutions and the possible ways of achieving them.

Techniques for developing ODR systems include legal knowledge based systems that provide legal advice to the disputing parties and also "systems that (help) settle disputes in an online environment" [6]. In this sense we can enumerate projects that make use of rule-based systems such as [25], negotiation support systems as in [26], [27] and [28], and others that look at game theory and heuristics [29]. In this paper, we consider the use of a Case-based Reasoning (CBR) [1] approach for the purpose of retrieving similar cases in order to advise the parties about the probable and possible outcomes and solution paths given former similar cases.

The so-called second generation of ODR systems is essentially defined by a more active role of technology [16]. It goes beyond putting the parties into contact and is used for idea generation, planning, strategy definition and decision making processes. The technologies used in this new generation of ODR systems will comprise not only the communication technologies used nowadays but also subfields of areas such as Artificial Intelligence, mathematics or philosophy: neural networks, intelligent agents, case-based reasoning, logical deduction, argumentation, methods for uncertain reasoning and learning methods. Thus being, the development of Second Generation ODR, in which an ODR system might act "as an autonomous agent" [16] is an appealing way for solving disputes.

In considering this possibility, we take in consideration the Katsh/Rifkin vision of the four parties in an ODR process: the two opposing parties, the third party neutral and the technology that works with the mediator or arbitrator [11]. But we must assume a gradual tendency to foster the intervention of software agents, acting either as decision support systems [3] or as real electronic mediators [16]. This latest role for software agents implies the use of artificial intelligence techniques such as case based reasoning and information and knowledge representation. "Models of the description of the fact situations, of the factors relevant for their legal effects allow the agents to be supplied with both the static knowledge of the facts and the dynamic sequence of events" [16].

Merely representing facts and events, whilst useful, is not sufficient for dispute resolution; the software agent, in order to perform actions of utility for the resolution of the dispute, also needs to know not only the terms of the dispute but also the rights or wrongs of the parties [16], and to foresee the legal consequences of the said facts and events. Thus we have to consider the issue of software agents really understanding law and to consider legal reasoning by software agents and its eventual legal responsibility: As [4] states, "are law abiding agents realistic?".

We need to consider whether agents can evaluate the position of the parties and present them with useful proposals, "taking into a consideration of which of the two parties would have a higher probability of being penalised or supported by a judicial decision of the dispute and, therefore, who would be more or less willing to make concessions in their claims" [16]. The ability to understand the position of the parties is vital for the successful involvement of software agents in the process. To do so, it is mandatory for the software agent to have the characteristics of consistency, transparency, efficiency and enhanced support for dispute resolution, in order to allow it to replicate "the manner in which decisions are made" and thus make the parties "aware of the likely outcome of litigation" [3]. That is to say, software agent intervention in an ODR procedure should take into account the alternatives, for the parties, to an ODR negotiated agreement. This kind of ODR environment involves much more than just transposing ADR ideas into ODR environments. It should actually proceed by being "guided by judicial reasoning", and getting disputants "to arrive at outcomes in line with those a judge would reach" [14]. Despite there being difficulties to overcome, the generalised use of software agents as decision support systems in a negotiation, is nevertheless a useful approach.

2   The Role of BATNA (Best Alternative to a Negotiated Agreement)

Principled negotiation is based on four fundamental principles: separate the people from the problem; focus on interests, not positions; invent options for mutual gain; insist on objective criteria [8, 21]. In interest-based negotiation, the disputants attempt to reconcile their underlying interests. Most negotiations are interest-based[1]. In this situation, disputing parties need to know their BATNA (or, the possible best outcome "along a particular path if I try to get my interests satisfied in a way that does not require negotiation with the other party"[15].

Whilst principled negotiation as an important concept, it must be supplemented with other approaches to negotiation. Justice or rights based negotiation (pointing out to the determination of who is – or who could be considered to be – right in accordance to norms or rules of behaviour) – should also be considered [8].

When taking a principled negotiation approach, we must understand the notion of a BATNA and what role it should play in ODR. "A precise notion of what constitutes a BATNA is not available" [6]. But "knowing one's BATNA may contribute to the acknowledgement that an agreement may be disadvantageous" [12].

Entering into negotiation or mediation is justified if the parties expect to get better results than those that could be obtained without the process. In order to evaluate this, one needs to know, at least what the best alternative to the negotiated agreement would be. Of course, parties will tend to enter into an agreement if they know that a possible settlement in ODR is undoubtedly better than her own BATNA [12]. This is an obvious case of interest in knowing one's BATNA. But the position of the parties may become much more unclear if they cannot foresee the possible results in case the negotiation / mediation fails. "If you are unaware of what results you could obtain if the negotiations are unsuccessful, you run the risk of entering into an agreement that you would be better off rejecting or rejecting an agreement that you would be better off entering into" [9]. That is to say, the parties, by determining their BATNA, would on one side become "better protected against agreements that should be rejected" and,

---

[1] One exception is Australian Family Law where the paramount interests of the children trump the interests of the divorcing parents. This is however not the case in US family law.

on the other side, they would be in a better condition to "reach an agreement that better satisfies their interests" [6].

A BATNA may also provide additional interesting features for the parties in the dispute procedure. For instance, it might also be used as a "way to put pressure on the other party", especially in dispute resolution procedures allowing the choice of going to court [6]. The important thing is that the choice of going to court, instead of continuing ADR or ODR, should be a "well-informed choice". And in ODR environments, either by the use of data mining techniques, semantic web technology or other adequate techniques possibly used to determinate the BATNA, the parties can foresee the possible outcome of the judicial dispute in the case of not reaching an agreement through ODR [3]. For that purpose, some technical possibilities have already been pointed out in literature. For instance, the use of a BATNA agent, an agent that has the knowledge necessary to compute the value of the BATNA, using Toulmin argument structures providing a "mechanism for decomposing a task into sub-tasks" has been pointed out [2]. Similarly, the possibility of the BATNA agent being modified in order to "include current case data and incorporate changes to law" is an important development [23]. The role of technology is becoming more appealing especially for the task of determining or establishing objective BATNAs [6].

3   How Understanding WATNAs Can Improve the ODR Process

No matter the alternative dispute resolution method chosen, parties will tend "to develop an overly optimistic view on their chances in disputes" [6]. This rather optimistic view may lead to differing attitudes taken by the parties, especially in their calculation of chances of success obtaining their goals in the dispute, and influencing the way disputants calculate their BATNA. In the course of the dispute, the parties may tend either to reject generous offers from the other parties, or to stand stubbornly fixed in some positions or even support "positions or options that are incorrect" [6]. This "optimistic overconfidence" [6] may lead the parties to miscalculate the possibilities of success in an eventual judicial decision.

It is important to reflect on the usefulness of the concept of a BATNA. On one side, a BATNA may be misevaluated through the above optimistic overconfidence of the parties. On the other side, there is no probabilistic measure for the correctness of BATNA. That is to say, the best alternative may not be the most probable one. And parties will certainly tend to underestimate the probabilities of an undesired result in judicial decision-making.

In many situations, the calculation of the possible outcomes of a judicial decision may become quite complex. One of the major reasons that disputants try to avoid litigation is the risks they might incur – in terms of legal costs and outcome – if they are unsuccessful [30]. In this situation it could certainly be useful, besides the BATNA, to consider a WATNA (Worst alternative to Negotiated Agreement) [8, 15, 20]. A WATNA intends to estimate the worst possible outcome along a litigation path [15]. It can be quite relevant in complementing principled negotiation with a justice or rights based approach and thus leading to a calculation of the real risks that parties will face in judicially determined litigation, imagining the worst possible outcome for the party. This calculation could prove interesting both for ADR and for

ODR. In the case of second generation ODR, it would be useful to develop a software agent to consider the whole space between the BATNA and WATNA. The larger the space, the greater the benefit in making, or accepting, a proposal.

Consider, for instance, from an organizational perspective, the relevance of labour rules in the functioning of professional virtual communities: as Willy Picard states, "the organizational structure of the population that may potentially execute activities may evolve, as some employees are promoted or are fired" [17]. In the case of a employee being fired, litigation will most likely occur. Under legal systems such as that of Portugal, a huge deal of legal parameters need to be considered:

(a) the antiquity of the worker in the company,
(b) supplementary work,
(c) night work,
(d) justified or unjustified absence to work,
(e) the possibility of a "just cause for dismissal" being declared by Court,
(f) the existence (or not) of a valid and legal procedure of dismissal,
(g) the possibility of dismissal being accepted without indemnities or
(h) of it being accepted but accompanied by indemnities that could range from a very low to a very high amount of money [7].

To dismiss a worker, the company needs to calculate the potential ensuing financial penalties. For the worker, the amounts involved are not irrelevant: being fired without good indemnities may be seen as a double sacrifice: not only would he lose his job but he could get no or little payment for his loss. But he might, on the other hand, receive adequate financial compensation. For the parties in a labour conflict, it can be said that the calculation of the possible results of litigation (or of the various possible outcomes for litigation) are vital.

In order to clearly understand the advantages of a proposed agreement, parties need to know not only their BATNA but also their WATNA (the worst alternative they may obtain in case they do not reach an agreement), and they certainly should consider the spectrum between their BATNA and their WATNA. Of course, the less space there is between BATNA and WATNA, the less dangerous it becomes for the party not to accept the agreement (unless, of course, their BATNA is really disadvantageous). A wider space between BATNA and WATNA would usually mean that it can become rather dangerous for the party not to accept the ODR agreement (except in situations when the WATNA is not undesirable for the party).

Of course, this consideration of the values appearing between the BATNA and the WATNA is related to the Zone of Possible Agreement proposed by Raiffa (1982) [19]. It is the zone where an agreement can be met that is acceptable to both parties. The consideration of the space between BATNA and WATNA has, in our vision, a clear risk oriented approach – the intention is to estimate the risks and, thus, to avoid them. And this vision may well push the possible agreement to a space not exactly coincident with the traditional ZOPA. And certainly it can even be considered here the existence of a MLATNA – most likely alternative to a negotiated agreement [20].

In terms of the system we are developing, it does not much matter what is the most likely outcome, which might be hard to estimate, but rather it is vital to foresee the real risks that the parties are facing. And the extreme value presented by WATNA may well force the parties to change the ideas they have about their BATNA and ZOPA.

We accept that this analysis is still in an early stage and that other relevant parameters should also be considered: for instance, the existence of metrics in order to measure the probabilities of each possible outcome. Nonetheless, judicial decisions, although having to be based on legal rules and reasoned from them, arise from a process in which it must be determined that some issues are true or false, or are considered as proved, partially proved, or not proved [18]. This characteristic of judicial decisions certainly makes it advisable for parties to consider not just a single value, in the case of judicial litigation, but rather a spectrum of values, situated between a BATNA and a WATNA.

4   UMCourt Architecture

On-line dispute resolution methods can provide easy, efficient, fast ways for resolving disputes. Labour disputes need to be quickly resolved. The judicial path (which, in countries such as Portugal, often leads to a judicial conciliation led by a Judge) is expensive and time consuming. First and second generation ODR [16], with agents performing relevant parts of the agreement procedure can be of inestimable use for the parties in a Portugese labour dispute.

UMCourt is a project being developed at University of Minho in the context of the TIARAC project (Telematics and Artificial Intelligence in Alternative Conflict Resolution) that intends to help parties involved in legal disputes. The current application relates to the domain of Portuguese labor law. UMCourt represents one of the first steps in Portugal to implement the ideas depicted in the previous sections [5].

It is based on the agent paradigm, which means that the resulting architecture is highly modular and expansible. This choice has not been a random one. Although we are currently addressing the specific domain of the Portuguese labour law, we are aware that by defining a few domain-dependent agents and reusing many of the core agents, it is possible to extend the platform to address other domains. To put this idea into practice, an extension to this architecture that addresses consumer protection law is now being developed, that uses much of the already defined architecture. The building blocks of this modular architecture are agents or groups of agents with well defined roles that, through their interactions, configure an intelligent system. Following the methodology proposed by [22], our work in this system began with a high level definition of the members of the architecture in terms of their roles. In this phase we have looked at existing agent-based architectures in the legal domain, namely at [2], and made the necessary improvements in order to adapt it to our needs. We therefore arrived at a configuration of four high-level agents with their roles shown in Table 1. The implementation of this architecture is based on a range of technologies chosen with the objective of making it a distributed, expansible and independent one [5].

The core of the architecture is the Jade (Java Agent Development Framework) platform[2]. JADE is a software framework that significantly facilitates the development of agent-based applications in compliance with the FIPA specifications. FIPA (Foundation for Intelligent Physical Agents) promotes standards that aim at the interoperability and compatibility of agents, specifically targeting the fields of agent communication,

---

[2] See http://jade.tilab.com/ last accessed January 2 2010.

Table 1. The four high-level agents and their main roles

High-level Agent	Description	Main Roles
Security	This agent is responsible for dealing with all the security issues of the system	Establish secure sessions with users
		Access levels and control
		Control the interactions with the knowledge base
		Control the lifecycle of the remaining agents
Knowledge Base	This agent provides methods for interacting with the knowledge stored in the system	Read information from the KB
		Store new information in the KB
		Support the management of files within the system
Reasoning	This agent embodies the intelligent mechanisms of the system	Compute the BATNA and WATNA values
		Compute the most significant outcomes and their respective likeliness
		Proactively compile and provide useful information based on the phase of the dispute resolution process
Interface	This agent is responsible for establishing the interface between the system and the user in a intuitive fashion	Define a intuitive representation of the information of each process
		Provide an intuitive interface for the interaction of the user with the system
		Provide simple and easy access to important information (e.g. laws) according to the process domain and phase

agent transport, agent management, agent architecture and applications. Of these FIPA categories, agent communication is the core category at the FIPA multi-agent system model and is the one that is more closely followed in our system. Our interest in this category is focused on specification 61, which defines the structure of the Agent Communication Language (ACL), .i.e., the structure that the messages exchanged between agents respect. An example of use of this standard is shown below.

Example of an ACL message from agent Coordinator to agent Retriever requesting the cases similar to 1263491000923, assuming the default settings.

```
Sender : ( agent-identifier
  :name Coordinator@davide-desktop:1099/JADE
  :addresses (sequence http://davide-desktop:7778/acc ))
Conversation-ID : 1263492569251
Reply-To : Coordinator@davide-desktop:1099/JADE
Ontology : CBR_LABOUR
Content : RETRIEVE_SIMILAR DEFAULT 1263491000923
```

Jade has also the advantage of dealing with all the issues of message transport and agent registry thanks to a wide number of services provided by the *ams* and *df* agents. This significantly simplifies the creation of new agents thus decreasing the development time and costs.

The interaction with the system can be performed in two ways: by means of a JSP based Guided User Interface (GUI) or with remote agents that interact directly with the agent platform. In the first case, the interface was designed so that the users could remotely interact with the system using any common web browser. Through the browser, the client sends requests to the server which interacts with the Jade platform, collects the answers and returns the HTML code to be shown in the browser of the user. By doing this, we not only make sure that the user can understand and interact with the system through an intuitive interface but also grant the security of the whole system. In the second case, Jade agents external to the platform can interact with the agents present by means of FIPA-ACL messages. In fact, using the Jade platform, sending and receiving remote messages becomes as easy as performing the task locally. This increases the expansibility and compatibility of the architecture, making sure that it can interact with other architectures with similar or complementary functionalities or even that automated agents representing the parties can interact.

The high level agents depicted in Table 1 have been submitted to a cut-down process in order to more precisely define their roles and make them more simple and refined. In this task we have defined several simpler agents, such as the coordinator, with the task of load balancing, the retriever which interacts with the KB agents in order to retrieve cases, among others. A simplified view of the architecture highlighting the main agents is presented in Figure 1. However, as this paper is centred on the concepts of BATNA and WATNA and the determination of the space in between, we will from now on focus on the *Reasoning* agent. The *Reasoning* agent was defined with some objectives in mind, being one of them to assist the parties in determining the possible outcomes. When the parties have knowledge about what could possibly happen, they can take more informed and, hopefully, rational decisions. As exposed before, in this paper we support the idea that it is vital to know the BATNA and WATNA values, as well as the most significant values contained in the interval

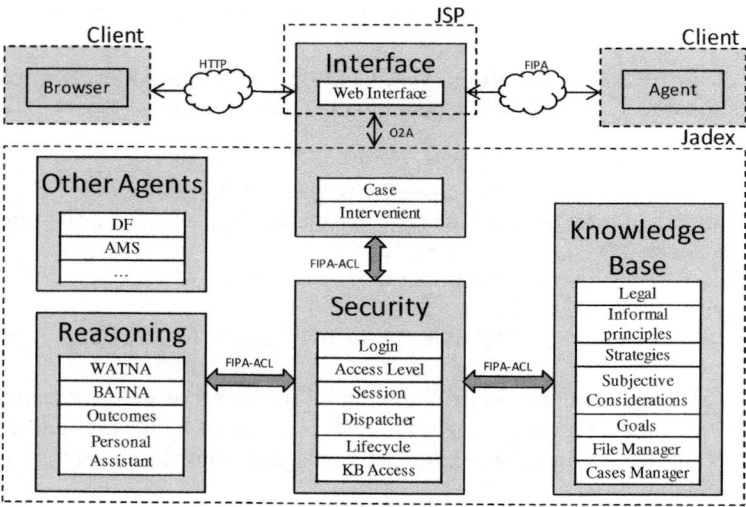

Fig. 1. A simplified view of the architecture

between them, so that an optimal global solution can be achieved. Indeed, the WATNA of one party is frequently close to the BATNA of the other so limiting the possible choices to these two values represents a serious drawback and the impossibility of reaching a global optimization. The optimal solution lies somewhere in the intersection of the possible solutions of each of the parties: the Zone of Possible Agreement (ZOPA) [13]. This zone represents all the solutions that can happen at the end of the process. The role of the *Reasoning* agent is, in the first place, to determine if an agreement is possible. If the ZOPA does not exist, then an agreement is not possible. However, if there is a ZOPA with a range of solutions, the objective of the agent is to determine which is the best option and suggest it to the parties, and then trying to work for a mutually favourable outcome from that point on. This agent has been subdivided into four simpler agents: *BATNA, WATNA, Outcomes* and *Personal Assistant.*

The *BATNA* and *WATNA* agents respectively compute the values of the BATNA and the WATNA. This calculus is based on mathematical formulae that are well defined in the Portuguese labour law and have been transported to the agents in the form of logical predicates. These are simple formulae that map a set of values of parameters such as antiquity or extra and night working hours to an economical value of indemnity. The first agent therefore analyzes the case and computes the value of the best legally possible outcome, i.e., it assumes that the employee is absolutely right in his allegations. In the other hand, the *WATNA* agent does the same, assuming that the employee cannot prove any of the arguments in his defence, determining the worst possible case according to the rules of the Portuguese labour law. The output of these two agents, according to what is defined in Portuguese labour law, is a pair of values: one quantifying the value of the indemnity and the other one stating if the employee looses the job or is reintegrated.

The *Outcomes* agent has as objective, as the name depicts, to compute the possible outcomes of a new case, which configures one important feature in online dispute resolution systems. For determining them, this agent uses a Case-based Reasoning model. Our conviction that CBR is an appropriate method for such a problem solving domain relies on the fact that law itself implements a very similar concept: the legal precedent [24]. This concept is defined by the Blacks Law dictionary as "an adjudged case or decision of a court, considered as furnishing an example or authority for an identical or similar case afterwards arising or a similar question of law."[31] A precedent, in the legal domain, can be sub-divided into two categories: the binding precedent which must be applied and the persuasive precedent, which is not mandatory but is relevant. This labeling has generally to do with the courts that decide on the case. If it is a higher court making the decision, it usually becomes a more persuasive precedent. By looking at all the significant past cases contained in the Knowledge Base, this agent is able to determine which outcomes are possible to occur, given the properties of the current case.

In order for this agent to correctly use the information contained in each case, a syntactic structure for the case has been defined. Each case contains the laws used by each party and all the remaining information that is requested by law, including a pair of values denoting the outcome: one quantifying the value of the indemnity and the other one stating if the employer looses the job or is reintegrated. All this data is stored in XML files but, in order to fasten the retrieval processes, the cases are indexed in a database by the laws that they address and the way that they are addressed, i.e., if these laws are used by the employee, employer or by a witness. This allows us to efficiently search for a case with given characteristics in the database and then retrieve it from its location in the file system to parse all the information.

Essentially, the process of estimating outcomes is as follows. The agent looks at the new case, specifically at the norms that are addressed by each party. Afterwards, it applies a template retrieval algorithm in order to narrow the search space. This can be performed, a priori in determining which type of cases have the possibility of being similar and which ones do not. In this sense, template retrieval works much like SQL queries: a set of cases, with given characteristics, is retrieved from the database. In the next step, a nearest neighbor algorithm is applied to this set of cases instead of applying it to all the cases in the case memory, a task that could be very time consuming as our nearest neighbor algorithm has linear complexity (Formula 1).

$$\frac{\sum_{i=1}^{n} W_i * fsim_i(Arg_i^N, Arg_i^R)}{\sum_{i=1}^{n} W_i} \tag{1}$$

In formula 1, our closest neighbor algorithm is shown. In this equation,

- n – number of elements to consider to compute the similarity;
- W_i – weight of element i in the overall similarity;
- Fsim – similarity function for element i;
- Arg – arguments for the similarity function representing the values of the element i for the new case and the retrieved case, respectively N and R.

We now discuss in greater detail the information of the case that is considered to be relevant for the computation of the similarity, i.e., the components. According to the scope of application, we consider three types of information: the objectives stated by each party in the beginning of the dispute, the norms addressed by each party and by the eventual witnesses and the date of the dispute. The norms addressed and the objectives are lists of elements, thus the similarity function consists in comparing two lists (equation 2). The similarity is higher when the two lists have a higher percentage of common members. As for the date, the similarity function verifies if the two dates are within a given time range, having a higher similarity when the two dates are closer.

$$fsim_{list} = \frac{|L_N \cap L_R|}{n}, n = \begin{cases} |L_N|, & |L_N| \geq |L_R| \\ |L_R|, & |L_N| < |L_R| \end{cases} \tag{2}$$

Once each case is associated with a value that denotes its similarity with another given case, we can perform more interesting and useful operations on the cases. One of these operations is to determine to which extent we want the cases to be similar. We do so by selecting a threshold. If our current selection results in few cases, we can lower the threshold resulting in a wider number of cases but with an expected smaller degree of similarity. On the other hand, if we have many cases, we can increase the threshold in order to get a more restricted set with a higher degree of similarity.

The most interesting aspect is that the agent can autonomously apply these operations in real-time, when choosing the cases to present to the user: if there are many cases, the agent will increase the threshold to select less and more similar cases and vice versa. For the agent to determine if it should change the value of the threshold, it generates a box-and-whisker diagram using the values of the indemnities of each case selected, and looks at the dispersion of the data. The dispersion of the data is determined by the Euclidian distance between the indemnities of each of the two consecutive cases. The agent decides to decrease the value of the threshold if the data is much dispersed and the other way around if the data is not very dispersed.

Having done this, the agent sorts the cases for each of the parties, starting with the WATNA, passing through all the intermediary cases and ending with the BATNA. The cases are sorted according to the numeric value of the indemnity and how favorable it is to each party. At this point, the parties have an intuitive picture of what may happen, including not only the best and worst case but all the intermediary cases that have happened in the past and may happen again, accompanied by the respective likeliness to occur. All this information is shown in Figure 2. This figure contains two axes, one for each party, with a direction for increasing satisfaction. Cases are here represented in these axes by the smaller rectangles and the Euclidian distance between the values of their indemnities determines how they are distributed and ordered in the axis of each party and therefore highlights the dispersion.

The likeliness of a given outcome is represented in Figure 2 by the colored curves which denote the area in which the cases are more likely to occur. We can see that the line is more distant from the axis when there are more cases that are concentrated. This denotes a higher likeliness for a case in this region. However, it is not only the amount of cases that is important. We also consider the type of case, i.e., if it is a case with a binding or a persuasive precedent or if it has been decided by a higher or a lower court. In some cases, we may even have groups of cases instead of single cases, as cases which are highly similar are grouped together into a single case with a weight that is proportional to the number of cases merged.

Still looking at Figure 2, we can see the range of possible outcomes for each of the parties in the form of the two big colored rectangles and the result of its intersection, the ZOPA. It is also possible to see each case and its position in the ordered axis of increasing satisfaction, in the shape of the smaller rectangles. As stated earlier, the cases are more likely to occur for each party when they are in the area where the colored lines are further away from the axis of that party. This is highlighted in the figure by the big dot. Therefore, the probable outcome of the dispute will probably be near the area where the two lines are closer. Looking at this information, the parties can have an approximate notion of the most likely outcome. Although a single solution is proposed by the system, parties can look at these cases in order to search for alternative solutions in search for a mutual agreement.

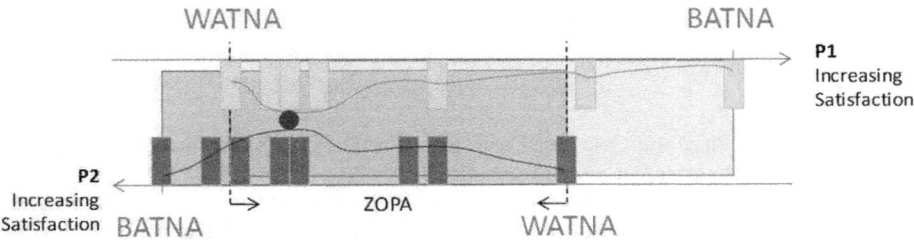

Fig. 2. The graphical representation of the possible outcomes for each party

Finally, the *Personal Assistant* agent is responsible for knowing the role of the user and adapting the results of the remaining agents according to that role, i.e., the employee will see the information in a different way than the employer or even than the witnesses. This agent will also be extended with more features, namely the adaptation of the interfaces and the remaining information that is presented to the users, including help information, according to the roles.

5 Conclusions

When using software agents in electronic contracting the possibility of disputes is prevalent. Thus there is a demand for fast and efficient ways for resolving the eventual disputes using ODR. For a second generation ODR, software agents are a useful tool to help the parties reach an agreement. The acceptance of a certain proposal by the parties in ODR must take into consideration relevant parameters, such as the BATNA.

Parties in ODR tend to adopt an over-optimistic view on the possible outcomes in the case of litigation. In many situations the calculation of the possible outcomes of litigation may become rather complex, with a huge range of possibilities to be evaluated. This is particularly clear in Portuguese labour legal cases. In these situations it may become interesting to consider not just the value of BATNA but also the value of the WATNA.

It is vital, to consider the space that lies between the BATNA and the WATNA. It would be advisable for parties to consider not just a single value but rather a spectrum of values, situated between a BATNA and a WATNA. For this purpose, we have developed an architecture supported by a JADE platform, allowing the user to interact with the system. Software agents are used to compute and provide the parties with the best and worst alternative to a negotiated agreement as well as the spectrum of possible outcomes between these two values and their likeliness.

The system presented here, by being based on the multi-agent paradigm ensures that it can be easily extended with the addition of new agents. To address the challenges of incorporating new agents, we rely on open standards and technologies. Moreover, by adopting a case-based approach, we achieve a system that can learn and adapt to new situations and changes in the law. In future work we intend to allow the system to estimate outcomes of dispute resolution processes based on CBR and other paradigms (e.g., neural networks, Adjusted Winner algorithm) in order to compare the performance and results of different approaches.

Acknowledgments. The work described in this paper is included in TIARAC - Telematics and Artificial Intelligence in Alternative Conflict Resolution Project (PTDC/JUR/71354/2006), which is a research project supported by FCT (Science & Technology Foundation), Portugal.

References

1. Aamodt, A., Plaza, E.: Case-based reasoning: Foundational issues, methodological variations, and system approaches. AI Communications 7(1), 39–59 (1994)
2. Abrahams, B., Zeleznikow, J.: A multi-agent architecture for online dispute resolution services. Expanding the horizons of ODR. In: Proceedings of the 5th International Workshop on Online Dispute Resolution (ODR Workshop 2008), Firenze, Italy, pp. 51–61 (2008)

3. Bellucci, E., Lodder, A., Zeleznikow, J.: Integrating artificial intelligence, argumentation and game theory to develop an online dispute resolution environment. In: ICTAI 2004 - 16th IEEE International Conference on Tools with Artificial Intelligence, pp. 749–754 (2004)
4. Brazier, F., Kubbe, O., Oskamp, A., Wijngaards, N.: Are Law abiding agents realistic? In: Proceedings of the workshop on the Law of Electronic Agents (LEA 2002), CIRSFID, pp. 151–155. University of Bologna (2002)
5. Carneiro, D., Novais, P., Andrade, F., Zeleznikow, J., Neves, J.: The Legal Precedent in Online Dispute Resolution, in Legal Knowledge and Information Systems. In: Governatori, G. (ed.) Proceedings of the Jurix 2009 - the 22nd International Conference on Legal Knowledge and Information Systems, Rotterdam, The Netherlands, pp. 47–52. IOS press, Amsterdam (2009) ISBN 978-1-60750-082-7
6. De Vries, B.R., Leenes, R., Zeleznikow, J.: Fundamentals of providing negotiation support online: the need for developping BATNAs. In: Proceedings of the Second International ODR Workshop, Tilburg, pp. 59–67. Wolf Legal Publishers (2005)
7. Fernandes, A.M.: Direito de Trabalho, Almedina (2005) (in Portuguese)
8. Fisher, R., Ury, W.: Getting To Yes: Negotiating Agreement Without Giving. Houghton Mifflin, Boston (1981) ISBN 0-395-31757-6
9. Goldberg, S.B., Sander, F.E., Rogers, N., Cole, S.R.: Dispute Resolution: negotiation, mediation and other processes. Aspen Publishers, New York (2003)
10. Goodman, J.W.: The pros and cons of online dispute resolution: an assessment of cyber-mediation websites. Duke Law and Technology Review (2003)
11. Katsh, E., Rifkin, J.: Online dispute resolution – resolving conflicts in cyberspace. Jossey-Bass Wiley Company, San Francisco (2001)
12. Klaming, L., Van Veenen, J., Leenes, R.: I want the opposite of what you want: summary of a study on the reduction of fixed-pie perceptions in online negotiations. In: Expanding the horizons of ODR, Proceedings of the 5th International Workshop on Online Dispute Resolution (ODR Workshop 2008), Firenze, Italy, pp. 84–94 (2008)
13. Lewicki, R., Saunders, D., Minton, J.: Zone of Potential Agreement. In: Negotiation, 3rd edn. Irwin-McGraw Hill, Burr Ridge (1999)
14. Muecke, N., Stranieri, A., Miller, C.: The integration of online dispute resolution and decision support systems. In: Expanding the horizons of ODR, Proceedings of the 5th International Workshop on Online Dispute Resolution (ODR Workshop 2008), Firenze, Italy, pp. 62–72 (2008)
15. Notini, J.: Effective Alternatives Analysis In Mediation: "BATNA/WATNA" Analysis Demystified (2005), http://www.mediate.com/articles/notini1.cfm (Accessed July 24, 2009)
16. Peruginelli, G., Chiti, G.: Artificial Intelligence in alternative dispute resolution. In: Proceedings of the Workshop on the law of electronic agents – LEA (2002)
17. Picard, W.: Support for Power in adaptation of social Protocols for Professional Virtual Communities. In: Camarinha-Matos, L., Afsarmanesh, H., Novais, P., Analide, C. (eds.) Establishing the Foundation of Collaborative Networks. IFIP International Federation for Information Processing, pp. 363–370. Springer, Heidelberg (2007) ISBN: 978-0-387-73797-3
18. Pimenta, J.C.: A Lógica da Sentença, Livraria Petrony (2003) (in Portuguese)
19. Raiffa, H.: The art and science of negotiation: how to resolve conflicts and get the best out of bargaining. The Belknap Press of Harvard University Press, Cambridge (1982)
20. Steenbergen, W.: Rationalizing Dispute Resolution: From best alternative to the most likely one. In: Proceedings 3rd ODR workshop, Brussels (2005)

21. Ury, W., Brett, J.M., Goldberg, S.B.: Getting Disputes Resolved: Designing Systems to Cut the Costs of Conflict. Jossey-Bass Publishers, San Francisco (1988)
22. Wooldridge, M., Jennings, N.R., Kinny. D.: The Gaia Methodology for Agent-Oriented Analysis and Design. In: Autonomous Agents and Multi-Agent Systems, vol. 3 (2000)
23. Zeleznikow, J., Abrahams, B.: Incorporating issues of fairness into development of a multi-agent negotiation support system. In: Proceedings of the 12th International Conference on Artificial Intelligence and Law, pp. 177–184. ACM, Barcelona (2009)
24. Zweigert, K., Kötz, H.: An Introduction to Comparative Law. Clarendon Press, Oxford (1998)
25. Waterman, D.A., Peterson, M.: Rule-based models of legal expertise. In: The Proceedings of the First National Conference on Artificial Intelligence. Stanford University, Stanford (1980)
26. Cáceres, E.: EXPERTIUS: A Mexican Judicial Decision-Support System in the Field of Family law. In: Francesconi, E.B.E., Sartor, G., Tiscornia, D. (eds.) Legal Knowledge and Information Systems, pp. 78–87. IOS Press, Amsterdam (2008)
27. Kersten, G., Noronha, S.: Negotiation via the World Wide Web: A Cross-cultural Study of Decision Making. Group Decision and Negotiation 8, 251–279 (1999)
28. Thiessen, E.M.: ICANS: An Interactive Computer-Assisted Multi-party Negotiation Support System. PhD Dissertation, School of Civil & Environmental Engineering, Cornell University, Ithaca, NY (1993)
29. Zeleznikow, J., Bellucci, E.: Family_Winner: integrating game theory and heuristics to provide negotiation support. In: Proceedings of Sixteenth International Conference on Legal Knowledge Based System, pp. 21–30 (2003)
30. Lodder, A., Zeleznikow, J.: Enhanced Dispute Resolution through the use of Information Technology. Cambridge University Press, Cambridge (2010)
31. Black, H.C.: Black's Law Dictionary. West Publishing Company, St. Paul (1990)

Thai Succession and Family Law Ontology Building Using Ant Colony Algorithm

Vi-sit Boonchom and Nuanwan Soonthornphisaj*

Department of Computer Science, Faculty of Science
Kasetsart University, Bangkok, Thailand
{g5184018,fscinws}@ku.ac.th

Abstract. Ontology building is a tedious job and a time consuming task for user. The quality of ontology plays an important role in information retrieval application. Therefore, we need the expertise from law users to extract their knowledge in term of ontology structure. To overcome these difficulties, we develop an algorithm using ant colony to automatically extends a seed ontology that was initially created by law expert. Two seed ontologies are constructed by experts which are succession law and family law ontology. These seed ontologies can be extended using Supreme Court sentences as a corpus. We used 2 datasets from civil Supreme Court sentences, which are succession law and family law.

Our ontologies are embedded with weight values that are the product of pheromone updating process of ant colony. We investigate the experimental results and found that the weighted ontology provides significant improvement on the retrieval process. Furthermore, the size of our ontologies obtained from TLOE is suitable since their retrieval performances are acceptable by users.

Keywords: Ontology Building, Ontology Expansion, Legal, Thai Law, Ant Colony Algorithm.

1 Introduction

Ontology can be used as a knowledge representation method to enhance the performance of the retrieval process. There are several domain specific ontologies such as e-government domain [1], e-learning domain [2], legal knowledge [3], etc. Unfortunately for the Thai law, we haven't had any ontology for law users. Therefore the Supreme Court sentences retrieval process cannot provide satisfied results to law user.

Studying Thai law needs Supreme Court sentences as a set of precedent cases to fulfill legal knowledge. The Supreme Court of Thailand has provided the search engine as a service to lawyers, legal students, judges and law researchers. Another advantage of Thai law ontology is that users can understand the overview of main legal knowledge concepts obtained from the ontology. These main concepts help user for self studying in order to gain the complete legal knowledge in the domain specific law.

This research proposes a new algorithm to extend a seed ontology and uses ant colony [4] to do weight updating on the links of ontology. The mechanism of our

---

* Corresponding author.

K. Nakakoji, Y. Murakami, and E. McCready (Eds.): JSAI-isAI, LNAI 6284, pp. 19–32, 2010.

proposed algorithm complys with the concept of Ant Colony algorithm. There are 26 ants (users) in the colony. The mission of each ant is to find the food (document). Ants traverse through the ontology to find an appropriated node in the ontology in order to attach the new found node. The pheromone from the ant is accumulated on the link, if it passes through by ant. On the other hand, the pheromone is evaporated from the link if that link is abandoned by the colony. We apply a ThaiLegalWordNet in the ontology expansion process to identify a concept or superclass in order to correctly connect the new conceptual node to the predefined ontology. Note that, the ThaiLegalWordNet covers 2 datasets of Thai civil Supreme Court sentences.

The contributions of this work are the Thai succession law ontology and family law ontology that can be applied in any applications and the retrieval algorithm that is suitable for the Supreme Court sentences search engine.

This paper is structured as follows. Section 2 describes the overview of other research using ontology. In section 3, we propose our framework and a new algorithm. Section 4 is about the experimental results and discussion. Finally, the conclusion and future work will be addressed in section 5.

2   Related Works

There are several methodologies for ontology construction such as dictionary, thesaurus, natural language processing or NLP, WordNet and specific domain dictionary, etc.

Nowadays, dictionary and thesaurus are widely used to ontology building. In 2005, Corcho, O. *et al.* developed a template for law experts to create a Spanish law ontology. They used thesaurous and dictionary as a tool for building a relationship diagram and used software called WebODE to construct the template in OWL format [5]. In 1997, Kurematsu, M. and Yamaguch, T. initialized legal ontology using the MRD dictionary. They found the most related legal concept that is extracted from an MRD. Their algorithm based on the comparison between the initial legal ontology and the best MRD correspondences legal concepts [8]. Then, in 2007, Zhang, X. and Xia, G. proposed the pattern to extract the terms and their relationship from the documents. The prototype is constructed based on the thesaurus that construction depends on the domain experts. Their algorithm makes a mapping and expands the prototype between the prototype and the documents using pattern on the documents [9].

NLP technique has more benefits for automatic ontology construction. In 2006, Despres, S. and Szulman, S. developed the TERMINAE method based on micro-ontology building as core ontology that is used in legal knowledge. Their algorithm aligned the legal micro-ontology with NLP tools and combined knowledge with modeling technique that kept link between them. Two or more ontology are brought into mutual agreement using ontology alignment method [11].

Furthermore, WordNet is the electronic lexical database that can be employed to many applications. In 2007, Hu, H. and Du, X. built a building bilingual ontology based on the WordNet and Chinese Classified Thesaurus (CCT). The matrixes of both ontologies are not in the same dimension. They use lattice method for transform to ontology matrix [10].

Moreover, specific domain dictionary can be created by domain expert and applied to ontology construction. In 2006, the first Korean law ontology was generated using natural language processing technique cooperated with Korean WordNet. The dataset was collected from law text books and research papers about law. After that, their algorithm found a set of word co-occurrences and hand on ability to construct a graph that connects these words together [7]. In addition, in 2009, the framework of Thai Law Ontology Expansion was proposed by Boonchom, V. and Soonthornphisaj, N. Their algorithm created seed ontology and ontology expansion via ThaiLegalWord-Net that was created by law experts [12].

Kayed, A. used text mining technology for law ontology construction from e-commerce law cases. He proposed a new algorithm to reduce a number of links connected between nodes in order to descrease the ontology complexity [6].

Behavior of ants is capable on finding the shortest path between food and nest. Each ant leaves pheromone while moving on the shortest path (chosen path). The chosen path is increased pheromone and on the other hand, the pheromone decrease as time passes due to evaporation [4].

3 The Framework of Thai Law Ontology Expansion (TLOE)

There are two main parts in TLOE which are a retrieval module and ontology expansion module (see Fig.1)

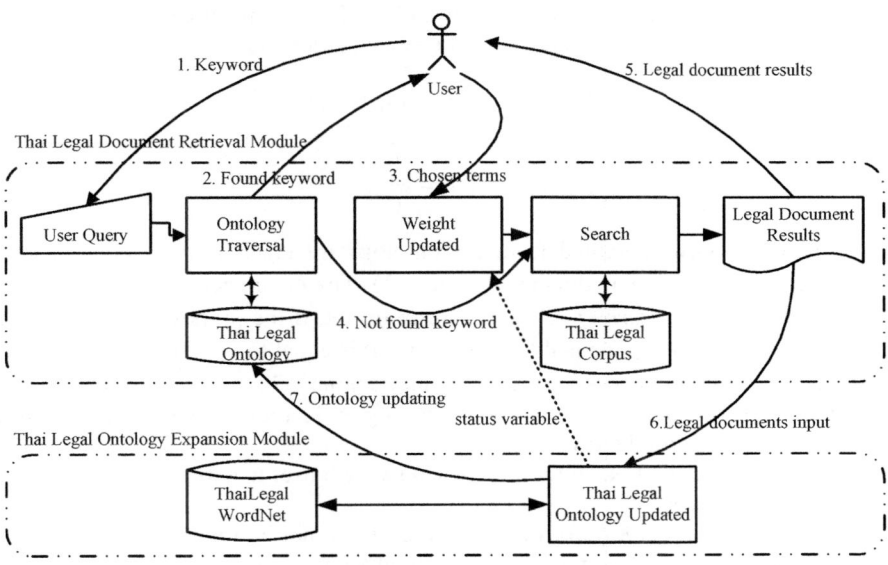

Fig. 1. The framework of TLOE

3.1 The Thai Legal Document Retrieval Algorithm

The process is started when users enter keyword to the Thai Legal Document Retrieval Module. This module proposes set of terms to user using ontology traversal. The structure of ontology consists of a term, an ID number and a pheromone value

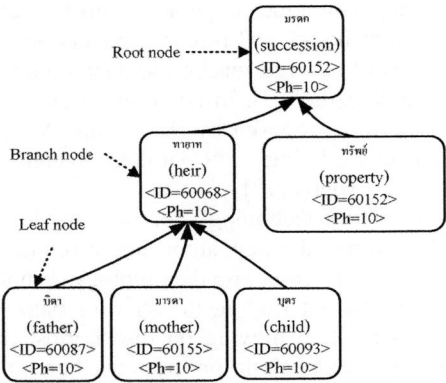

Fig. 2. Three kinds of nodes and three parameters in ontology

(see Fig. 2). When the node is found, the algorithm determines the type of node whether it is a root node or a branch node. The OntologyTraversal function converts a keyword to ID number and use it in the next process. Then the algorithm explores the links attached with that node and proposes those words to users.

Once TLOE gets information about the selected words from each user (ant). The pheromone updating process is performed using equation (1).

$$\tau(term_k) = \tau(term_k) + (\xi \cdot \Delta\tau) \tag{1}$$

$$\Delta\tau = \frac{1}{f} \tag{2}$$

Given a set of terms obtained from the ontology traversal ($term_k$), the pheromone level $\tau(term_k)$ is calculated using equation (1). The parameter ξ is an indicator to determine whether the $term_k$ is chosen by user or not. If the proposed term is chosen, the value of parameter ξ is equal to 1 otherwise it is equal to -1. This parameter simulates the accumulation and evaporation of pheromone. The notation $\Delta\tau$ in (2) represents the updated value for pheromone increasing or decreasing. The value of   is the total number of proposed terms.

Then TLOE system uses all terms selected by each user to expand the query and do searching in the corpus.

In case that the input keyword from each user can not be found in the ontology or found as a leaf node, TLOE will directly use that keyword to search in the corpus.

$$DocScore_j = \sum_{i=1}^{n} \left(\tau(term_k) \times tf_{ij} \right) \tag{3}$$

Given n chosen terms obtained from user, the searching module calculates a term frequency (tf_{ij}) for each term. The document ranking process is performed using *DocScore* in equation (3).

Table 1. Thai Legal Document Retrieval Algorithm

```
Algorithm: ThaiLegalDocRetrieval
Input: User keyword, status
Begin
  If foundConcept(keyword,Ontology) Then
    Terms←OntologyTraversal(keyword,Ontology)
    For each t_i∈Terms
      QExpand←ChosenTerm(t_i)
      Ph_i←WeightUpdated(QExpand,Ontology)
        If status<6 then
          If Ph_i<= 0 && term_i = LeafNode Then
            DeleteNode(term_i, Ontology)
        End If
      For each Doc_k∈Corpus
        DocScore_k←Ph_i*TF_i
    LegalDocRanked←Sort(DocScore,Corpus)
  End If
End.
Output: LegalDocRanked
```

Table 1 show the OntologyTraversal function is started to explore all links and outputs a set of terms. The ChosenTerm procedure displays a set of terms for user to filter out some terms that may not relate to the user requirement. After that, the WeightUpdated procedure is performed on a set of chosen words. The algorithm considers the *Ph* value of *term_i*. If the value is less than or equal to 0 and the *term_i* is a leaf node, then the DeleteNode module deletes *term_i*. The retrieval step is done by calculating the score for each document in corpus and the final results are shown to user. Note that this retrieval module gets the signal from the ontology expansion algorithm via the parameter call status. The status value informs the module about the convergence condition found by the ontology expansion. If the value of status exceeds the limit (5), the pheromone evaporation on that like will be discarded. Otherwise the node deletion might be performed if the pheromone level is completely evaporated and that term is the leaf node.

3.2 Thai Legal Ontology Expansion Algorithm

The ontology expansion algorithm gets a set of ranked documents from the retrieval module. *Max* variable is the total number of document sentences from the corpus. Given the first ranked document, the word segmentation is performed to get a set of terms, *term_i*. For each *term_i*, the algorithm traverses through the ontology to find *term_i*. If it is not found, the algorithm will explore the ThaiLegalWordNet to get the synonym or the superclass of the *term_i*. The superclass concept is the general meaning of the *term_i* that will be used to join its concept to the ontology. In the case of new node is added to the ontology only one node per document, our algorithm consider for five times and stop to expanded ontology suddenly (see Table 2).

Table 2. Ontology expansion algorithm

```
Algorithm: OntologyExpansion
Input: LegalDocRanked ←{Doc₁, Doc₂,…, Docₙ}, Ontology,
       Threshold ←Max, status
Begin
If status<6 then
   For each Docₖ ∈ RankedDoc
     Term ← WordSegmentation(Docₖ)
     For each termᵢ ∈ Term
       OntologyTraversal(termᵢ)
       If notFound(termᵢ)
         Superclassᵢ ← Explore(ThaiLegalWordNet(termᵢ))
         OntologyTraversal(Superclassᵢ)
         If Found(Superclassᵢ)
           JoinConcept(termᵢ,Ontology)
             If CountNewNode(JoinConcept)=1
               status=status+1
     CountDoc++
     If CountDoc > Threshold
       Return Ontology
End.
Output: OntologyUpdated
```

For example, the term 'F' cannot be found in the ontology. Therefore, the algorithm explores the ThaiLegalWordNet and found that the superclass of 'F' is 'B'. Therefore, node 'F' is connected to node 'B' (see Fig. 3).

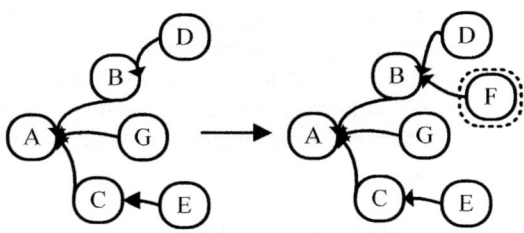

Fig. 3. An example ontology

In case that, the superclass of the $term_i$ is not found, the $term_i$ is discarded. The ontology expansion process convergences when there is only one new node found in the retrieved document.

3.3 The Seed Ontology

We store the initial ontology structure in XML format since it supports the traversal process (see Fig.4).

```
<?xml version="1.0" encoding="windows-874"?>
<chunks>  <chunk>
    <class><term>มรดก (succession)</term>
      <ph>10</ph>
      <id>60152</id></class>
    <subclass><term>ทายาท (heir)</term>
      <ph>10</ph>
      <id>60068</id></subclass>
    <subclass><term>ทรัพย์ (property)</term>
      <ph>10</ph>
      <id>60068</id></subclass>
</chunk>  </chunks>
```

Fig. 4. Illustrates an example of succession ontology in the XML file

There are two seed ontologies which are succession law and family law ontology. The initial succession law ontology consists of 19 nodes and 18 links (see Fig.5). In addition, the initial family law ontology have 7 nodes and 6 links (see Fig.6).

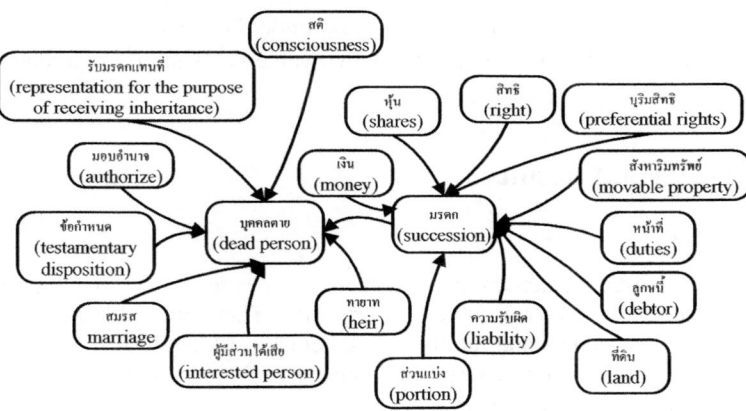

Fig. 5. The seed of succession law ontology

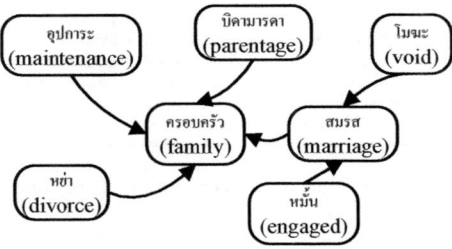

Fig. 6. The seed of family law ontology

3.4   The Structure of ThaiLegalWordNet Dictionary

ThaiLegalWordNet Dictionary covers 2 main Thai Civil Law: succession and family law. There are 282 main concepts obtained from law experts.

Table 3. Nodes of the seed ontology and meaning

Concept_Name	Concept _ID	Concept _Val	Sup_ID
มรดก (succession)	60152	10	60000
ทายาท (heir)	60068	10	60152
ทรัพย์ (property)	60065	10	60152
บิดา (father)	60087	10	60068
บุตร (child)	60093	10	60068
มารดา (mother)	60155	10	60068
...	...	...	...

Consider Table 3, Concept _Name is the name of term, Concept_ID is the concept identifier of each term, Concept_Val is the pheromone value of each term and Sup_ID is the concept identifier of superclass node. For example, the concept ID of property (ทรัพย์) is 60065. Its initial pheromone value is 10 and its superclass is 60152 (succession).

4   Experimental Results

4.1   Data Set

The Supreme Court sentences repository was collected from the Supreme Court search engine web site (www.supremecourt.or.th). There are 1,303 sentences related to the Thai succession law and family law (article No. 1435-1755).

A word segmentation procedure is applied for each court sentence (see Fig.7). We obtain 126,996 terms in the corpus.

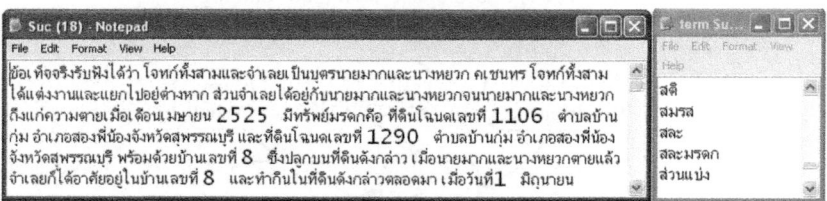

Fig. 7. An Example of Supreme Court Sentences and the words obtained from word segmentation procedure

4.2 Results

In order to simulate the concept of ant colony in the ontology expansion algorithm, a set of keywords are given to 26 law experts. Each expert randomly picks up the keywords and retrieves the set of court sentences using ThaiLegalDocRetrieval algorithm. The pheromone updating process is performed according to the chosen words obtained by each expert. Note that each expert can arbitrarily select the words those are the result of the ontology traversal module. The ontology expansion process is activated when the ThaiLegalDocRetrieval algorithm is finished.

We setup 2 experiments to investigate the contribution of ant colony in ontology expansion module. We found that the pheromone updating process has effect on ontology expansion efficiency in terms of convergence speed and total processing time.

$$\text{precision} = \frac{\text{number of relevant documents retrieved by algorithm}}{\text{number of retrieved documents}} \tag{4}$$

In order to evaluate the performance of the extended ontologies, we measure the precision value obtained from the LegalDocRanked module using equation (4).

The performance measured on Thai succession Law data set is shown in Table 4. From the table, weighted ontology is the output of TLOE which uses Ant Colony's strategy. Non-weighted ontology is obtained from the baseline algorithm without Ant Colony (there is no pheromone updating). The experiment shows that TLOE outperforms the baseline algorithm in term of precision value. The average retrieval performance of TLOE is 89% which is higher than that of the baseline algorithm. Note that the ontology obtained from TLOE is shown in Table 5.

Table 4. The precision values measured on documents retrieval process

Keyword	Precision	
	Non-weighted ontology	TLOE algorithm
มรดก (succession)	0.67	0.82
ทายาท (heir)	0.62	0.87
ผู้จัดการมรดก (administrator of an estate)	0.70	0.90
พินัยกรรม (wills)	0.76	0.90
มิให้รับมรดก (exclusion for the succession)	0.74	0.92
รับมรดกแทนที่ (representation for the purpose of receiving inheritance)	0.79	0.91
สมรส (marriage)	0.71	0.89
ส่วนแบ่ง (portion)	0.73	0.88
Average	0.72	0.89

Table 5. The extended nodes and meaning of succession ontology

Keyword	Extended nodes
มรดก (succession)	ความรับผิด (liability) เงิน (money) ทายาท (heir) ที่ดิน (land) บุริมสิทธิ (preferential rights) ลูกหนี้ (debtor) ส่วนแบ่ง (portion) สังหาริมทรัพย์ (movable property) สิทธิ (right) หน้าที่ (duties) หุ้น (shares) ทรัพย์ (property)
ทายาท (heir)	ตา (grandfather) น้า (uncle) ผู้จัดการมรดก (administrator of an estate) ทรัพย์ (property) น้อง (young brother or young sister) บุตร (child) พี่ (brother or sister) ย่า (grandmother) อา (uncle) บิดา (father) บิดามารดาเดียวกัน (parents) ลำดับ (order) มารดา (mother) มิให้รับมรดก (exclusion for the succession) ยาย (grandmother) ครอบครอง (possess) วิกลจริต (insane person) ผู้รับพินัยกรรม (legatee) สละมรดก (renunciation of an estate) ผู้เยาว์ (a minor) ชำระหนี้ (performance of the obligation) ความประพฤติ (behavior) ผู้สืบสันดาน (descendants)
ผู้จัดการมรดก (administrator of an estate)	ทรัพย์ (property) พินัยกรรม (wills) ศาล (court) บัญชี (account) จำหน่าย (dispose of) อำนาจ (authorize) ตัวแทน (agent) ผู้รับพินัยกรรม (legatee) ลาออก (resign) เสียงข้างมาก (majority of votes) คนล้มละลาย (bankrupt person) ไร้ความสามารถ (incompetence) วิกลจริต (insane person) นิติกรรม (juristic act) เจ้าหนี้ (creditor) ทุจริต (bad faith) ประมาท (negligently) พนักงานอัยการ (public prosecutor) มูลนิธิ (foundation) ความประพฤติ (behavior) บำเหน็จ (remuneration) บรรลุนิติภาวะ (sui juris)
พินัยกรรม (wills)	ทำพินัยกรรม (make wills) พยาน (witness) ทรัพย์ (property) ผู้จัดการมรดก (administrator of an estate) ลงลายมือชื่อ (sign) ผู้เขียน (writer) อำเภอ (district) ลบ (erasure) ผู้รับพินัยกรรม (legatee) มูลนิธิ (foundation) โมฆะ (void) ข้อกำหนด (testamentary disposition) รายงาน (report) หนังสือ (written evidence) เติม (add) แต่ง (change) วาจา (oral) แก้ไข (modification) ผู้สืบสันดาน (descendants) ขีด (cancelled) เอกสารฝ่ายเมือง (public document) โอนเปลี่ยนแปลง (transfer) เอกสารเขียน (written document)
มิให้รับมรดก (exclusion for the succession)	ปิดบัง (conceal) ยักย้าย (divert) ฉ้อฉล (fraudulent) พินัยกรรม (wills) ผู้รับพินัยกรรม (legatee) ความประพฤติ (behavior) พนักงานเจ้าหน้าที่ (competent official)
รับมรดกแทนที่ (representation for the purpose of receiving inheritance)	บุตร (child) ผู้สืบสันดาน (descendants) บิดา (father) มารดา (mother) ผู้ตาย (dead person) ทรัพย์ (property)
สมรส (marriage)	สินสมรส (marriage portion) ภริยา (wife) สามี (husband) สินเดิม (the separate property)
ส่วนแบ่ง (portion)	ทรัพย์ (property) ทายาท (heir) มรดก (succession) สมรส (marriage) ฟ้อง (prosecute) ภริยา (wife) สามี (husband)

Considering the experiments done on Family Law dataset (see Table 6, 7), we found that the average retrieval performance of TLOE is 86% which is higher than that of the non-weighted ontology obtained from the baseline algorithm.

Table 6. The precision values measured on documents retrieval process

Keyword	Precision	
	Non-weighted ontology	TLOE algorithm
ครอบครัว (family)	0.62	0.87
อุปการะ (maintenance)	0.70	0.90
บิดามารดา (parentage)	0.76	0.90
หย่า (divorce)	0.74	0.92
สมรส (marriage)	0.79	0.91
หมั้น (engaged)	0.71	0.89
โมฆะ (void)	0.73	0.88
Average	0.69	0.86

Table 7. The extended nodes and meaning of family ontology

Keyword	Extended nodes
ครอบครัว (family)	สมรส (marriage) หย่า (divorce) บิดามารดา (parentage) อุปการะ (maintenance)
อุปการะ (maintenance)	สามี (husband) ภริยา (wife) บิดา (father) มารดา (mother) บุตร (child)
บิดามารดา (parentage)	บุตร (child) มรดก (succession) สิทธิ (right) หน้าที่ (duties) ปกครอง (guardianship) บุตรบุญธรรม (adoption) จดทะเบียน (entry in the register)
หย่า (divorce)	ยินยอม (consent) จดทะเบียน (entry in the register) ชู้ (adulterer) ชั่ว (badly) ทำร้าย (harm) หมิ่นประมาท (insult) เหยียดหยาม (look down upon) จำคุก (imprison) แยกกันอยู่ (be separated) อุปการะ (maintenance) แบ่ง (partition of the property) ทรัพย์ (property) ละทิ้ง (abandon) ค่าทดแทน (compensate)
สมรส (marriage)	หมั้น (engaged) สินสอด (bride-price) อายุ (age) จดทะเบียน (entry in the register) ยินยอม (consent) ความสัมพันธ์ (relation) ทรัพย์ (property) หนี้ (debt) โมฆะ (void) สิ้นสุด (end)
หมั้น (engaged)	โมฆะ (void) อายุ (age) ยินยอม (consent) โมฆียะ (voidable) ของหมั้น (marriage portion) สัญญา (contract)
โมฆะ (void)	ญาติ (relatives) มีคู่ (partner) ยินยอม (consent) มรดก (succession) ค่าทดแทน (compensate) อุปการะ (maintenance)

The objective of TLOE is the build an ontology by extending the nodes from a seed ontology created by law expert. Therefore it is an important issue to discuss about the quality of the extended ontology. We set up another experiment to investigate the retrieval performance of the extended ontology obtained from TLOE using the convergence criteria as the stop condition for ontology expansion module (see Table 2). TLOE finishes the ontology expansion task when only one new node is obtained from the retrieved document. Note that the numbers of new nodes obtained from the retrieved document are decreased until the last retrieved document. It infers that there is no new concept found in the retrieved document after the size of ontology is increased to a certain number of nodes. To prove our assumption, we compare the retrieval performance between 2 ontologies, (the TLOE ontology with and without convergence). For the no convergence option, TLOE will extend the node continuously until no document left in the corpus.

From Table 8 and 9, we found that the performance of TLOE using convergence option is comparable. TLOE with convergence option gets the same performance as TLOE without convergence option in both data sets. Figure 8 illustrates the example of the extended family law ontology.

Table 8. The appropriate number of extended nodes of Succession Law ontology

Keyword	TLOE with convergence		TLOE without convergence	
	Number of Nodes	Precision	Number of Nodes	Precision
มรดก (succession)	100	0.82	121	0.82
ทายาท (heir)	111	0.87	122	0.87
ผู้จัดการมรดก (administrator of an estate)	71	0.85	115	0.90
พินัยกรรม (wills)	95	0.90	120	0.90
มิให้รับมรดก (exclusion for the succession)	110	0.92	121	0.92
รับมรดกแทนที่ (representation for the purpose of receiving inheritance)	104	0.91	121	0.91
สมรส (marriage)	98	0.89	108	0.89
ส่วนแบ่ง (portion)	92	0.88	103	0.88
Average	98	0.88	116	0.89

Table 9. The appropriate number of extended nodes of Family Law ontology

Keyword	TLOE with convergence		TLOE without convergence	
	Number of Nodes	Precision	Number of Nodes	Precision
ครอบครัว (family)	47	0.87	54	0.87
อุปการะ (maintain)	42	0.90	57	0.90
บิดามารดา (parentage)	45	0.90	55	0.90
หย่า (divorce)	55	0.90	59	0.92
สมรส (marriage)	50	0.91	57	0.91
หมั้น (engaged)	44	0.89	53	0.89
โมฆะ (void)	40	0.86	54	0.88
Average	46	0.89	56	0.90

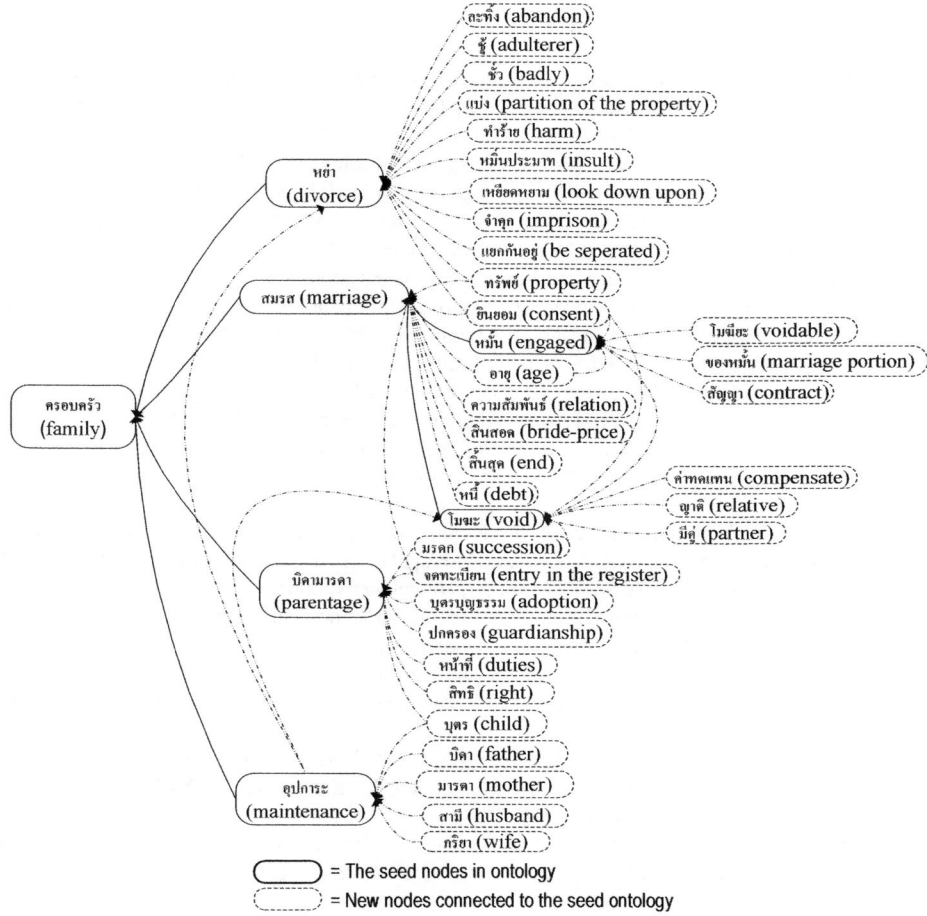

Fig. 8. An example of extended family law ontology

5   Conclusions

This research proposes two algorithms, the Supreme Court retrieval algorithm and the ontology expansion algorithm. We use ant colony algorithm as a mechanism in order to increase or decrease weight in term of ant pheromone and ant behaviours.

The pheromone level of each word is increased when law users select that word to expand the query. Moreover, the pheromone level effects the retrieval performance compare to the non-weighted ontology. In the near future, we plan to extend our work to other areas of Thai law.

Acknowledgments

This work has been supported by Office of the Higher Education Commission, Thailand.

References

1. Perez, A.G., Rodriguez, F.O., Terrazas, B.V.: Legal Ontologies for the Spanish e-Government. In: Marín, R., Onaindía, E., Bugarín, A., Santos, J. (eds.) CAEPIA 2005. LNCS (LNAI), vol. 4177, pp. 301–310. Springer, Heidelberg (2006)
2. Henze, N., Dolog, P., Nejdl, W.: Reasoning and Ontologies for Personalized E-learning in the Semantic Web. Educational Technology & Society 7(4), 82–98 (2004)
3. Benjamins, V.R., Contreras, J., Casanovas, P., Ayuso, M., Becue, M., Lemus, L., Urios, C.,: Ontologies of Profesional Legal Knowledge as the Basic for Intelligent IT Support for Judges. Artificial Intelligence and Law 12, 359–378 (2006)
4. Abachizadeh, M., Tahani, M.: An Ant Colony Optimization Approach to Multi-objective Optimal Design of Symmetric Hybrid Laminates for Maximum Fundamental Frequency and Minimum Cost. Springer Struct. Multidisc. Optim. 37, 367–376 (2009)
5. Corcho, O., Lopez, M.F., Perez, A.G., Cima, A.L.: Building Legal Ontologies with Methontology and WebODE. Law and the Semantic Web, 142–157 (2005)
6. Kayed, A.: Building e-Laws Ontology: New Approach. In: Meersman, R., Tari, Z., Herrero, P. (eds.) OTM-WS 2005. LNCS, vol. 3762, pp. 826–835. Springer, Heidelberg (2005)
7. Soonhee, H., Youngim, J., Aesun, Y., Kwon, H.C.: Building Korean Classifier Ontology Based on Korean WordNet. In: Sojka, P., Kopeček, I., Pala, K. (eds.) TSD 2006. LNCS (LNAI), vol. 4188, pp. 261–268. Springer, Heidelberg (2006)
8. Kurematsu, M., Yamaguch, T.: A Legal Ontology Refinement Support Environment Using a Machine-Readable Dictionary. Artificial Intelligence and Law 5, 119–137 (1997)
9. Zhang, X., Xia, G.: A Methodology for Domain Ontology Construction Based on Chinese Technology Documents. Research and Prcatice Issue of Enterprise Information Systems II 2, 1301–1310 (2007)
10. Hu, H., Du, X.: Building Bilingual Ontology from WordNet and Chinese Classified Thesaurus. In: Zhang, Z., Siekmann, J.H. (eds.) KSEM 2007. LNCS (LNAI), vol. 4798, pp. 649–654. Springer, Heidelberg (2007)
11. Despres, S., Szulman, S.: TERMINAE Method and Integration Process for Legal Ontology Building. In: Ali, M., Dapoigny, R. (eds.) IEA/AIE 2006. LNCS (LNAI), vol. 4031, pp. 1014–1023. Springer, Heidelberg (2006)
12. Boonchom, V., Soonthornphisaj, N.: Thai Succession Law Ontology Building Using Ant Colony Algorithm. In: Proceeding of the Third International Workshop on Jurisinformatics (JURISIN), pp. 27–37. Campus Innovation Center, Tokyo (2009)

Reflective Visualization of the Agreement Quality in Mediation

Yoshiharu Maeno[1], Katsumi Nitta[2], and Yukio Ohsawa[3]

[1] Social Design Group, Bunkyo-ku, Tokyo 112-0011, Japan
maeno.yoshiharu@socialdesigngroup.com
[2] Tokyo Institute of Technology, Yokohama-shi, Kanagawa 226-8502, Japan
[3] The University of Tokyo, Bunkyo-ku, Tokyo 113-8563, Japan

Abstract. Training for mediators is a complex issue. It is generally effective for trainees to reflect on their past thinking, speaking, and acting. We present a text processing method which aids mediation trainees in reflecting on how they reached an agreement from their dialogue. The method is an improved variant of the Data Crystallization algorithm, which visualizes the inter-topic associations which foreshadow the intentional or unintentional subsequent development of topics far apart in time. We demonstrate how the dialogues which differ in the agreement quality affects the topological characteristics of the associations.

1   Introduction and Background

Resolving a conflict between parties having opposing opinions is an important social requirement. Mediation is a form of alternative dispute resolution, which refers to a rather private and confidential extrajudicial process. Mediation aims at assisting disputants in reaching an agreement on a disputed matter. Companies often hire mediators in an attempt to resolve a dispute with workers' unions. Mediation is different from arbitration where an arbitrator imposes a solution on the disputants. Rather, a mediator uses appropriate skills to improve the dialogue between the disputants and find solution.

Information technologies are applied to assist mediators and disputants in reaching a good agreement. For example, software agents are used in many related works in analyzing and aiding the mediation process. A software agent [1] is a piece of software that acts on behalf of a user. An intelligent software agent is capable of adapting to the environment by choosing problem solving rules, and of learning by trial and error, or by generalizing the given examples. The software agent analyzes the mediation and aids mediators or disputants in making decisions. Case based reasoning is a powerful technique for the software agents. This technique is the process of solving new problems based on the solutions of similar past problems. The process formalizes four steps: retrieving the relevant cases from a knowledge base, reusing the retrieved case to a new problem, revising the retrieved solution to a new situation, and retaining the new problem and its solution to the knowledge base.

K. Nakakoji, Y. Murakami, and E. McCready (Eds.): JSAI-isAI, LNAI 6284, pp. 33–44, 2010.

Besides, logic programming and ontology are frequently used as techniques to implement the case based reasoning. Logic programming [16] is the use of logic as a language for declarative and procedural representation. Prolog remains one of the most commonly used logic programming languages today. It has been applied to the fields of theorem proving, expert systems, games, automated answering systems, and ontology. Building ontology is an essential task in analyzing the knowledge base. Ontology refers to a formal representation of a shared conceptualization of a particular domain. It includes a set of individuals which are the basic objects, classes which are the collections of things, attributes which describe the aspects of the individuals and classes, and relations in which the individuals and classes can be related to one another.

On the other hand, education and training for mediator trainees (improvement of mediator trainees' human skills) become a complex issue because the mediator's skills range widely from the ability to remain neutral, the ability to move the disputants from the impasse points, to the ability to evaluate the strength and weakness of the disputants correctly. Appropriate means are, therefore, necessary to education and training. The idea of reflection can be a clue in the situation when we need to improve a skill which can not be defined clearly and taught by trainers.

Reflection in cognitive science [17] and computer-mediated communication [20] means the ability to recognize and understand oneself, discover something unexpected, and create something new [7], [19], [18]. *Particularly, visualization of the past utterances, decision-making, and actions is one of the most practical tools to aid the trainees in reflection.* Reflective visualization and verbalization are proven effective in helping a person become aware of his or her unconscious preferences [9], [6]. We expect that such reflective visualization is also promising in education and training for mediation trainees. Utterances are relevant and convenient information records for the trainees to reflect on. They are essential inputs to negotiation log analysis [13] and online agent based negotiation assistant systems [21], [22]. Similarly, mediators and disputants can reflect on the quality of the agreement they made by looking back the way how the dispute was resolved in a dialogue.

In this paper, we explore a text processing method for reflective visualization and apply it to a mediation case. It is an improved variant of the Data Crystallization algorithm [15] in which a graph-structured diagram evolves to explore unknown structures with the introduction of dummy variables. The Data Crystallization algorithm has also been studied in [10], [11], and [14]. The method derives temporal topic clusters and inter-topic associations from the recorded utterance texts, and draws the clusters and associations on a graph-structured diagram. *The inter-topic associations foreshadow the intentional or unintentional subsequent development of topics far apart in time.* Two dialogue examples in mediating a dispute on cancelling a purchase transaction at an online auction site demonstrate how the difference between the agreement quality affects the topological characteristics of the associations.

2   Method

2.1   Dialogue

The dialogue d is a time sequence of the recorded utterance texts u_t from a mediator and disputants. It is represented by eq.(1) formally. The subscript t means the time when the utterance is observed. We do not use the absolute time from the beginning of mediation. Instead, the i-th utterance from the beginning is associated with an integer time $t = i$ approximately. In eq.(1), T is the number of utterances in mediation.

$$d = (u_0, \ldots, u_t, \ldots, u_{T-1}). \tag{1}$$

A recorded utterance text is a set of words w_i which appear in the sentences in an utterance. It is in the form of eq.(2). The number of words in an utterance text u_t is $|u_t|$.

$$u_t = \{w_i\} \ (0 \leq i < |u_t|). \tag{2}$$

The utterances are analyzed morphologically while assembling a dialogue. Morphology is the identification, analysis and description of structure of words. Verbs are changed into un-conjugated forms. Nouns are changed into un-inflected forms. Besides, irrelevant words are deleted. They are articles, prepositions, pronouns, and conjunctions. Periods are not words. For example, the first utterance of a mediator, *Thank you for agreeing in attempting to solve the dispute by mediation. Are you ready for starting mediation?*, becomes $u_0 = \{$agree, attempt, be, dispute, mediation, solve, start, thank, ready$\}$. A word may appear in many utterance texts. On the other hand, a word which appears multiple times in an utterance appears only once in the set of words in eq.(2).

2.2   Graphical Diagram

A graph-structured diagram [15], [9] is employed here to represent the dialog d visually. Two characteristic structures are extracted from the time sequence pattern of word appearance in d. The first structure is a temporal topic cluster. It is a group of words whose time sequence pattern of appearance is similar. The cluster is drawn as a sub-graph including nodes representing words and links representing strong similarity between words. The second structure is an inter-topic association. The ability to extract the inter-topic association is the strength of our proposed method described in 2.3 and 2.4. The inter-topic association corresponds to an utterance which can be a trigger to move from a temporal topic cluster to another. It does not necessarily mean a temporally adjacent relationship between 2 clusters. Rather, it may foreshadow the intentional or unintentional subsequent development of topics indicated by clusters which are far apart in time. The inter-topic association is drawn as a set of links between multiple temporal topic clusters. The set of links has a label pointing to a trigger utterance.

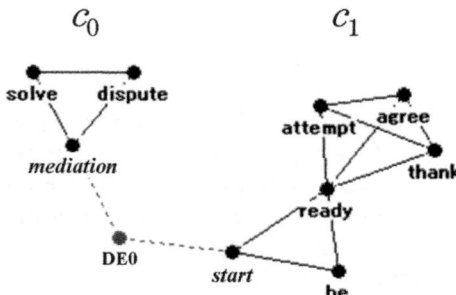

Fig. 1. Example of a graph-structured diagram which visualizes temporal topic clusters (c_0 and c_1), and inter-topic associations (DE0) found in the recorded utterance texts in a dialog. The nodes *mediation* and *start* are gateway words of the clusters.

Fig. 1 shows an example of a graph-structured diagram which represents temporal topic clusters and inter-topic associations in the recorded utterance texts in a dialog. The cluster c_0 includes 3 words and c_1 includes 6 words. The association is a link between c_0 and c_1 which is labeled as DE0 pointing to u_0. The gateway word is the word in a cluster to which the link representing an inter-topic association is connected.

The gateway word has strong associations with the words in other clusters as well as the words in the cluster to which it belongs. It is interpreted as a trigger (or a switch) which plays a role to ignite the subsequent development from a cluster to another. For example, the word *mediation* is the gateway word to c_0 for DE0 (u_0).

2.3   Temporal Topic Cluster

Every word w_i which appears in $\boldsymbol{d}$ is classified into temporal topic clusters. The number of clusters C is a granularity parameter which can be adjusted so that visualization can assist mediation trainees in reflecting their dialogues most effectively. As the granularity becomes finer, C increases, that is, the number of words in a cluster decreases and the time difference between neighbor clusters decreases. As the granularity becomes coarser, C decreases and the time difference increases. At the beginning of reflection, if the trainees want to grasp the rough sketch of the subsequent development of topics, coarser granularity visualization may be appropriate. Large number of clusters is not necessary for this purpose. After that, if they want to detail the subsequent development between the topics of particular interests, finer granularity visualization may be appropriate.

At first, a simple measure is introduced to characterize the time sequence pattern of word appearance. The characteristic time of a word is defined by eq.(3). It is the average of the time when a word appears.

$$a(w_i) = \frac{\sum_{t=0}^{T-1} t B(w_i \in u_t)}{\sum_{t=0}^{T-1} B(w_i \in u_t)}. \tag{3}$$

The function $B(s)$ is define by eq.(4).

$$B(s) = \begin{cases} 1 & \text{if the statement } s \text{ is true} \\ 0 & \text{otherwise (false)} \end{cases}. \tag{4}$$

The similarity between 2 words is defined by eq.(5). The function min is included to avoid divergence of the similarity when the characteristic time is very close. Two words appear closely in time if the similarity is large. Eq.(5) measures the degree of similarity in temporal appearance pattern while the Jaccard coefficient used in text analysis [8] measures the degree of co-occurrence.

$$I(w_i, w_j) = \min\left(\frac{1}{|a(w_i) - a(w_j)|}, 1\right). \tag{5}$$

Then, a clustering algorithm for discrete objects is applied for given C. The k-medoids algorithm is a simple example [5]. A medoid is an object that is the closest to the center of gravity in a cluster. Its principle is similar to that of the k-means algorithm [4] for continuous numerical variables where the center of gravity is updated repeatedly according to the expectation-maximization method [3]. The distance between words is evaluated by the similarity in eq.(5). Initially, the words are classified into clusters at random in the k-medoids algorithm. The cluster into which a word w_j is classified is denoted by $c(w_j)$. It is given by eq.(6).

$$c(w_j) = \text{random interger} \in [0, C - 1]. \tag{6}$$

The medoid $w_{\text{med}}(c_k)$ of a cluster c_k is also assigned at random. It is given by eq.(7).

$$w_{\text{med}}(c_k) = \text{random word} \in c_k \ (0 \leq k < C). \tag{7}$$

The clusters into which words are classified and the medoids are updated repeatedly. The cluster into which a word is classified is updated according to eq.(8). The operator arg in eq.(8) means that $c(w_j)$ is the cluster which gives the largest $I(w_{\text{med}}(c_k), w_j)$. of all the clusters.

$$c(w_j) = \arg \max_{c_k} I(w_{\text{med}}(c_k), w_j). \tag{8}$$

The medoid is updated according to eq.(9). The operator arg in eq.(9) means that the medoid is the word w_j classified into c_k, which maximizes $M(c_k, w_j)$.

$$w_{\text{med}}(c_k) = \arg \max_{w_j \in c_k} M(c_k, w_j) \ (0 \leq k < C). \tag{9}$$

The quantity $M(c_k, w_j)$ in eq.(9) is given by eq.(10). The operator $\wedge$ means logical AND.

$$M(c_k, w_j) = \sum_{w_l \in c_k \wedge w_l \neq w_j} I(w_l, w_j). \tag{10}$$

After the medoids are determined, the clusters into which words are classified are updated according to eq.(8) again. Eq.(8), (9), and (10) are calculated repeatedly until their value converges. The characteristic time of a cluster is defined by eq.(11). The time when a topic cluster c_k appears is evaluated by the time when its medoid word appear approximately.

$$a(c_k) = a(w_{\mathrm{med}}(c_k)) \tag{11}$$

2.4   Inter-topic Association

After extracting temporal topic clusters, every utterance is next assigned a score which measures the degree of being an inter-topic association [15], [10], [9]. The score $s(u_t)$ of the utterance u_t is calculated by eq.(12).

$$s(u_t) = \max_{w_i \in u_t} \sum_{c_k} \max_{w_j \in c_k \wedge w_j \neq w_i} I(w_i, w_j). \tag{12}$$

The utterances which are assigned large value of the score are extracted to draw on a diagram. The utterance which has the l-th largest score is given by eq.(13).

$$U(\boldsymbol{d}, l) = \arg \max_{u_i \neq U(\boldsymbol{d},m) \text{ for } \forall m < l} s(u_i). \tag{13}$$

A gateway word of a cluster is selected when $U(\boldsymbol{d}, l)$ is drawn as a link between clusters on a graph. It is given by eq.(14). The operator arg means that the gateway word $w_{\mathrm{gtw}}(l, c_k)$ of a cluster c_k for the utterance of the l-th largest score is the word $w_j \in c_k$ which maximizes $I(w_i, w_j)$.

$$w_{\mathrm{gtw}}(l, c_k) = \arg \max_{w_j \in c_k w_i \in U(\boldsymbol{d},l)} I(w_i, w_j). \tag{14}$$

A set of links are drawn between the gateway words $\{w_{\mathrm{gtw}}(1, c_k)\}$ $(0 \leq k < C)$ for the utterance assigned the largest value of the score $(u_t = U(\boldsymbol{d}, 1))$ on a diagram. The label DEt indicating u_t is attached to the links. Similarly, a set of links are drawn between the gateway words $\{w_{\mathrm{gtw}}(l, c_k)\}$ $(0 \leq k < C)$ for the utterance assigned the l-th largest value of the score $(u_t = U(\boldsymbol{d}, l))$.

3   Extended Example

3.1   Mediation Case

The method described in 2 is applied to dialogues recorded in a mediation training program. The disputed matter in the program is on a purchase transaction at an online auction site. Two groups of three mediation trainees played mediator and disputant roles. Their utterances until the dispute is resolved were recorded and assembled into two dialogues.

Disputed Matter. The disputed matter is on cancelling a purchase transaction between two persons (seller disputant and buyer disputant) at an online auction site. A seller disputant offered a car muffler for bid at the auction site. The seller disputant provided bidders with photographs of the muffler and showed them its condition and vendor information. A buyer disputant won the bid 7 days later. The buyer disputant paid for 20,000 yens two days later, and the seller disputant sent the muffler to the buyer disputant. The transaction at the auction site completed.

The disputed matter consists of many sub-matters. After two and half months, the buyer disputant asked the seller disputant whether the muffler is made of stainless steel or aluminum-plated steel. The seller disputant answered that the muffler is made of aluminum-plated steel as the photographs at the auction site had indicated. But, all the mufflers found at the muffler vender's web catalogue were made of stainless steel at the time of bidding. A muffler made of stainless steel is expensive, but excellent in quality. The muffler vender's hallmark can not be found on the muffler which the seller disputant sent to the buyer disputant. The muffler vender used to place the vender's hallmark on the products at the time of bidding. The buyer disputant became disappointed at this. The buyer disputant requested that the purchase transaction be cancelled and the paid money be returned. The seller disputant rejected the buyer disputant's request. The buyer disputant assigned a low rating score to the seller disputant at the auction site. The seller disputant did similarly in return. These low rating scores had the effect of making them untrustworthy at the auction site.

Resolution. The buyer disputant asked the seller disputant to attempt to resolve their dispute by mediation. The seller disputant agreed. With the aid of a mediator, the disputants talked about the undisclosed facts on each side in both of the two dialogues.

The seller disputant had bought the muffler at the same auction site before. The seller disputant was not aware that the muffler vender did not supply mufflers made of aluminum-plated steel at the time of bidding. The seller disputant investigated the muffler after the buyer requested cancellation. The seller disputant found that it was a custom-made muffler from the vendor. The seller disputant was embarrassed by the low rating score, which harmed the seller disputant's business reputation. The seller disputant could not agree on the request that the paid money be returned because the seller disputant happened to have little money at the time of mediation. Instead, the seller disputant had a number of mufflers, which might be used to make an agreement. The buyer disputant had not tried to check the quality of the sent muffler for a long time. The buyer disputant had trouble with a car because of the insufficient quality of the sent muffler, and wanted to settle the trouble by all means.

Finally, they reached an agreement although the buyer disputant's original request on returning the paid money did not survive the mediation. In the two dialogues, however, the two groups of the trainees reached different agreements. In dialogue 1, the agreement was to substitute the muffler for one made of stainless steel which the seller disputant possesses and delete both of the low rating scores at the online auction site. In dialogue 2, the agreement was to substitute the disputed muffler for one that cost 70 % as much. They discussed about the low rating scores at the online auction site but did not make an agreement about them.

In both dialogues, the mediators succeeded in assisting the disputants in reaching an agreement on a disputed matter. The mediator in dialogue 2, however, failed to make an agreement on the disputed sub-matter of the low rating scores. It was not beneficial to both of the disputants. The agreement in dialogue 1 looks better than that in dialogue 2 in that it incorporates beneficial compromises on most of the disputed sub-matters. How is this intuitive interpretation of the agreement quality visualized in terms of the difference between the diagrams? It is demonstrated next.

3.2   Visualization

Figure 2 shows the diagram drawn by the method from the dialogue 1. The number of temporal topic clusters is $C = 12$. They are placed clockwise from c_0 to c_{11}. The black nodes in the clusters are labeled with Japanese words which they represent. The inter-topic associations found within the whole $0 \le t < T (= 54)$ utterances are labeled by the red nodes DEt. The number of the associations which have a non-zero score $s(u_t) > 0$ and are drawn on a diagram is 24 $(t = 0, \cdots, 23)$. Figure 3 shows the diagram drawn by the method from the dialogue 2. The number of temporal topic clusters is $C = 12$. They are placed clockwise from c_0 to c_{11}. The black nodes in the clusters are labeled with Japanese words which they represent. The inter-topic associations found within the whole

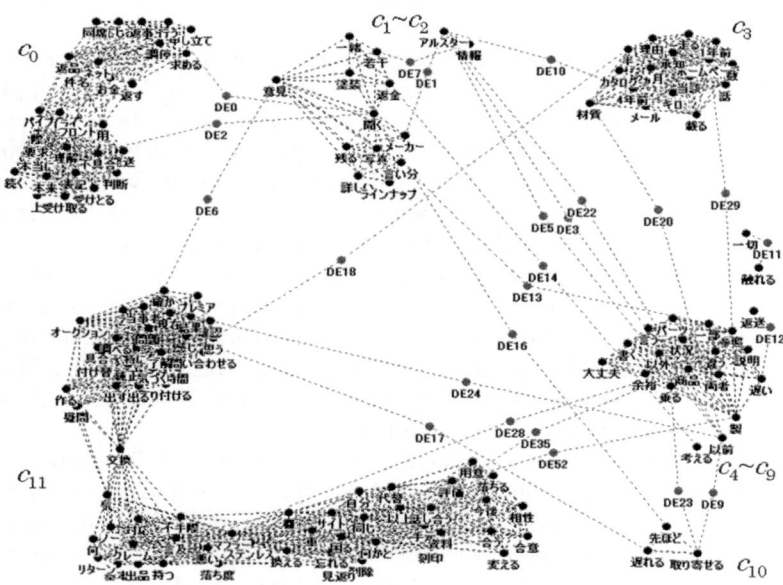

Fig. 2. Diagram drawn by the method from dialogue 1. The number of temporal topic clusters is $C = 12$. The inter-topic associations found within the whole $0 \le t < T (= 54)$ utterances are labelled by the red nodes DEt.

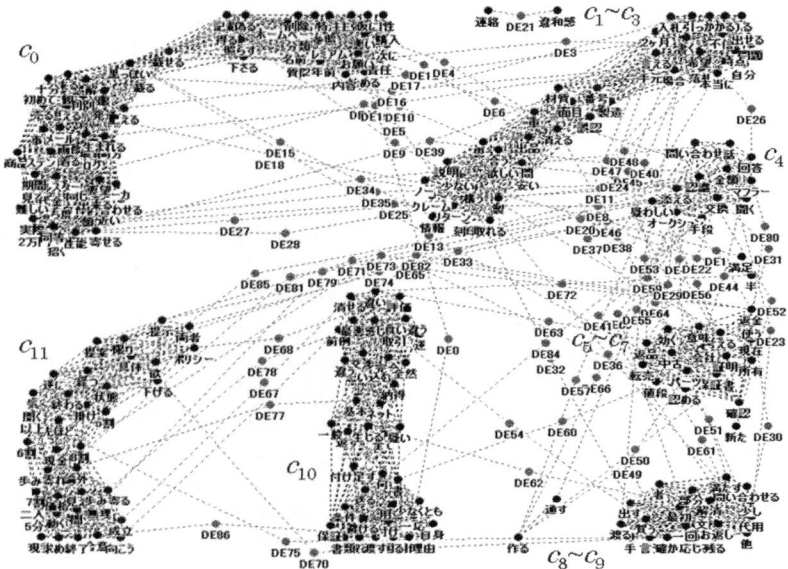

Fig. 3. Diagram drawn by the method from dialogue 2. The number of temporal topic clusters is $C = 12$. The inter-topic associations found within the whole $0 \leq t < T(= 87)$ utterances are labelled by the red nodes DEt.

$0 \leq t < T(= 87)$ utterances are labeled by the red nodes DEt. The number of the associations which have a non-zero score $s(u_t) > 0$ and are drawn on a diagram is 81 $(t = 0, \cdots, 80)$.

There are inter-topic associations between most of the clusters in Figure 2 while there are few inter-topic associations between the early-stage clusters (c_0 to c_4) and the late-stage clusters (c_5 to c_{11}). To analyze quantitatively, let us count the number of the inter-topic associations between individual clusters and the concluding cluster c_{11}. The concluding cluster at the end of the mediation summarizes the content of the agreement. Table 1 shows the result.

Table 1. Number of the inter-topic associations between individual clusters and the concluding cluster c_{11}

	c_0	c_1	c_2	c_3	c_4	c_5	c_6	c_7	c_8	c_9	c_{10}
Dialogue 1	0	1	0	1	0	0	0	2	0	1	0
Dialogue 2	0	0	0	0	6	0	4	0	2	0	5

The Gini coefficient can be calculated from Table 1. The Gini coefficient is a measure of statistical dispersion. It is commonly used in economics as a measure of inequality of income or wealth. The Gini coefficient can range from 0 to 1. A small Gini coefficient indicates a more equal distribution while a large Gini coefficient indicates more unequal distribution. The value of 0 corresponds to perfect equality. The value of 1 corresponds to perfect inequality. The Gini coefficient is 0.56 for the dialogue 1 and 0.65 for the dialogue 2. The inter-relationship between the temporal clusters in dialogue 2 suffers from more inequality. The concluding cluster in dialogue 2 is less related to the sub-matters in the early-stage clusters than that in dialogue 1. This may result in the defect in the agreement quality of the dialogue 2.

This result demonstrates that the agreement quality and the visual structure of the diagrams may be dependent. This implication is relevant because the diagrams provide the mediation trainees with a clue to assess the agreement quality. The number of the associations between the concluding cluster and the early-stage clusters is of particular interest. A list of the problems posed in the dispute and the preliminary opinions on them from the disputants are usually presented in the early-stage clusters. The agreement in the concluding clusters is supposed to solve all the problems in a successful mediation. For this reason, very weak inter-relationship between the early-stage clusters and the concluding clusters can be a sign of bad agreement quality. The method can be an effective means in education and training to visualize such inequality and present it to mediator trainees for reflection on how they reached an agreement and how good or bad the way they discussed and decided was.

4 Discussion

The agreement which the disputant reached in mediation is often different from the best solution from the viewpoints of many similar past mediation cases. It is, therefore, more important to understand the intention in the agreement toward which the disputants are reaching (and possibly foresee and adjust it) than to know the best solution of the disputed matter. Our study focuses on the reflective visualization of mediators' human skills, rather than the information technology based mediation assistance. Analyzing the subsequent development of topics in the dialogue is essential for this purpose. The utterances are noisy and fluctuating information, which are not suitable for machines to understand by means of the knowledge base and ontology (above mentioned purely computational information technologies). Rather, a human-computer interacting process is potentially advantageous in combining text processing methods with experts' opinion and trainees' reflection through visual interfaces. The method in this paper emphasizes the practical usefulness of visualization in strengthening the mediation trainees' intuitive understanding of the vast amount of texts in the dialogue.

Although the topology of the inter-topic associations is a big clue to understand the dialogue, the graph-structured diagrams such as Figure 2 and Figure 3 may be difficult to interpret for those who are not familiar with such mathematical constructs Instead, short easy-to-understand text-based summary information on the quality of the agreement and other characteristics of the utterances is convenient to the trainees. Furthermore, a single digit, score for a mediator and disputants, may be simple but very useful information. With such information, trainees' reflection on their utterances with the aid of the information would remain very relevant and essential to improve the skills of the trainees. The score needs to include a clue to reflect on how and why the disputants have reached better or worse agreements in the dialogue.

The method is more suitable to analyze the Internet based text records than real time face-to-face conversations. This is mainly because speech recognition is not a mature technology. Negotiation between two parties through on-line chats [2] is an attractive field of application. For this purpose, the method needs to be improved to analyze streaming data (accumulating records of utterances) rather than a database (a complete dataset from the beginning to the end of the dialogue). The graph-structured diagrams will have to be updated utterance by utterance. A big change in the diagram during a single utterance affects badly the understanding of a mediator and disputants. They are likely to be confused. Gradual but noticeable change in the diagram would aid the trainees in understanding the effect of an utterance on the subsequent development of topics. Such visualization techniques are within the scope for future works.

5 Conclusion

Training for mediators is a complex issue. We present a text processing method which is a promising tool to address such an issue. The method aids mediation trainees in reflecting on how they reached an agreement from their dialogue. The strength of the method is the ability to visualize the inter-topic associations which foreshadow the intentional or unintentional subsequent development of topics indicated by temporal topics clusters far apart in time. The method is applied to a mediation case where a dispute between a seller and a buyer at an online auction site is resolved. The result demonstrates that the agreement quality and the visual structure of the diagrams which our method outputs may be dependent.

This implication is relevant because the diagrams may provide the mediation trainees with a clue to access the agreement quality from the associations between the concluding cluster and the early-stage clusters. With this in mind, in the future, we may be able to design an effective computer-aided training tool which visualizes mediation trainees' dialogue in real time, aids them in reflecting on how they are making a discussion and how they are reaching an agreement, and strengthens their ability to foresee the conclusion to which they are approaching and set it in the right direction.

References

1. Abbas, S., Sawamura, H.: Mining impact on argument learning environment and arguing agent. In: Proceeding of the International Workshop on Juris-informatics, Tokyo (2009)
2. Andrade, F., Novais, P., Carneiro, D., Zeleznikow, J., Neves, J.: Using BATNAs and WATNAs in online dispute resolution. In: Proceeding of the International Workshop on Juris-informatics, Tokyo (2009)
3. Dempster, A.: Maximum likelihood from incomplete data via the EM algorithm. Journal of the Royal Statistics Society 39, 1–38 (1977)

4. Duda, R.O., Hart, P.E., Stork, D.G.: Pattern classification. Wiley-Interscience, Hoboken (2000)
5. Hastie, T., Tibshirani, R., Friedman, J.: The elements of statistical learning: Data mining, inference, and prediction. Springer, Heidelberg (2001)
6. Ishii, N., Miwa, K.: Interactive processes between mental and external operations in creative activity: A comparison of experts' and novices' performance. In: Proceedings of the Creativity and Cognition Conference, Loughborough (2002)
7. Larkin, J., Simon, H.A.: Why a diagram is (sometimes) worth ten thousand words? Cognitive Science 11, 65–99 (1987)
8. Liben-Nowell, D., Kleinberg, J.: The link prediction problem for social networks. Journal of American Society of Information Science and Technology 58, 1019–1031 (2007)
9. Maeno, Y., Ohsawa, Y.: Reflective visualization and verbalization of unconscious preference. International Journal of Advanced Intelligence Paradigms (2010) (in press)
10. Maeno, Y., Ohsawa, Y.: Human-computer interactive annealing for discovering invisible dark events. IEEE Transactions on Industrial Electronics 54, 1184–1192 (2007)
11. Maeno, Y., Ohsawa, Y.: Analyzing covert social network foundation behind terrorism disaster. International Journal of Services Sciences 2, 125–141 (2009)
12. Maeno, Y.: Node discovery problem for a social network. Connections 29, 62–76 (2009)
13. Miura, T., Katagami, D., Nitta, K.: Analysis of negotiation logs using diagrams. Japanese Society for Information and Systems in Education Research Report 22, 33–38 (2007)
14. Nitta, K., Zeze, K., Miura, T., Katagami, D.: Scenario extraction using word clustering and data crystallization. In: Proceeding of the International Workshop on Juris-informatics, Tokyo (2009)
15. Ohsawa, Y.: Data crystallization: chance discovery extended for dealing with unobservable events. New Mathematics and Natural Computation 1, 373–392 (2005)
16. Sato, K.: A formalization for burden of production in logic programming. In: Proceeding of the International Workshop on Juris-informatics, Tokyo (2009)
17. Schön, D.A.: The reflective practitioner: How professionals think in action. Basic Books, New York (2006)
18. Suwa, M., Tversky, B.: Constructive perception: Metacognitive skill for coordinating perception and conception. In: Proceedings of the Annual Conference of the Cognitive Science Society, Boston (2003)
19. Suwa, M., Tversky, B.: Constructive perception: An expertise to use diagrams for dynamic interactivity. In: Proceedings of the Annual Conference of the Cognitive Science Society, Fairfax (2002)
20. Thurlow, C., Lengel, L., Tomic, A.: Computer mediated communication. Sage Publications Ltd., Thousand Oaks (2004)
21. Yasumura, Y., Oguchi, K., Nitta, K.: Negotiation strategy of agents in the Monopoly game. In: Proceedings of the IEEE International Workshop on Robot and Human Interactive Communication, San Francisco (2003)
22. Yuasa, M., Yasumura, Y., Nitta, K.: A tool for animated agents in network-based negotiation. In: Proceedings of the IEEE International Symposium on Computational Intelligence in Robotics and Automation, Banff (2001)

Implementing Temporal Defeasible Logic
for Modeling Legal Reasoning

Guido Governatori[2], Antonino Rotolo[1], and Rossella Rubino[1]

[1] CIRSFID, University of Bologna, Italy
[2] NICTA, Queensland Research Laboratory, Australia

Abstract. In this paper we briefly present an efficient implementation of temporal defeasible logic, and we argue that it can be used to efficiently capture the the legal concepts of persistence, retroactivity and periodicity. In particular, we illustrate how the system works with a real life example of a regulation.

1 Introduction

Defeasible Logic (DL) is based on a logic programming-like language and, over the years, proved to be a flexible formalism able to capture different facets of non-monotonic reasoning (see [4]). Standard DL has a linear complexity [22] and has also several efficient implementations (e.g., [8,3,6,21]).

DL has been recently extended to capture the temporal aspects of several phenomena, such as legal positions [18], norm modifications (e.g., [16]), and deadlines [13]. The resulting logic, called Temporal Defeasible logic (TDL), has been developed to model the concept of temporal persistence within a non-monotonic setting and, remarkably, it preserves the nice computational properties of standard DL [15]. In addition, this logic distinguishes between permanent and transient (non-permanent) conclusions, which makes the language suitable for applications. In the legal domain, typically we have two types of effects. The first type is where normative effects may persist over time unless some other and subsequent events terminate them (example: "If one causes damage, one has to provide compensation"). For the second type we have that normative effects hold at a specific time on the condition that the antecedent conditions of the rules hold and with a specific temporal relationship between the antecedent of the rule and the effect (example: "If one is in a public building, one is forbidden to smoke", that is, if one is a public building at time t, then at time t one has the prohibition to smoke).

This paper illustrates how [23,24] Java implementation of TDL can be fruitfully applied in the legal domain by discussing the concepts of normative persistence, retroactivity, and periodicity. While the idea of persistence is of paramount importance for modelling the type of normative effects mentioned above, it is still an open question whether we really need to explicitly distinguish between persistent and non-persistent rules, as done in TDL: indeed, it may be the case that persistent effects are simulated by suitable sets of non-persistent rules. Our answer, however, is negative, since the introduction of persistent rules allows for a more efficient computation. As we will see, also the concepts retroactivity and periodicity can be feasibly handled within TDL.

The layout of the paper is as follows. Section 2 briefly presents TDL. Section 3 outlines [23,24]'s Java implementation of TDL. Section 4 tests and validates this implementation

K. Nakakoji, Y. Murakami, and E. McCready (Eds.): JSAI-isAI, LNAI 6284, pp. 45–58, 2010.

against some interesting cases of normative persistence, retroactivity and periodicity: Section 4.1 discusses some aspects of the concept of persistence and its relation with that of retroactivity and shows that our system allows for a very efficient computation of conclusions; Section 4.2 illustrates how to handle the regulation on road traffic restrictions of the Italian town of Piacenza, which specifies normative deadlines and periodical provisions. Some conclusions end the paper.

2   Temporal Defeasible Logic

The language of propositional TDL is based on the concept of *temporal literal*, which is an expression such as l^t (or its negation, $\neg l^t$), where l is a literal and t is an element of a discrete totally ordered set $\mathcal{T}$ of instants of time $\{t_1, t_2, \dots\}$: l^t intuitively means that l holds at time t. Given a temporal literal l the complement $\sim l$ is $\neg p^t$ if $l = p^t$, and p^t if $l = \neg p^t$.

A *rule* is an expression $lbl : A \hookrightarrow^x m$, where lbl is a unique label of the rule, A is a (possibly empty) set of temporal literals, $\hookrightarrow \in \{\rightarrow, \Rightarrow, \rightsquigarrow\}$, m is a temporal literal and x is either π or τ signaling whether we have a *persistent* or *transient* rule. *Strict rules*, marked by the arrow $\rightarrow$, support indisputable conclusions whenever their antecedents, too, are indisputable. *Defeasible rules*, marked by $\Rightarrow$, can be defeated by contrary evidence. *Defeaters*, marked by $\rightsquigarrow$, cannot lead to any conclusion but are used to defeat some defeasible rules by producing evidence to the contrary. A *persistent* rule is a rule whose conclusion holds at all instants of time after the conclusion has been derived, unless interrupting events occur; *transient* rules establish the conclusion only for a specific instant of time. Thus $ex_1 : p^5 \Rightarrow^\pi q^6$ means that if p holds at 5, then q defeasibly holds at time 6 and continues to hold after 6 until some event overrides it. The rule $ex_2 : p^5 \Rightarrow^\tau q^6$ means that, if p holds at 5, then q defeasibly holds at time 6 but we do not know whether it will persist after 6. Note that we assume that defeaters are only transient: if a persistent defeasible conclusion is blocked at time t by a transient defeater, such a conclusion no longer holds after t unless another applicable rule reinstates it. Furthermore, as we will see, according to the proof conditions for TDL defeaters cannot be used to prove positive conclusions directly, thus, in this respect the distinction between transient and persistent defeaters is irrelevant. In addition, due to the skeptical nature of defeasible logic, negative conclusions can be considered persistent by default, in the sense that if no reason to prove a conclusion is given for an instant t, the conclusion is deemed as not provable at that instant. Thus, also in this case the distinction between persistent and transient defeaters is irrelevant. Finally, given the intended interpretation of rules and their applicability conditions, the effects of defeaters are, essentially, those of transient rules.

We use some abbreviations. Given a rule r and a set R of rules, $A(r)$ denotes the antecedent of r while $C(r)$ denotes its consequent; R^π denotes the set of persistent rules in R, and $R[\psi]$ the set of rules with consequent ψ. R_s, R_{sd} and R_{dft} are respectively the sets of strict rules, defeasible rules, and defeaters in R.

There are in TDL three kinds of features: facts, rules, and a superiority relation among rules. Facts are indisputable statements, represented by temporal literals. The superiority relation describes the relative strength of rules, i.e., about which rules can overrule which other rules. A *TDL theory* is a structure $(F, R, \prec)$, where F is a finite set of facts, R is a finite set of rules and $\prec$ is an acyclic binary superiority relation over R.

TDL is based on a constructive inference mechanism based on tagged conclusions. Proof tags indicate the strength and the type of conclusions. The strength depends on whether conclusions are indisputable (the tag is Δ), namely obtained by using facts and strict rules, or they are defeasible (the tag is ∂). The type depends on whether conclusions are obtained by applying a persistent or a transient rule: hence, conclusions are also tagged with π (persistent) or τ (transient).

Provability is based on the concept of a derivation (or proof) in a TDL theory D. Given a TDL theory D, a *proof* P from D is a finite sequence of tagged temporal literals such that: (1) each tag is either $+\Delta^\pi$, $-\Delta^\pi$, $+\partial^\pi$, $-\partial^\pi$, $+\Delta^\tau$, $-\Delta^\tau$, $+\partial^\tau$, or $-\partial^\tau$; (2) the proof conditions *Definite Provability* and *Defeasible Provability* given below are satisfied by the sequence P^1.

The meaning of the proof tags is a follows:

- $+\Delta^\pi p^{t_p}$ (resp. $+\Delta^\tau p^{t_p}$): we have a definite derivation of p holding from time t_p onwards (resp. p holds at t_p);
- $-\Delta^\pi p^{t_p}$ (resp. $-\Delta^\tau p^{t_p}$): we can show that it is not possible to have a definite derivation of p holding from time t_p onwards (resp. p holds at t_p);
- $+\partial^\pi p^{t_p}$ (resp. $+\partial^\tau p^{t_p}$): we have a defeasible derivation of p holding from time t_p onwards (resp. p holds at t_p);
- $-\partial^\pi p^{t_p}$ (resp. $-\partial^\tau p^{t_p}$): we can show that it is not possible to have a defeasible derivation of p holding from time t_p onwards (resp. p holds at t_p).

The inference conditions for $-\Delta$ and $-\partial$ are derived from those for $+\Delta$ and $+\partial$ by applying the Principle of Strong Negation [5]. For space reasons, in what follows we show only the conditions for $+\Delta$ and $+\partial$.

Definite Provability
If $P(n+1) = +\Delta^x p^{t_p}$, then
1) $p^{t_p} \in F$ if $x = \tau$; or
2) $\exists r \in R_s^x[p_p^{t'}]$ such that
$\quad \forall a^{t_a} \in A(r): +\Delta^y a^{t_a} \in P[1..n]$

Defeasible Provability
If $P(n+1) = +\partial^x p^{t_p}$, then
1) $+\Delta^x p^{t_p} \in P[1..n]$ or
2) $-\Delta^x \sim p^{t_p} \in P[1..n]$ and
2.1) $\exists r \in R_{sd}^x[p_p^{t'}]$ such that
$\quad \forall a^{t_a} \in A(r): +\partial^y a^{t_a} \in P[1..n]$, and
2.2) $\forall s \in R^y[\sim p^{t_{\sim p}}]$ either
$\quad$ 2.2.1) $\exists b^{t_b} \in A(s), -\partial^y b^{t_b} \in P[1..n]$ or
$\quad$ 2.2.2) $\exists w \in R^y[p^{t_{\sim p}}]$ such that
$\qquad \forall c^{t_c} \in A(w): +\partial^y c^{t_c} \in P[1..n]$ and
$\qquad s \prec w$

where (for both proof conditions) (a) $y \in \{\pi, \tau\}$; (b) if $x = \pi$, then $t_p' \leq t_{\sim p} \leq t_p$; (c) if $x = \tau$, then $t_p' = t_{\sim p} = t_p$.

Consider the conditions for definite provability. If the conclusion is transient (if $x = \tau$), the above conditions are the standard ones for definite proofs in DL, which are just monotonic derivations using forward chaining. If the conclusion is persistent ($x = \pi$), p can be obtained at t_p or, by persistence, at any time t_p' before t_p. Finally, notice that facts lead to strict conclusions, but are taken not to be persistent.

Defeasible derivations run in three phases. In the first phase we put forward a supported reason (rule) for the conclusion we want to prove. Then in the second phase we

---

[1] Given a proof P we use $P(n)$ to denote the n-th element of the sequence, and $P[1..n]$ denotes the first n elements of P.

consider all (actual and potential) reasons against the desired conclusion. Finally in the last phase, we have to rebut all the counterarguments. This can be done in two ways: we can show that some of the premises of a counterargument do not obtain, or we can show that the counterargument is weaker than an argument in favour of the conclusion. If $x = \tau$, the above conditions are essentially those for defeasible derivations in DL. If $x = \pi$, a proof for p can be obtained by using a persistent rule which leads to p holding at t_p or at any time t'_p before t_p. In addition, for every instant of time between the t'_p and t_p, p should not be terminated. This requires that all possible attacks were not triggered (clause 2.2.1) or are weaker than some reasons in favour of the persistence of p (clause 2.2.2). Consider the following theory.

$$(F = \{a^{t_1},\ b^{t_3},\ c^{t_3},\ d^{t_4}\},$$
$$R = \{r_1 : a^{t_1} \Rightarrow^{\pi} e^{t_1}, r_2 : b^{t_3} \Rightarrow^{\pi} \neg e^{t_3}, r_3 : c^{t_3} \rightsquigarrow^{\tau} e^{t_3}, r_4 : d^{t_4} \Rightarrow^{\tau} \neg e^{t_4}\},$$
$$\succ = \{r_3 \succ r_2, r_1 \succ r_4\})$$

At time t_1, r_1 is the only applicable rule; accordingly we derive $+\partial^{\pi} e^{t_1}$. At time t_2 no rule is applicable, and the only derivation permitted is the derivation of $+\partial^{\pi} e^{t_2}$ by persistence. At time t_3 both r_2 and r_3 are applicable, but r_4 is not. If r_2 prevailed, then it would terminate e. However, it is rebutted by r_3, so we derive $+\partial^{\pi} e^{t_3}$. Finally at time t_4, rule r_4 is applicable, thus we derive $+\partial^{\tau} \neg e^{t_4}$ and $-\partial^{\pi} e^{t_4}$, which means that r_4 terminates e. Notice that, even if r_4 is weaker than r_1, the latter is not applicable at t_4, thus it does not offer any support to maintain e.

3   The Implementation

The system implementing TDL consists of three elements: (a) a parser, which loads sets of rules, stored either in plain text format or in RuleML format, and generates a corresponding TDL theory; (b) a Graphical User Interface for selecting defeasible theories, and for visualizing conclusions and the execution time of the algorithm; (c) an inference engine which implements the algorithm of [15] to compute conclusions of the generated TDL theory [23,24].

The parser translates sets of rules in plain text or RuleML formats to generate a corresponding theory to be processed by the inference engine. The plain text format is mostly useful to handle rulesets in a simple presentation syntax easily understandable by human users. RuleML[2] is an open, general and vendor neutral XML/RDF dialect for the representation of rules. The ability to handle sets of rules written in an interchange format is valuable insofar as it allows one to exchange rules across different rule engines and different languages (e.g., W3C RIF [20], R2ML[3] and LKIF [10]) just using XSLT transformations to translate from one language to another language. Furthermore, the use of XML based languages makes possible interactions of our implementation and Semantic Web technology. For example, it is possible to use the RDF loaders of [3,8,14] to load RDF stores as sets of facts. At the same time, it permits the integration of a non-monotonic reasoner and OWL reasoners [2], thus enabling the use of (legal) ontologies,

---

[2] http://www.ruleml.org
[3] http://rewerse.net/I1

paving thus the way for the use of TDL in expressive and powerful Semantic Web applications, as well as in SOA based applications.

The Graphical User Interface allows the user to select a set of rules in RuleML or plain text format and to decide the time interval within which to compute their conclusions. Rules are then elaborated by assigning a unique label to each rule and the signs $+/-$ to the literals according to whether they are positive or negative: the rules are visualized accordingly.

The inference engine of the system implements in Java the algorithm for TDL developed by Governatori and Rotolo in [15][4]. The algorithm computes the extension of any TDL theory D, where the concept of extension is defined as follows: if HB_D is the Herbrand Base for D, the extension of D is the 4-tuple $(\Delta^+, \Delta^-, \partial^+, \partial^-)$, where $\#^\pm = \{p^t | p \in HB_D, D \vdash \pm\#^x p^t, t \in \mathscr{T}\}$, $\# \in \{\Delta, \partial\}$, and $x \in \{\pi, \tau\}$. Δ^+ and Δ^- are the positive and negative definite extensions of D, while ∂^+ and ∂^- are the positive and negative defeasible extensions. The extension of a theory D contains the set of all possible conclusions (and their type) that can be derived from the theory. For example, given a theory D, ∂^+, the positive defeasible extension, is the set of the (temporalised) literals that can be proved defeasibly from the theory D, similarly for the other elements of the extension.

The computation of the extension of a TDL theory runs in three steps [15][5]:

(i) in the first step the superiority relation is removed by creating an equivalent theory where $\prec = \emptyset$; any fact a^t, too, is removed by replacing it with a rule $\rightarrow^\tau a^t$;

(ii) in the second step the theory obtained from the first phase is used to compute the definite extension;

(iii) in the third step the theory from the first step and the definite extension are used to generate the theory to be used to compute the defeasible extension.

The Java class implementing the algorithm is TDLEngine. This has, as its main attributes, the theory (a set of rules and atoms), the theory conclusions, the time interval within which to compute these conclusions, the execution time of the algorithm, and a log manager.

The methods of the class TDLEngine are of two types: (i) those that are proper of the algorithm; (ii) those that are functional to the algorithm execution. Here, we will only describe the former ones.

It is worth noting that the computation makes use of time intervals to give a compact representation for sets of contiguous instants. The algorithm works both with proper intervals such as $[t, t']$, i.e., intervals with start time t and end time t', and punctual intervals such as $[t]$, i.e., intervals corresponding to singletons.

Following the idea of [22], the computation of the definite and defeasible extensions is based on a series of (theory) transformations that allow us (1) to assert whether a literal is provable or not (and the strength of its derivation) (2) to progressively reduce

---

[4] See `http://www.defeasible.org/implementations/TDLJava/index.html` for the full code and the Javadoc documentation. See [23] for more details.

[5] Governatori and Rotolo [15] proved that, given a TDL theory D, the extension of D can be computed in linear time, i.e., $O(|R| * |H_D| * |\mathscr{T}_D|)$, where R are the rules of D and $\mathscr{T}_D$ is the set of distinct instants occurring in D. It is also shown that the proposed algorithm is correct.

and simplify a theory. The key ideas depend on a procedure according to which, once we have established that a literal is positively provable we can remove it from the body of rules where it occurs without affecting the set of conclusions we can derive from the theory. Similarly, we can safely remove rules from a theory when one of the elements in the body of the rules is negatively provable. The methods of TDLEngine for this purpose are computeDefiniteConclusions and computeDefeasibleConclusions.

The method computeDefiniteConclusions works as follows. At each cycle, it scans the set of literals of the theory in search of temporal literals for which there are no rules supporting them (namely, supporting their derivation). This happens in two cases: (i) there are no rules for a temporal literal l^t or (ii) all the persistent rules having the literal in their head are parametrized by a greater time than t. For each of such temporal literals we add them to the negative definite extension of the theory, and remove all rules where at least one of these literals occurs. Then, the set of rules is scanned in search of rules with an empty body. In case of a positive match we add the conclusion of the rule to the positive definite extension (with an open ended interval for a persistent rule and with a punctual interval otherwise). Finally we remove such temporal literals matching the newly added conclusions from the body of rules. The cycle is repeated until (1) there are no more literals to be examined, or (2) the set of strict rules is empty, or (3) no addition to the extension happened in the cycle.

The method computeDefeasibleConclusions is more complex. As regards the scanning of the set of literals of the theory–in search of temporal literals for which there are no rules supporting them–the procedure is basically the same of computeDefiniteConclusions (with the difference that when we eliminate a rule we update the state of the extension instead of waiting to the end as in the case of the definite extensions). Then we search for rules with empty body. Suppose we have one of such rules, say a rule for l^t, and we know that the complement of l, i.e., $\sim l$, cannot be proved at t. So we add $(\sim l, [t])$ to ∂^-. At this stage we still have to determine whether we can insert l in ∂^+ and the instant/interval associated to it. We have a few cases. The rule for l is a defeater: defeaters cannot be used to prove conclusions, so in this case, we are done. If the rule is transient, then it can prove the conclusion only at t, and we have to see if there are transient rules for $\sim l^t$ or persistent rules for $\sim l^{t'}$ such that $t' \leq t$. If there are we have to wait to see if we can discard such rules. Otherwise, we can add $(l, [t])$ to ∂^+. Finally, in the last case the rule is persistent. What we have to do in this case is to search for the minimum time t' greater or equal to t in the rules for $\sim l$, and we can include $(l, [t, t'])$ in ∂^+.

The method computeDefeasibleConclusions basically calls three subroutines: proved, discard, and persistence.

The subroutine corresponding to persistence updates the state of literals in the extension of a theory after we have removed the rules in which we know at least one literal in the antecedent is provable with $-\partial^x$. Consider, for example, a theory where the rules for p and $\neg p$ are: $r : \Rightarrow^\pi p^1$, $s : q^5 \Rightarrow^\tau \neg p^{10}$, $v : \Rightarrow^\pi \neg p^{15}$. In this theory we can prove $+\partial^\pi p^t$ for $1 \leq t < 10$, no matter whether q is provable or not at 5. Suppose that we discover that $-\partial^x q^5$. Then we have to remove rule s. In the resulting theory from this transformation can prove $+\partial^\pi p^t$ for $1 \leq t < 15$. Thus we can update the entry for l from $(l, [1, 10])$ to $(l, [1, 15])$.

Secondly, `discard` adds a literal to the negative defeasible extension and then removes the rules for which we have already proved that some literal in the antecedent of these rules is not provable. The literal is parametrised by an interval. Then it further calls `persistence` that updates the state of the extension of a theory.

Third, `proved` allows to establish if a literal is proved with respect to a given time interval I. As a first step, it inserts a provable literal in the positive defeasible extension of the theory. Then it calls `discard` with the complementary literal. The next step is to remove all the instances of the literal temporalised with an instant in the interval I from the body of any rule. Finally, the rule is removed from the set of rules.

4   Validation and Testing in the Legal Domain

The main contribution of this paper is the validation in the legal domain of [23,24]'s Java implementation of TDL.

We are still at a preliminary stage for a general testing the system. In particular, we have not yet done a systematic performance evaluation using tools generating scalable test defeasible logic theories: this study is a matter of further research.[6] In this section we test and validate the implementation (and the logic) with regard to three complex temporal phenomena occurring in the legal domain, i.e., persistence, retroactivity, deadlines and periodicity. To test persistence and retroactivity we have generated some synthetic theories modelling various features of the logic and we have examined the difference of computing persistence directly or via computation for all instants in an interval. The test for deadlines and periodicity are based on a theory encoding a real life scenario.

4.1   Persistence and Backward Persistence in Legal Reasoning

We have generated some theory types (exemplified below), and for each of them we have instantiated the set of rules and computed the conclusions in the interval $[0, 100]$.

- **Persistence**
  - Rules: $\Rightarrow^\pi a^0$
  - Output: $(a, [0, 100])$
- **Backward Persistence Persistence**
  - Rules: $a^x \Rightarrow^\pi a^{x-1}, \Rightarrow^\tau a^{100}$
  - Output: $(a, [0, 100])$
- **Backward Persistence Lazy**
  - Rules: $a^{100} \Rightarrow^\pi a^0, \Rightarrow^\tau a^{100}$
  - Output: $(a, [0, 100])$
- **Persistence via Transient**
  - Rules: $a^x \Rightarrow^\tau a^{x+1}, \Rightarrow^\tau a^0$
  - Output: $(a, [0, 100])$
- **Backward Persistence Transient**
  - Rules: $a^x \Rightarrow^\tau a^{x-1}, \Rightarrow^\tau a^{100}$
  - Output: $(a, [0, 100])$
- **Backward Opposite Persistence**
  - Rules: $a^{100} \Rightarrow^\pi \neg a^0, \Rightarrow^\tau a^{100}$
  - Output: $(a, [100]), (\neg a, [0, 99])$

Our preliminary experiments, reported in Table 1, were performed on an Intel Core Duo (1,80 GHz) with 3 GB main memory. The results about the execution time are fully alligned with the theoretical result about the linear computational complexity of TDL (see footnote 5), with minor variations mostly due start-up time.

---

[6] Some first non-systematic tests on large theories are encouraging: random-generated theories with about 100,000 rules, 1,000,000 of atoms, and 100 instants of time show computation times comparable to the result of [21] for theories of similar size. See [23] for a discussion.

Table 1. Performances on Persistence

Theory	Rules	Atoms	Time	Execution time
Backward Persistence Transient	101	1	[0,100]	1110 ms
Backward Persistence Persistence	101	1	[0,100]	984 ms
Backward Opposite Persistence	2	1	[0,100]	15 ms
Backward Persistence Lazy	2	1	[0,100]	32 ms
Persistence via Transient	101	1	[0,100]	1250 ms
Persistence	1	1	[0,100]	16 ms

Arguably, one the most distinctive features of legal reasoning is that some normative effects persist over time unless some other and subsequent events terminate them, while other effects hold on the condition and only while the antecedent conditions of the rules hold. This is reflected in TDL in the distinction between persistent and transient rules. However, we may have different ways the logic can handle this notion of persistence. One could argue that persistence is not necessary since the logic is meant to give a response for a query (i.e., whether a conclusion holds) given a start time and an instant in which to evaluate a formula and the time line is taken as a discrete total order, thus persistence could be simulated using transient rules. In particular for each (persistent) literal a, one could introduce a set of rules $a^t \Rightarrow a^{t+1}$, for t in the fixed time interval.

Unfortunately, the approach using transient rules suffers from two main limitations: (1) the approach is not efficient: compare, in Table 1, the execution times for 'Persistence' and 'Persistence via Transient'. Indeed, persistence allows us to adopt a concise and computationally efficient encoding for the representation of the phenomenon. (2) fixing the time interval based on the start time and the evaluation time of a conclusion is not enough. For example, it is not enough to consider the interval $[0, 100]$ if one wants to know if a holds at time 100. Consider for example the theory $a^1 \Rightarrow^\pi b^{10}, b^{99} \Rightarrow^\tau c^{1000}, c^{1000} \Rightarrow^\tau a^{100}$. In this theory, limiting to the generation of the transient rules in $[0, 100]$ does not lead to the right results. Using the 'persistence via transient' method we have to generate all rules in the interval $[0, 1000]$, with a consequent, useless increase of the execution time.

The second problem reported above is due to the possibility of having rules where the time labelling the conclusion precedes some of the times in the antecedent (i.e., retroactivity). In legal reasoning it is not unusual to obtain conclusions about the past.[16] This means that a norm is introduced at a particular time, but its normative effects must be considered at times preceding the validity. In fact, this is typical, e.g., of taxation law. A common example is the introduction of norms whose validity is retroactive: for instance, it is possible to claim a tax benefit from a date in the past[7]. Even though trivial

---

[7] Consider for example Section 165–55 of the Australian Goods and Services Tax Act 1999 prescribing: "For the purpose of making a declaration under this Subdivision, the Tax Commissioner may: (a) treat a particular event that actually happened as not having happened; and (b) treat a particular event that did not actually happen as having happened and, if appropriate, treat the event as: (i) having happened at a particular time; and (ii) having involved particular action by a particular entity; and (c) treat a particular event that actually happened as: (i) having happened at a time different from the time it actually happened; or (ii) having involved particular action by a particular entity (whether or not the event actually involved any action by that entity."

cases of this phenomenon are captured by single rules whose conclusions hold at times preceding some of the times of the antecedents, we should be able to detect retroactivity also in other scenarios, where normative effects are in fact applied retroactively to some conditions as a result of complex arguments that involve more rules, such as $a^{10} \Rightarrow^\pi b^{10}$ and $b^{10} \Rightarrow^\pi c^0$. This problem is of great importance not only because the designer of a normative system may have the goal to state retroactive effects in more articulated scenarios of taxation law, but also because she should be able to check whether such effects are not obtained when certain regulations regard matters for which retroactivity is not in general permitted. This is the case of criminal law, where the principle *Nullum crimen, nulla poena sine praevia lege poenali* is valid.

Modelling retroactivity is challenging if it is combined with the notion of persistence. In our synthetic experiments we focused on some types of backward persistence, where conclusions persist from times which precede the ones when rules leading to such conclusions apply. As expected, all cases of backward persistence where conclusions are re-used to derive persistent literals ('Backward Persistence Transient' and 'Backward Persistence Persistence') are more computationally demanding, while the other cases, where literals persist by default, are comparable to standard persistence (the last row in Table 1). In the theories 'Backward Persistence Transient/Persistence' we perform a regression from one instant to the previous instant, and the we compute the conclusions, thus the difference between persistence and transient is not relevant. On the contrary, in 'Backward Persistence Lazy' and 'Backward Opposite Persistence' we first go back in the past and then we set persistence: for this reason, the performance of the system is here in line with that of the case 'Persistence'. However, while it is not investigated in this paper, to address the problem of (persistent) regression, one could define a logic, where conclusion are persistent in the past, and this can be achieved with the same mechanism we use to deal with persistence in the future (all one need is a discrete linearly order set of instants).

4.2   A Real-Life Scenario: Road Traffic Restrictions of Piacenza

In this paper we report our test with some real-life scenarios [23]. One of them is particularly significant. It formalizes the regulation on road traffic restrictions of the Italian town of Piacenza. This case illustrates how the logic and its implementation behave in handling the concepts of deadline and periodical provision. As regards the former concept, Piacenza regulation contains several instances of so-called maintenance obligations, which state that a certain condition must obtain during all instants before the deadline [13]. On the other hand, the regulation includes an example of periodical provision, stating that a certain legal effect periodically occurs because of intermittent character of the pre-conditions of this effect.

The road traffic restrictions regulation of the town of Piacenza consists of the following rules:

N1. From 1 October 2008 to 1 March 2009 all vehicles Euro 0, diesel Euro 1 are prohibited from circulating in the city centre.
N2. The prohibition of clause N1 is suspended in occasion of public holidays.
N3. From 7 January 2009 to 31 March 2009 vehicles diesel Euro 2 without particulate filters are prohibited from circulating in the city centre.

N4. From 8 January 2009 to 31 March 2009, on all Thursdays all vehicles are prohibited from circulating in the city centre.

N5. All vehicles Euro 4 and Euro 5, and the vehicles diesel Euro 3 with particulate filter are permitted to circulate in the city centre.

N6. WARNING: on the occasion of the snowfall of 7 January 2009, traffic restrictions scheduled for Thursday 8 January 2009 do not apply.

For the representation of the above regulation in TDL we assume that 1/10/2008 corresponds to instant 1, while the other dates are associated to integers according to the time granularity Day.

The regulation above corresponds to the following TDL rules:

$$r_1 : \;\Rightarrow^\pi trafficBlock^1 \qquad\qquad r_9 : e2 \Rightarrow^\pi \neg restricted^{180}$$
$$r_2 : \;\Rightarrow^\pi \neg trafficBlock^{151} \qquad r_{10} : \;\Rightarrow^\pi circulate^x$$
$$r_3 : \;\Rightarrow^\pi \neg restricted^1 \qquad\qquad r_{11} : restricted^x, trafficBlock^x \Rightarrow^\pi \neg circulate^x$$
$$r_4 : e0 \Rightarrow^\pi restricted^1 \qquad\qquad r_{12} : thursday^x \Rightarrow^\tau \neg circulate^x$$
$$r_5 : e1 \Rightarrow^\pi restricted^1 \qquad\qquad r_{13} : thursday^x \Rightarrow^\tau circulate^{x+1}$$
$$r_6 : e3 \Rightarrow^\pi restricted^1 \qquad\qquad r_{14} : festivity^x \Rightarrow^\tau \neg trafficBlock^x$$
$$r_7 : e3, filter \Rightarrow^\pi \neg restricted^1 \qquad r_{15} : festivity^x \Rightarrow^\pi trafficBlock^{x+1}$$
$$r_8 : e2, \neg filter \Rightarrow^\pi restricted^{99} \qquad r_{16} : snowfall^x \Rightarrow^\tau \neg circulate^{x+1}$$

The meaning of the propositions in the above set of rules is given in Table 2.

Table 2. Meaning of Predicates in the Piacenza Traffic Regulation

Temporal literal	Meaning
$trafficBlock^t$	the traffic block restrictions are in force at time t
$e0, e1, e2, e3, e4, e5$	type of vehicles according to Euro classification
$filter$	whether a vehicle is equipped with a particulate filter or not
$restricted^t$	whether a vehicle is subject to traffic restrictions a time t
$circulate^t$	whether a vehicle is permitted to circulate in the city centre at time t
$thursday^t$	it evaluates to true if t is a Thursday
$festivity^t$	it evaluates to true is t is a gazetted holiday
$snowfall^t$	whether a snowfall happened at time t

In the above set of rules, rules 1–9 are single rules (i.e., instances of rules), while rules 10–16 are schemata, where the temporal variables have to be instantiated. For the superiority relation we have[8]

$$r_4, r_5, r_6 \succ r_3, \quad r_7 \succ r_6, \quad r_2 \succ r_{15}, \quad r_{11} \succ r_{10}, \quad r_{11} \succ r_{13}, \quad r_{14} \succ r_{15}, \quad r_{16} \succ r_{11}, r_{12}$$

Rules r_{12} and r_{13} could have been simply replaced by rules with empty antecedents and whose conclusion is the literals temporalised with the day number, similarly for rules r_{14} and r_{15}.

---

[8] For rule schemata, we use the superiority to mean that each instance of a schema is superior to any instance of a second schema.

The above set of rule exhibits several typical features of reasoning with time and norms.[9] Norms (and their effects) have a time of efficacy. This can be expressed by deadlines. Thus for example, we can consider the first norm, N1, whose efficacy is from 1 October 2008 to 1 march 2009. This is represented by rule r_1, saying that from time 1 the traffic block restrictions are in force, and they will stay in force until they are interrupted or terminated. Where the termination is represented by rule r_2: the traffic restrictions are no longer effective from day 151. However, traffic block conditions can be suspended in the circumstances defined by rule r_{12} (norm N4) and rule r_{16} (Warning, N6). Notice that these two norms suspend the general traffic block restrictions, but do not terminate the efficacy of the norm. In temporal defeasible logic, we have that we can use rule r_2 to derive that *trafficBlock* holds from time 1, so it persists till a rule for $\neg trafficBlock$ becomes applicable. At that time, we conclude $\neg trafficBlock$; a 'fresh' conclusion takes precedence over a conclusion persisting from the past. However, the new conclusion is transient. After, we have to reinstate the *trafficBlock* conclusion. This is done, by a rule, i.e., r_{15}, with the same antecedent of the suspending rule, i.e., r_{14}.

The traffic regulation shows a very common feature of normative systems, norms can be enacted at different time and can have different validity time. In the regulation at hand norms N1–N4 are all enacted at the same time. However, norms N2 and N3 are in force (i.e., they can produce normative effects), only after day 99 and day 100.

Rule r_{11} could be instantiated for all days, and then we could reinstate the prohibition to enter in the city centre many times. However, it is not necessary to instantiate it for all days, all we have to do is to instantiate it for days just after turning points for the temporal literals $restricted^t$, $trafficBlock^t$, and $circulate^t$.[10] Thus, for our example, given that we know the gazetted public holidays we can instantiate

$$r_{10}^{87} : \Rightarrow^\pi circulate^{87}$$
$$r_{10}^{88} : \Rightarrow^\pi circulate^{88}$$
$$r_{14}^{87} : festivity^{87} \Rightarrow^\tau \neg trafficBlock^{87}$$
$$r_{14}^{87} : festivity^{87} \Rightarrow^\tau trafficBlock^{88}$$
$$r_{11}^{87} : restricted^{87}, trafficBlock^{87} \Rightarrow^\pi \neg circulate^{87}$$
$$r_{11}^{88} : restricted^{88}, trafficBlock^{88} \Rightarrow^\pi \neg circulate^{88}$$

The intuition of rule r_{17}, which corresponds to the generalisation of norm N6, is similar to the case of the suspension of the traffic block conditions. However, there is a caveat with this rule: We do not know in advance whether there will be a snowfall on a particular day. Thus it is not possible to generate in advance the relevant instances. However, the norm will produce its effect only when we have a fact $snowfall^t$. Thus we can generate dynamically such instances, when the relevant facts are given. We can think of this case as the introduction of the norm at the time of the snowfall. This further illustrate another important aspect to consider when reasoning with time and norms: typically one has to consider at least three temporal dimensions: The time of validity of the norms (when a norm is enacted), the time of force of the norm (when a norm can produce an

---

[9] For a comprehensive discussion of requirements for conceptual representation of norms using rules see [11].

[10] A turning point is an instant in time –day, in the granularity of the example–, where a change of provability of a literal could happen.

effect), and the time of efficacy (when a norm produces an effect). A proper model to handle this should consider the view-point at which we look at a system of norm. Thus if I ask whether I can drive my e0 car in the city center of Piacenza on January 8, then, if I ask on January 1, the answer is no, since the norm N5 is not valid, but if I ask during the snowfall of January 7, then the answer is yes, since the norm N5, rule r_{16} become valid, in force and effective.

The model of TDL presented in this paper, and thus the implementation cannot directly handle this, it needs to instantiate the rule dynamically. In [17,16] we have extended TDL to cover the three temporal dimensions. However, currently is it not know, if it is possible to implement the extended TDL efficiently and maintaining good computational properties.

4.3   Validation of TDL

The material presented in this section is a first step towards the empirical validation of TDL as a formalism suitable for modelling and reasoning with and about norms and regulations. The results so far are promising; in particular the synthetic experiment shows:

1. The computation model behind TDL and the infrastructure to consider time does not produce a substantial overhead over standard DL;
2. The introduction of persistence leads to a substantial speed-up in computation, and reduction of the complexity of rule-set. In other term it allows for concise encoding of the formal representation of norms and regulations;
3. The implementation is able to handle large rulesets;
4. Retroactivity does not pose particular concerns both from a conceptual point of view and computationally.

On the other hand, the experiment where we encoded the Traffic Restriction Regulation of Piacenza in TDL shows that TDL is able to handle different phenomena common in norms and regulations: deadlines, interruption of the efficacy of norms, and periodical norms, as well as exceptions, and derogations. Exceptions and derogation follows immediately from the basic properties of standard DL with immediate adaptation to the temporal case.

The analysis of the scenario suggests that TDL is appropriate as computational model for regulations involving temporal references. We notice that some normative aspects require several rules for their representation. However, these constructions exhibit clear and regular patterns for the rules needed to capture them. Thus syntactic forms can be defined for them[11]. This will hidden the apparent complexity of these construction for people without expertise in defeasible logic and in general formal methods.

5   Conclusions

There are two mainstream approaches to reasoning with and about time: a point based approach, as TDL, and an interval based approach [1]. Notice, however, that TDL is

---

[11] For a discussion about patterns for modelling deadline and norms see [13], and for periodicity, see [19].

able to deal with constituents holding in an interval of time: an expression $\Rightarrow a^{[t_1, t_2]}$, meaning that a holds between t_1 and t_2, can just be seen as a shorthand of the pair of rules $\Rightarrow^\pi a^{t_1}$ and $\leadsto^\tau \neg a^{t_2}$.

Non-monotonicity and temporal persistence are covered by a number of different formalisms, some of which are quite popular and mostly based on variants of Event Calculus or Situation Calculus combined with non-monotonic logics (see, e.g., [25,26]). TDL has some advantages over many of them. In particular, while TDL is sufficiently expressive for many purposes, it is possible in TDL to compute the set of consequences of any given theory in linear time to the size of the theory. To the best of our knowledge, no logic covering a set concepts comparable to what TDL covers is so efficient (see [9] for a comprehensive list of complexity results for various forms of the Event Calculus).

Temporal and duration based defeasible reasoning has been also developed by [7,19]. [19] focuses on duration and periodicity and relationships with various forms of causality. In particular, [7] proposed a sophisticated interaction of defeasible reasoning and standard temporal reasoning (i.e., mutual relationships of intervals and constraints on the combination of intervals). In these cases no complexity results are available, but these systems cannot enjoy the same nice computational properties of TDL, since both are based on more complex temporal structures.

On account of the feasibility of TDL, in this paper we reported on [23,24]'s Java implementation of this logic. In particular, we tested and validated the implementation (and the logic) with regard to three complex temporal phenomena occurring in the legal domain, i.e., persistence, retroactivity, deadlines and periodicity. Our results are encouraging and so we plan to extend the system in order to handle time plus deontic operators (see [21]) and to add temporal parameters, such as in expressions like $(a^t \Rightarrow^\pi b^{t'})^{t''}$, where t'' stands for the time when the norm is in force [16].

Acknowledgements

NICTA is funded by the Australian Government as represented by the Department of Broadband, Communications and the Digital Economy and the Australian Research Council through the ICT Centre of Excellence program.

References

1. Allen, J.: Towards a general theory of action and time. Artificial Intelligence 23 (1984)
2. Antoniou, G.: A nonmonotonic rule system using ontologies. In: Proc. RuleML 2002. CEUR Workshop Proceedings, vol. 60 (2002)
3. Antoniou, G., Bikakis, A.: DR-Prolog: A system for defeasible reasoning with rules and ontologies on the semantic web. IEEE Transactions on Knowledge and Data Engineering (2), 233–245 (2007)
4. Antoniou, G., Billington, D., Governatori, G., Maher, M.J.: Representation results for defeasible logic. ACM Transactions on Computational Logic 2, 255–287 (2001)
5. Antoniou, G., Billington, D., Governatori, G., Maher, M.J.: Embedding defeasible logic into logic programming. Theory and Practice of Logic Programming 6, 703–735 (2006)
6. Antoniou, M.R., Maher, M.J., Rock, A., Antoniou, G., Billington, D., Miller, T.: Efficient defeasible reasoning systems. International Journal of Artificial Intelligence Tools 10 (2001)

7. Augusto, J., Simari, G.: Temporal defeasible reasoning. Knowledge and Information Systems 3, 287–318 (2001)
8. Bassiliades, N., Antoniou, G., Vlahavas, I.: A defeasible logic reasoner for the Semantic Web. International Journal on Semantic Web and Information Systems 2, 1–41 (2006)
9. Cervesato, I., Franceschet, M., Montanari, A.: A guided tour through some extensions of the event calculus. Computational Intelligence 16(2), 307–347 (2000)
10. ESTRELLA Project. The reference LKIF inference engine. Deliverable 4.3, European Commission (2008)
11. Gordon, T.F., Governatori, G., Rotolo, A.: Rules and norms: Requirements for rule interchange languages in the legal domain. In: Governatori, et al. (eds.) [12]
12. Governatori, G., Hall, J., Paschke, A. (eds.): RuleML 2009. LNCS, vol. 5858. Springer, Heidelberg (2009)
13. Governatori, G., Hulstijn, J., Riveret, R., Rotolo, A.: Characterising deadlines in temporal modal defeasible logic. In: Orgun, M.A., Thornton, J. (eds.) AI 2007. LNCS (LNAI), vol. 4830, pp. 486–496. Springer, Heidelberg (2007)
14. Governatori, G., Pham, D.: A semantic web based architecture for e-contracts in defeasible logic. In: Adi, A., Stoutenburg, S., Tabet, S. (eds.) RuleML 2005. LNCS, vol. 3791, pp. 145–159. Springer, Heidelberg (2005)
15. Governatori, G., Rotolo, A.: Temporal defeasible logic has linear complexity. In: Proceedings NMR 2010. CEUR Workshops Proceedings (2010)
16. Governatori, G., Rotolo, A.: Changing legal systems: Legal abrogations and annulments in defeasible logic. The Logic Journal of IGPL (forthcoming)
17. Governatori, G., Rotolo, A., Riveret, R., Palmirani, M., Sartor, G.: Variants of temporal defeasible logic for modelling norm modifications. In: Proc. ICAIL 2007, pp. 155–159 (2007)
18. Governatori, G., Rotolo, A., Sartor, G.: Temporalised normative positions in defeasible logic. In: ICAIL 2005, pp. 25–34. ACM Press, New York (2005)
19. Governatori, G., Terenziani, P.: Temporal extensions to defeasible logic. In: Orgun, M.A., Thornton, J. (eds.) AI 2007. LNCS (LNAI), vol. 4830, pp. 476–485. Springer, Heidelberg (2007)
20. Hawke, S.: Bringing order to chaos: RIF as the new standard for rule interchange. In: Governatori, et al. (eds.) [12], p. 1, http://www.w3.org/2009/Talks/1105-ruleml/
21. Lam, H.-P., Governatori, G.: The making of SPINdle. In: Governatori, et al. (eds.) [12]
22. Maher, M.: Propositional defeasible logic has linear complexity. Theory and Practice of Logic Programming 1, 691–711 (2001)
23. Rubino, R.: Una implementazione della logica defeasible temporale per il ragionamento giuridico. PhD thesis, CIRSFID, University of Bologna (2009)
24. Rubino, R., Rotolo, A.: A Java implementation of temporal defeasible logic. In: Governatori, et al. (eds.) [12]
25. Shanahan, M.: Solving the Frame Problem: A Mathematical Investigation of the Common Sense Law of Inertia. MIT Press, Cambridge (1997)
26. Turner, H.: Representing actions in logic programs and default theories: A situation calculus approach. Journal of Logic Programming 31(1-3), 245–298 (1997)

Evaluating Cases in Legal Disputes as Rival Theories

Pontus Stenetorp[1] and Jason Jingshi Li[2,3]

[1] Graduate School of Information Science and Technology,
The University of Tokyo, Tokyo, Japan
pontus@is.s.u-tokyo.ac.jp
[2] College of Engineering and Computer Science,
The Australian National University, Canberra, Australia
[3] Canberra Research Laboratory,
NICTA, Canberra, Australia
jason.li@anu.edu.au

Abstract. In this paper we propose to draw a link from the quantitative notion of coherence, previously used to evaluate rival scientific theories, to legal reasoning. We evaluate the stories of the plaintiff and the defendant in a legal case as rival theories by measuring how well they cohere when accounting for the evidence. We show that this gives rise to a formalized comparison between rival cases that account for the same set of evidence, and provide a possible explanation as to why judgements may favour one side over the other. We illustrate our approach by applying it to a known legal dispute from the literature.

Keywords: legal argument, legal justification, theory construction, coherence.

1 Introduction

In legal disputes each side present their case before the court, outlining the issues, positions, and arguments taken with respect to the issues. The "story" is supported by evidence, which is sometimes explicitly sought by the judge as burden of proof. Each side must explain how the evidence fits their story, though there may be elements of their story that for some reason cannot be verified by evidence or empirical testing.

Similarly in the philosophy of science, rival, possibly incompatible scientific theories must also account for all observations, but empirical testing cannot always be used to differentiate or rank theories, as they make the same empirical claims. One possible measure to evaluate theories is how *coherent* a theory is in accounting for a given set of observations.

We propose to draw parallels from this notion of coherence to legal reasoning, where we view the cases of the plaintiff and the defendant in a legal dispute as rival theories, and evaluate the cases by measuring how coherent the stories are in their account for the evidence. Intuitively, just as a good scientific theory uses

K. Nakakoji, Y. Murakami, and E. McCready (Eds.): JSAI-isAI, LNAI 6284, pp. 59–72, 2010.

only a few credible postulates to explain a large body of evidence, a good "story" in a legal case must account for the evidence using only a few minor assumptions. This provide an alternative view on how the cases can be evaluated and decided, and provides a possible explanation as to why judgements may favour one side over the other, in a formal and structured manner.

The notion of coherence in regard to legal justification is well explored by Hage [3] and Amaya [1]. Equally, there is abundant literature on abductive reasoning with respect to the evidence and the burden of proof by Prakken et al. [10] and Satoh et al. [11]. However, the existing literature on coherence is mostly concerned with how a decision can cohere with current law and cases, whereas we are interested in the overall picture of how the "story" of the plaintiff/defendant coheres with the evidence.

In this paper we introduce the notion of *coherence* in the form proposed by Kwok et al. [5, 6] for evaluating scientific theories. We then propose a possible scheme based on the previous work by Kwok et al. [5, 6] for evaluating the coherence of cases in a legal dispute. This is followed by an example of applying our theory to an actual legal dispute previously formalized by Prakken [8]. In conclusion, we discuss what is implied by our coherence measure and possible future directions of this work.

2   Coherence of Theories

Traditionally in the philosophy of science, coherence has always been a criterion in evaluating the quality of scientific theories. The extent of coherence of a theory depends on informal, qualitative notions such as "brevity", "predictive scope" and "tightness of coupling" of the components of the theory.

Kwok et al. [5] proposed a quantitative measure of coherence based on the average utilization of formulas in accounting for observations. Their later work, Kwok et al. [6], better mirrored scientific practice by introducing input and output sets. The proposed measure facilitates the testing of theories with experiments that have varying inputs and outputs, where each theory is expressed as a set of clauses.

When performing a scientific experiment we provide a certain input I and observe a certain output O. It is then the objective of a theory T to explain how the input leads to the output. For example, in an experiment verifying the theory of gravity, the input being an object dropped in vacuum, the output would be the measured velocity of the object some time after the drop and the theory provides a link between the input and output. In this section we summarize the approach developed by Kwok et al. [6].

Definition 1 (Support Sets). *Given an input set I, an output set O, a subset of the theory T being Γ. Then, Γ is a I-relative support set of O if:*

1. $\Gamma \wedge I \models O$ and
2. Γ is minimal (wrt set inclusion).

Support sets are the building blocks of the coherence measure. They are the formulas that account for a particular observation for a given input. We denote $S(T, I, O)$ to be the family of all I-relative support sets for O.

Definition 2 (Utility of a formula). *Given an input set I, an output set O and a theory T. For a formula $\alpha \in T$, the utility of α with respect to T, I, O is given by:*

$$U(\alpha, T, I, O) = \frac{|\{\Gamma : \alpha \in \Gamma \text{ and } \Gamma \in S(T, I, O)\}|}{|S(T, I, O)|} \quad if \quad S(T, I, O) \neq \emptyset$$

The support sets give rise to the definition of the utility of a formula of the theory. Informally, this is the relative frequency of occurrence of formula α in the support sets of T, I, O. This reflects the contribution of α in T to account for the pair (I, O).

Definition 3 (Coherence of a Theory). *The* coherence *of a theory T with formulas $\{\alpha_1, \ldots, \alpha_n\}$ with respect to input observations $I = \{I_1, \ldots, I_m\}$ and output observations $O = \{O_1, \ldots, O_m\}$ is:*

$$C(T, I, O) = \frac{1}{mn} \sum_{i=1}^{n} \sum_{j=1}^{m} U(\alpha_i, T, I_j, O_j)$$

The coherence of a theory is the average utility of the theory's formulas in accounting for all the observations from possibly multiple experiments. This measure has been shown by Kwok et. al. [6] to demolish Craig's trick as shown by Craig [2], where empirical observations are simply added to the theory as exceptions. They showed that such handling of exceptions results in the formulation of highly incoherent theories, since the measure favours theories in which a small subset of the theory accounts for a large body of evidence.

3   Evaluating Legal Cases

In this section, we draw a link from scientific disputes between rival theories to ordinary legal disputes between the cases of the plaintiff and the defendant. We treat the "stories" given from both sides as rival theories, each of which can be tested against the evidence presented to the court. Just as scientific theories can be tested over multiple experiments, a case in a legal dispute can be tested by multiple pieces of evidence and testimonies.

In the following sections we will show how the analogy can be made. Our approach enables us to evaluate the coherence of a case by measuring how well the components of the case are utilized when accounting for the evidence and testimonies presented to the court. We assume that all the components of the sets mentioned in the following subsections are in clausal form.

3.1   Inputs

The presentation of evidence and testimonies can be viewed as experiments testing the theory. Hence the input of the experiments are the relevant laws and the mutually accepted state of affairs that are necessary for the theory to entail the output. Laws themselves can not be disputed although their validity for a certain case may very well be questioned.

A mutually accepted state of affairs is a state of affairs that is presented by some participant of the trial, but is not contested by any other participant. This notion is the same as the view on common knowledge given in Walton and Macagno [12]. It can thus be deemed that a mutually accepted state of affairs can be regarded as a fact from which one may draw conclusions or aid arguments, even though the truth of such a state of affairs is never proven explicitly to the court. We relax the requirement on inputs to include laws and facts that are not used in deriving the output, as they do not affect the utility of any component of the theory, and hence have no effect on measuring coherence.

3.2   Outputs

The outputs are the evidence and testimonies presented to the court. They pose the main problem for any theory since it must explain how an output can be explained using the theory itself and the inputs mentioned in the previous section.

Both the plaintiff and the defendant must account for the observations in the output such as why certain DNA is present at the crime scene, why the witness x testify that y loaned equipment to z. Without explaining such circumstances a case may not be considered to fulfil the requirements of the court.

3.3   Theory

The theory in the case of a legal dispute is the "story" that is told by one of the sides. It may contain several components to explain the evidence presented to the court, and why the desired outcome holds for the plaintiff or the defendant. The requirement to hold the sought outcome becomes obvious: if one considers the "story", it must in some way justify the conclusion the side hopes for or nothing can be gained from the trial.

The two theories, while arguing for different outcomes, will have to take into account the same laws and mutually accepted state of affairs, which in our framework corresponds to the inputs. But they must also account for the same evidence and testimonies put forth to the court, in our framework the outputs. Making the assumption that they both account for all the facts they are clearly rival theories that account for the same set of evidence but must somehow be differentiated regarding how well they do so.

3.4   Support Sets and Coherence

For our measure of coherence we need to observe the I-relative support set for the outputs O that is a subset of the theory that accounts for a particular piece of evidence O_i given input I_i. The relative frequency of a component of the theory appearing in the support sets give rise to the utility of the component, where coherence is measured as the average utility of the components over all the given evidence.

Intuitively, the support set for a given piece of evidence is how the plaintiff/defendant explain that piece of evidence with respect to the mutually accepted state of affairs and relevant laws presented before the court. The coherence of the theory measures how well the overall "story" of the plaintiff and the defendant explains all the evidence. Any theory that assumes freely without proper support will thus be punished with lower coherence than a theory which can utilize a small amount of assumptions in combination with the inputs to account for a larger body of evidence.

4   Example

As an example of how our notion of coherence can be applied in order to evaluate rival legal cases between a plaintiff and a defendant, we apply it to a known legal dispute first formalized by Prakken [8].

Unlike Prakken [8] we simplify the case into sets of clauses to form the basis of the judgement rather than utilizing an argumentation framework to focus on the process of the trial. We will attempt to stay consistent with as much of the previous formalization as possible, differing only when our notion of coherence and our focus on the judgement rather than the process forces us to do so.

4.1   The Dispute

The legal case formalized by Prakken [8] is a Dutch civil case from 1978, concerning the ownership of a moveable good, a large tent. The owner of the tent, Mr. van der Velde, put the tent up for sale at the price of 850 Gulden (approx. 380 Euro). Mr. Nieborg, who was a friend of Mr. van der Velde, said that he was interested in buying the tent but could not afford it. Mr. van der Velde made the tent available to Mr. Nieborg, who in return helped Mr. van der Velde to paint his house. Also, Mrs. Nieborg helped Mrs. van der Velde with her domestic work for some time.

Later, Mr. Nieborg claimed that he and his wife had performed enough work for Mr. and Mrs. van der Velde to cover the cost of the tent, thus implicitly claiming that he had now become the legitimate owner of the tent. This angered Mr. van der Velde since he perceived the work performed by Mr. and Mrs. Nieborg as an expression of gratitude for allowing them to use the tent as a loan. He immediately demanded that Mr. Nieborg would return the tent. When his demands were not met, Mr. van der Velde, with assistance, threw Mr. Nieborg's son, who was the person currently occupying the tent, out of the tent and took possession of it.

Some time later, Mr. van der Velde sold the tent to a Mr. van der Weg. Mr. van der Weg paid for the tent by performing work, which was similar to the work performed earlier by Mr. and Mrs. Nieborg, for Mr. van der Velde. Mr. Nieborg took his case against Mr. van der Weg to court within a period of time which was less than three years after the repossession of the tent carried out by Mr. van der Velde, a fact that should be noted due to implications in regard to Dutch law.

We present the cases of both Mr. Nieborg and Mr. van der Weg in clausal form. In order to make our clauses as brief as possible, we abbreviate the names of the people as presented in Table 1. We also abbreviate the relevant points in time as presented in Table 2. Both sets of abbreviations conform to those used by Prakken [8]. To make the feel of the running text more natural, we will still make use of the full names and points in time.

Table 1. Abbreviations for the participants of the trial

Surname	Abbreviation	Role
Mr. Nieborg	N	Plaintiff
Mr. van der Weg	vdW	Defendant
Mr. van der Velde	vdV	Witness
Mr. Sluis	S	Witness
Mr. Galtema	G	Witness

Table 2. Abbreviations for the points in time relevant to the trial

Point in time	Event
t_1	N held the tent
t_2	Violent events between vdV and N
t_3	Time of the trial

4.2   Inputs – Relevant Laws, Mutually Accepted State of Affairs and Consequences

We will in this section present the relevant laws, mutually accepted state of affairs and consequences, all of which will be presented in clausal form as well as informally in the running text. These clauses will serve as input to be utilized by the respective theories in order to derive the evidence and testimonies which we will refer to as output. This derivation will be done using the theory itself and a subset of the input clauses. The clauses are partially those found in Prakken [8], but with some additions.

One difference to Prakken [8] is that we only consider the law, mutually accepted state of affairs and consequences that are relevant to the final juridical judgement. This difference is due to our focus on the final judgement rather than the process of the trial itself as is the case with an argument framework, thus we can disregard of a more general law that is later refuted in favour of one which for our case applies more specifically. We will point out these special cases when presenting the clauses.

All laws formalized here were in effect at the time of the trial. Also, just as in Prakken [8] we do not motivate the notion of persistence of ownership that Mr. Nieborg implicitly uses to justify that he indeed is still the owner of the tent at the time of the trial, since according to him no change in ownership has taken place since he took possession of the tent from Mr. van der Velde.

$$Hold(N, Tent, t_2) \tag{1}$$

$$Hold(vdW, Tent, t_3) \tag{2}$$

The first two clauses, 1 and 2, concerns the holder of the tent at different points in time. Clause 1, that Mr. Nieborg held the tent at the time it was taken from him. Clause 2, that Mr. van der Weg now holds the tent. Both parties agrees upon these state of affairs.

$$Loan(x, y) \rightarrow TestimonyLoan(z, x, y) \wedge (x \neq z) \tag{3}$$

$$FalseTestimonyLoan(x) \rightarrow TestimonyLoan(x, y, z) \wedge (x \neq y) \tag{4}$$

$$Violence(x, y) \rightarrow TestimonyViolence(z, x, y) \wedge (y \neq z) \tag{5}$$

$$FalseTestimonyViolence(x) \rightarrow TestimonyViolence(x, y, z) \wedge (x \neq z) \tag{6}$$

The next set of clauses, clause 3 to clause 6, lays forth the logic concerning testimonies, it should be noted that we assume a primitive notion of lying to simplify our set of clauses. It should also be noted that these clauses have no temporal components, as we previously saw for clauses 1 and 2, the reasoning behind this is that the testimonies and events in our particular case do not need any temporal components due to them occurring only once. These four clauses can not be refuted logically and have to be accepted by all parties.

Clause 3 and 5 simply state that if a person x borrowed an item y or violence was inflicted by a person x towards person y, then a third person z can deliver a testimony of the event. Clause 4 and 6 provide an alternate mode for explaining each testimony. If a witness is lying, then he would deliver the same testimony that a witness who had observed the events would have delivered.

$$Hold(x, y, t) \wedge \neg Loan(x, y) \rightarrow Possess(x, y, t) \tag{7}$$

Clause 7 is a formalization of Dutch law 590 BW. The loan condition is a simplification of the actual text that states that the holder may not be holding it for another person, this change is made to make the clause simpler since holding the item can be derived from a loan in our specific case. The loan condition is an exception added to the more general version of the law that lacks this condition, but since for our case the loan condition is relevant, we observe the more specific law. This constraint makes it possible to disregard the notion of law precedence in Prakken [8] and is justified by us only observing the judgement.

$$Possess(x, y, t) \rightarrow GoodFaith(x, y, t) \tag{8}$$

$$Possess(x, y, t) \land GoodFaith(x, y, t) \land Owner(z, y, t') \land (x \neq z)$$
$$\land\ InvoluntaryLoss(z, y, t') \land (t' - t) < 3\ years$$
$$\rightarrow \neg Owner(x, y, t) \tag{9}$$

Clause 8 is a formalization of Dutch law 589 BW which states that a possessor is presumed to be a possessor of good faith. Clause 9 is a formalization of Dutch law 2014 BW regarding the possession of a good. It covers the special case that x can not be the owner of y if it has occurred an involuntary loss of y from a previous owner z at a time t'. It also contains the restriction that a maximum of three years must have passed since the time of the involuntary loss t' and the current point in time t.

4.3   Evidence Presented

This section covers all irrefutable evidence presented to the court, these are facts that must be accounted for by any theory in order for a case to be considered valid. This can be done by utilizing a subset of a theory, by calling upon law and/or mutually accepted state of affairs, as presented in section 4.2. The set of all evidence clauses is referred to as the output set O.

$$O_1 = TestimonyLoan(vdV, N, Tent) \tag{10}$$

$$O_2 = TestimonyLoan(G, N, Tent) \tag{11}$$

$$O_3 = TestimonyLoan(S, N, Tent) \tag{12}$$

$$O_4 = TestimonyViolence(vdV, vdV, N) \tag{13}$$

$$O_5 = TestimonyViolence(G, vdV, N) \tag{14}$$

$$O_6 = TestimonyViolence(S, vdV, N) \tag{15}$$

Clauses 10 to 15 are all testimonies delivered to the court. That the testimonies took place is irrefutable, but the fact of them taking place has to be explained by each theory presented in the next two sections 4.4 and 4.5.

To simplify our clauses we have taken the liberty of stating that the testimonies of violence implied violence towards Mr. Nieborg. In reality the involved party was Mr. Nieborg's son. We have also done the same regarding the testimony of the tent being a loan, what was presented in reality was that Mr. Nieborg expressed gratitude towards Mr. van der Velde for being able to hold the tent for a limited time. This was observed by the witnesses, who gave testimonies to that effect. As described by Prakken [8] the violence towards Mr. Nieborg's son counts as violence towards Mr. Nieborg when legally proving that the loss was involuntary and the expression of gratitude observed by the three witnesses counts as the possession of the tent being perceived as a loan. We have once again simply left out these conclusions and replaced them with the results relevant to the judgement.

4.4   The Plaintiff's Case

To make a case, each party x must construct an input set I_x which consists of input subsets I_{xi} corresponding with the observation O_i presented in the previous section. The clauses of each subset I_{xi} are clauses from section 4.2, being formalizations of mutually accepted state of affairs and law.

Each party x also needs to produce a theory T_x representing his "story", that together with the input subset I_{xi} will explain the corresponding observation O_i. A theory may consist of any clauses, as long as it satisfies the previously mentioned condition to satisfy each O_i by using I_{xi} as input.

$$T_{P1} = Violence(vdV, N, t_2) \tag{16}$$

$$T_{P2} = FalseTestimonyLoan(vdV) \tag{17}$$

$$T_{P3} = FalseTestimonyLoan(G) \tag{18}$$

$$T_{P4} = FalseTestimonyLoan(S) \tag{19}$$

$$T_{P5} = \neg Loan(N, Tent) \tag{20}$$

The first clause of the plaintiff's theory T_P, clause 16, is an acceptance of the violent events when the possession of the tent was revoked by Mr. van der Velde since this plays in his favour. However, he is forced to add clauses 17 to 19 since he is unwilling to accept that his work was an expression of gratitude, which was how it was perceived by the witnesses. Not calling the testimonies false would render it impossible for him to claim previous possession, using law 590 BW with its special case (clause 7) and law 589 BW (clause 8). Thus being able to revoke the current hold of the tent by van der Weg using law 2014 BW (clause 9) with its exception which is his own goal and his theory must thus account for this. Clause 20 is included since it is a requirement for him to be able to use law 2014 BW (clause 9) with the special case applied.

$$I_{P1} = \{FalseTestimonyLoan(x) \rightarrow TestimonyLoan(x, y, z) \land (x \neq y)\} \tag{21}$$

$$I_{P2} = \{FalseTestimonyLoan(x) \rightarrow TestimonyLoan(x, y, z) \land (x \neq y)\} \tag{22}$$

$$I_{P3} = \{FalseTestimonyLoan(x) \rightarrow TestimonyLoan(x, y, z) \land (x \neq y)\} \tag{23}$$

$$I_{P4} = \{Violence(x, y) \rightarrow TestimonyViolence(z, x, y) \land (y \neq z)\} \tag{24}$$

$$I_{P5} = \{Violence(x, y) \rightarrow TestimonyViolence(z, x, y) \land (y \neq z)\} \tag{25}$$

$$I_{P6} = \{Violence(x, y) \rightarrow TestimonyViolence(z, x, y) \land (y \neq z)\} \tag{26}$$

Since it is an important point, we will once again stress that the input set I_x, unlike the theory set T_x, has the restriction that it can only consist of laws and mutually accepted state of affairs. This has significant consequences for our notion of coherence, which we will observe in the coming sections.

For his input set I_P the plaintiff alternates between clause 4 that implies that a testimony is a lie and clause 5 that implies that a testimony is accurate. This in combination with his theory T_P is enough to prove each O_i using the corresponding I_{Pi}, thus completing his task.

4.5   The Defendant's Case

The defendant's case is very similar to that of the plaintiff which we presented in the previous section. But the minute differences will have effects on how it interacts with our notion of coherence.

$$T_{D1} = Violence(vdV, N, t_2) \tag{27}$$

$$T_{D2} = Loan(N, Tent) \tag{28}$$

In his theory T_D, the defendant has no need to discredit the violent events taking place since they are neutral towards his goal of ownership when interacting with the laws contained in I. This is done by concurring with the violent events, just as the plaintiff did in clause 16 which corresponds to the defendant's clause 27.

In order to fulfil his goal of ownership the defendant simply has to assume that the testimonies regarding the loan are accurate, as is done in clause 28. This will make his theory capable of justifying his ownership of the tent since the attempts by the plaintiff to claim ownership using law 2014 BW (clause 9) with its exception, since law 590 BW with its exception (clause 7) is not applicable if the plaintiff was given the tent on loan.

$$I_{D1} = \{Loan(x, y) \rightarrow TestimonyLoan(z, x, y) \wedge (x \neq z)\} \tag{29}$$

$$I_{D2} = \{Loan(x, y) \rightarrow TestimonyLoan(z, x, y) \wedge (x \neq z)\} \tag{30}$$

$$I_{D3} = \{Loan(x, y) \rightarrow TestimonyLoan(z, x, y) \wedge (x \neq z)\} \tag{31}$$

$$I_{D4} = \{Violence(x, y) \rightarrow TestimonyViolence(z, x, y) \wedge (y \neq z)\} \tag{32}$$

$$I_{D5} = \{Violence(x, y) \rightarrow TestimonyViolence(z, x, y) \wedge (y \neq z)\} \tag{33}$$

$$I_{D6} = \{Violence(x, y) \rightarrow TestimonyViolence(z, x, y) \wedge (y \neq z)\} \tag{34}$$

Just like the plaintiff, the defendant alternates between two clauses when constructing his input set I_D, in this case clauses 5 and 3, both implying that the testimonies of the witnesses are true and are thus indications of a loan and a violent event taking place. He has thus also fulfilled his obligations.

4.6   Calculation of Coherence

We will now proceed to calculate the coherence of the "stories" given by the plaintiff and defendant using our measure of coherence introduced in section 3.

For the six observations $O : \{O_1, \ldots, O_6\}$, the plaintiff's theory T_P contains five clauses, whereas the the defendant's theory T_D contains two clauses. The support sets for the evidence from both sides are as noted in table 3. We remind the reader that a support set is the subset of a theory that is utilized to account for a given observation.

Table 3. Support sets for the observations

Observation	Plaintiff support set	Defendant support set
O_1	$\{T_{P5}, T_{P2}\}$	$\{T_{D2}\}$
O_2	$\{T_{P5}, T_{P3}\}$	$\{T_{D2}\}$
O_3	$\{T_{P5}, T_{P4}\}$	$\{T_{D2}\}$
O_4	$\{T_{P5}, T_{P1}\}$	$\{T_{D1}\}$
O_5	$\{T_{P5}, T_{P1}\}$	$\{T_{D1}\}$
O_6	$\{T_{P5}, T_{P1}\}$	$\{T_{D1}\}$

We remind ourselves that as described in section 4.4 and 4.5, T_P and T_D are comprised as shown in equation 35 and 36.

$$T_P = \{T_{P1}, T_{P2}, T_{P3}, T_{P4}, T_{P5}\} \tag{35}$$

$$T_D = \{T_{D1}, T_{D2}\} \tag{36}$$

The summation of the utility of each of the components of the plaintiff's theory T_P over all observations are as shown in equation 37 to 41. These reflect how much each theory component contributed in accounting for all the evidence.

$$\sum_{j=1}^{6} U(T_{P1}, T_P, I_{Pj}, O_j) = 3 \tag{37}$$

$$\sum_{j=1}^{6} U(T_{P2}, T_P, I_{Pj}, O_j) = 1 \tag{38}$$

$$\sum_{j=1}^{6} U(T_{P3}, T_P, I_{Pj}, O_j) = 1 \tag{39}$$

$$\sum_{j=1}^{6} U(T_{P4}, T_P, I_{Pj}, O_j) = 1 \tag{40}$$

$$\sum_{j=1}^{6} U(T_{P5}, T_P, I_{Pj}, O_j) = 6 \tag{41}$$

We then average the sum of all the utility of the components over the size of the theory and the number of evidence to derive the coherence measure. The calculation is described in equation 42.

$$C(T_P, I, O) = \frac{1}{6} \times \frac{1}{5} \times (3 + 1 + 1 + 1 + 6) = 0.4 \tag{42}$$

For the case of the defendant, there are only two parts to his story. The summation of the utility of each of the components in the defendant's theory over all observations are as follows.

$$\sum_{j=1}^{6} U(T_{D1}, T_D, I_{Dj}, O_j) = 3 \tag{43}$$

$$\sum_{j=1}^{6} U(T_{D2}, T_D, I_{Dj}, O_j) = 3 \tag{44}$$

The coherence of the defendant's theory can be derived as is done in equation 45.

$$C(T_D, I, O) = \frac{1}{6} \times \frac{1}{2} \times (3 + 3) = 0.5 \tag{45}$$

Our coherence measure shows that in this event where the explanations from both sides are considered equally valid, the defendant had provided a more coherent theory to account for the evidence.

The example illustrates a possible application of our notion of coherence in a legal dispute. Both sides were capable of producing a story explaining the evidence, but one did so better than the other as it provided a simpler account for the evidence. This is reflected in the higher coherence value derived from the explanations of the defendant compared to that of the plaintiff. In the following section we will discuss the merits and the shortcomings of this approach in evaluating cases in legal disputes, and identify possible lines of future work.

5   Discussion and Future Work

We proposed an approach to evaluate cases in a legal dispute as two rivaling scientific theories. The theories are measured by how well they account for the evidence. The proposed measure of coherence rewards simple theories that account for a large body of evidence, while punishing frivolous theories that regard much of the evidence as exceptions. We gave an example of a known legal dispute from the literature, and showed how the case fits into our framework. In the example, the side that lost due to insufficient evidence also had the less coherent theory.

We note that a key difference between scientific theories and cases in legal disputes is that scientific theories are evaluated primarily on how well they account for the evidence, whereas cases in legal disputes are ultimately concerned with proving a case in order to attain a goal. However, as the case is based on evidence, the quality of the theory in accounting for the evidence is still crucial when proving the case. We argue that our notion of coherence gives rise to an important measure to the quality of a case, and allow rival cases to be compared in a quantitative manner.

Our framework provides only a preliminary and approximate model for evaluating cases with respect to the given evidence. We intentionally chose to simplify the example to illustrate our goal of assessing the coherence of rival theories between the plaintiff and the defendant. In more complicated real-life examples, not all evidence is treated equally, some would be considered worthy of more merit than others, and some would be contradictory. The quality of explanations in accounting for a single piece of evidence can also be subjected to debate. The differing merits of evidence and explanations can be modelled by allocating weights to the evidence and also components of the theory in a way similar to the proposal made by Li et al. [7]. The merit of cases would then be dependent to the weights associated with each evidence and their explanations. This would be one possible line of future work, but it is beyond the scope of the current paper.

Our measure does not take into account that parts of a chain of conclusion interacts and the final conclusion in such a chain depends on the probability of the chain as a whole. This fact has been noted in Keppens [4]. Our measure does, however, capture the effect of corroboratory evidence as described in Walton and Reed [13], since a theory that can account for corroborating accounts of the same observation using only a few clever assumptions in combination with law and common knowledge will be considered "better" according to our measure. Contradictory evidence could be handled by argument-based version of extended logic programming with defeasible priorities such as the one proposed by Prakken and Sartor [9]. Incorporating these features into the evaluation of coherence is essential for extending our proposal to more complex legal disputes.

Our evaluation of coherence reflects the importance of the choice of evidence in a legal dispute. As the evidence in a legal dispute is essential for our calculation of coherence of the respective cases, different selection of evidence can change the theories, thus leading to different outcomes in accordance with our coherence measure. Therefore, intelligent allocation of the *burden of proof* is necessary to collect the relevant evidence in proving the cases, while avoiding material that may not necessarily relate to the case.

Acknowledgments. We would like to thank the referees for their comments, which helped improve this paper considerably. We would also like to thank Mattias Frånberg, Carl Johan Gustavsson, Luke McCrohon and Goran Topić for proof-reading this paper. This work was partially supported by Grant-in-Aid for Specially Promoted Research (MEXT, Japan), and NICTA. NICTA is funded by the Australian Government's Backing Australia's Ability initiative, in part through the Australian Research Council.

References

1. Amaya, A.: Formal Models of Coherence and Legal Epistemology. Artif. Intell. Law 15, 429–447 (2007)
2. Craig, W.: On Axiomatizability Within a System. The Journal of Symbolic Logic 18, 30–32 (1953)
3. Hage, J.C.: Formalizing Legal Coherence. In: Proc. of 8th International Conference on Artificial Intelligence and Law, pp. 22–31 (2001)
4. Keppens, J.: Towards qualitative approaches to Bayesian evidential reasoning. In: Proceedings of the 11th international conference on Artificial intelligence and law, pp. 17–25 (2007)
5. Kwok, R.B.H., Nayak, A.C., Foo, N.: Coherence Measure Based on Average Use of Formulas. In: Lee, H.-Y. (ed.) PRICAI 1998. LNCS, vol. 1531, pp. 553–564. Springer, Heidelberg (1998)

6. Kwok, R.B.H., Foo, N., Nayak, A.C.: The Coherence of Theories. In: Proceedings of the 18th Joint International Conference on Artificial Intelligence (IJCAI 2003), Acapulco, Mexico (August 2003)
7. Li, J.J., Kwok, R.B.H., Foo, N.: Coherence of Theories Dependencies and Weights. In: Trends in Logic, vol. 28, pp. 297–318. Springer, Berlin (2009)
8. Prakken, H.: Formalising Ordinary Legal Disputes: A Case Study. Artif. Intell. Law 16, 333–359 (2008)
9. Prakken, H., Sartor, G.: A dialectical model of assessing conflicting arguments in legal reasoning. Artif. Intell. Law 4, 331–368 (1996)
10. Prakken, H., Sartor, G.: Formalising Arguments About the Burden of Persuasion. In: Proc. of ICAIL 2007, pp. 97–106 (2007)
11. Satoh, K., Tojo, S., Suzuki, Y.: Abductive Reasoning for Burden of Proof. In: Proceedings of the 2nd International Workshop on Juris-informatics (JURISIN 2008), Asahikawa, Japan, pp. 93–102 (2008)
12. Walton, D., Macagno, F.: Common Knowledge in Legal Reasoning About Evidence. In: International Commentary on Evidence, vol. 3(1). The Berkeley Electronic Press (2005)
13. Walton, D., Reed, C.: Evaluating Corroborative Evidence. In: Argumentation, vol. 22, pp. 531–553. Springer, Berlin (2008)

Law-Aware Access Control: About Modeling Context and Transforming Legislation

Michael Stieghahn and Thomas Engel

University of Luxembourg, 6, rue R. Coudenhove-Kalergi, L-1359 Luxembourg
{michael.stieghahn,thomas.engel}@uni.lu

Abstract. Cross-border access to a variety of data defines the daily business of many global companies, including financial institutions. These companies are obliged by law and need to fulfill security objectives specified by legislation. Therefore, they control access to prevent unauthorized users from using data. Security objectives, for example confidentiality or secrecy, are often defined in the widespread eXtensible Access Control Markup Language that promotes interoperability between different systems.

In this paper, we show the necessity of incorporating the requirements of sets of legislation into access control. To this end, we describe our legislation model, various types of contextual information, and their interrelationship. We introduce a new policy-combining algorithm that respects the different precedence of laws of different controlling authorities. Finally, we demonstrate how laws may be transformed into policies using the eXtensible Access Control Markup Language.

1 Introduction

Although research on access control has been a topic of interest for years, the new field of *Legal Engineering* [4], in combination with access control, is of increasing importance. In times of an ongoing global financial crisis, an increasing demand for regulation of financial markets exists. Currently used remote desktop solutions, such as Citrix XenApp, VNC, or NX Nomachine, provide the convenience of a known desktop environment for their users. Such solutions are necessary because traveling employees of global working companies need access to data stored on the servers of their company. However, such remote desktop solutions do not dynamically restrict access to information that is necessary to fulfill a certain task but give full access to data. Similarly, today's access control systems (e.g. access control lists (ACL) and role-based access control (RBAC)) lack the possibility of including legal constraints in their access decisions. Nevertheless, deciding whether an access to specific data under a given context is legal is an indispensable factor for many companies.

We illustrate the necessity for a law-aware access control that incorporates legislation in an international banking application scenario using the following example, which is derived from results of interviews with bank consultants:

Example 1. A consultant travels by plane from country S to a customer located in a country D_1. The legislation of S comprises laws regarding bank secrecy and data protection. The destination country D_1 has a law that concedes the right to privacy; however, it has a restriction of this privacy that allows the border security to check mobile

K. Nakakoji, Y. Murakami, and E. McCready (Eds.): JSAI-isAI, LNAI 6284, pp. 73–86, 2010.

devices regarding their content. Therefore, airport security potentially checks the mobile device[1] and so, to avoid disclosure of confidential information, such data cannot be stored on the device. However, when meeting the customer, the consultant needs to access the data of the customer. Since bank secrecy and privacy can prohibit the use of a remote desktop solution in country D_1, the necessary data has to be transferred in advance to the device after the consultant has left the airport. An active connection possibly reveals a link between a customer and a bank. This breaks bank secrecy. Thus, a remote desktop solution might be the right choice, if the consultant and the customer could instead meet in country D_2, where the legal restrictions are not as strict as in D_1.

The legislation of a country litigates for everyone located within the country. However, accessing data such as confidential customer-related data or strategic information that is hosted in another country introduces the problem of being subject to at least two sets of legislation. The legislation of different countries may vary in respect of bank secrecy, data security, data privacy, cryptography, etc. Therefore, the access control system has to ensure a law-compliant access.

Various approaches, which extend RBAC by a variety of notions of context to overcome its limitations regarding dynamically changing situations, have been widely studied.

Bertino et al. introduce in [1] temporal authorization in a discretionary access control (DAC) system to combine authorization together with start and an expiration time. This approach supports temporal constraints, as we know from the time defined by the task when the data access is required and from the legislation at which time the access is legal. However, the time alone does not reveal whether or not an access to data is legal.

Strembeck and *Neumann* present in [11] an approach to enforce contextual constraints in a dynamic RBAC that checks the current values of contextual attributes for predefined conditions. In their approach, permissions can be associated with context constraints.

Damiani et al. define in [2] the spatially-aware access-control model GEO-RBAC. It enhances RBAC with spatial- and location-based information to model objects, user positions, and roles that are activated based on the position of the user. In add a physical position, users are also assigned a logical and device-independent position. However, binding the activation to roles based on the location information of the user is not sufficient when cross-border data access to confidential data is necessary. Thus, the location information can be used to serve two purposes: first, the location information for the start point and the end point of a connection is used to select the observable set of legislation, and second, it can localize a data access to a specific location to fulfill law-compliance.

Ungureanu and *Minsky* [13] and *Serban et al.* [8] describe a mechanism called Law-Governed Interaction (LGI) that regulates the activities of the participants in an e-commerce transaction. LGI allows participants, who are combined to a so-called open group of distributed heterogeneous agents, to interact with each other with confidence that this interaction is policy-compliant. The policies are called the law of the open group. In contrast to our solution, the term "location" means that laws are defined

---

[1] As happened recently:
http://www.theregister.co.uk/2009/08/11/ripa_iii_figures/,
http://www.theregister.co.uk/2009/11/24/ripa_jfl/

globally but enforced locally. Therefore, location is restricted to a group, a membership in a group, and contracts between participants, such that laws exist that are only valid for certain groups and not globally for all participants. Our solution uses location in the sense of a real location (a specific country or city as well as the proximity of a specific user). We also do not need a means for binding laws to certain users but bind instead to a location itself, because laws are enforced on the basis of the location.

1.1 Approach and Contribution

This paper reports our ongoing research to develop a law-aware access control system. We extend the approach introduced in [9], where we used a logic-based implementation, and in [10], where we used the eXtensible Access Control Markup Language (XACML). In this paper, we demonstrate how the widely used eXtensible Access Control Markup Language can be used to enhance an access control system. Today, XACML has become a *de facto* standard for access control policies. It is widely used to define policies that regulate access to data by providing a standard for access permissions as well as for access requests and their responses. Our contribution is to use the eXtensible Access Control Markup Language to incorporate legislation into access decisions by enriching policies with legal constraints. Those constraints are based on different types of context and their interrelations. By including legislation directly into access decisions, lawfulness can be ensured. To prevent overregulation, our approach guarantees that the access restrictions are only as strict as it is obliged by the legislation of the source and destination country.

1.2 Difference between a Health Care scenario and a Banking Scenario

Bank secrecy obliges financial institutions to secure data, which was given, for example, to provide a service, with appropriate security measures. For a data access from a location different from the head office, a financial institution has to guarantee that a data storage and a data processing is at least as secure as for a local access. Granting access from the outside may open unknown security holes and, therefore, this may compromise the security of this data. Applying the same access rights as for a local access is not sufficient to ensure the security of data, because usually a user is able to access more data than needed for a particular task. Financial institutions, however, have to ensure the security of accessed data and the best practice for the security is to limit access to necessary data. This does not imply that the risk of data leakage itself is decreased, but it minimizes the amount of data that may be lost.

Common examples of access control systems that deal with sensitive data are emergency scenarios, healthcare scenarios, and banking scenarios. A bank scenario differs from the two other scenarios in a fundamental way. The first priority in a bank scenario is to secure data against unauthorized access, data disclosure, and data loss. If this can be fulfilled, the second priority, the service to the customer, may be applied. In contrast, a healthcare scenario and, to an even greater degree, an emergency scenario rates safety over security, because saving life is more important than data security. If no service can be provided due to security reasons, it is inconvenient in the first case, but unacceptable in the second case. Therefore, an access control system for financial environments rather denies than grants access to data in a case of uncertainty.

1.3   Organization of the Paper

The remainder of this paper is organized as follows: In Section 2, we briefly describe the eXtensible Access Control Markup Language and how an XACML system decides about access. In Section 3, we describe our legislation model. To this end, we specify the different types of context information that are needed to incorporate legislation into access decisions. We introduce a new policy-combining algorithm that respects the precedence of different sets of legislation, which may overrule each other. Then, we describe how laws may be manually transformed into XACML policies. Finally, we describe briefly how an access decision is made by a system that follows our approach. Section 4 concludes the paper and outlines future work.

2   XACML

The eXtensible Access Control Markup Language is a declarative access control policy language designed to support authorization systems. XACML is implemented in XML to provides a processing model, describing how to interpret the policies and, as a second part, a request / response context language.

XACML [5] policies are structured as a tree of sub-policies (Fig. 1). Each tree has a defined target and a set of leaves containing a set of rules. A target defines certain conditions to determine whether this policy is applicable to a request. It is specified by four properties: a subject, a resource, an action, and an environment. *Subject* defines a user or process that requests access to a *resource*, which might be a file, a system, or a service. An operation on a resource is defined as *action*. *Environment* defines a set of attributes, which are necessary to decide about access, but which are not related to a specific subject, an action, or an environment. Attributes are features of a subject, a resource, an action or an environment. Rules define how to process a target and consist of Boolean expressions, which are interpreted and executed by the Policy Decision Point (PDP). Rules consist of a target, an effect, and conditions. The latter describe the state of the attributes of the target to satisfy the rule, whereas effect specifies how to proceed (e.g. *permit* or *deny*) if the conditions are satisfied. The response to the request is structured as follows: decision, status, and obligation. There are four possible decisions: *permit*, *deny*, *not applicable*, or *indeterminate*. Not applicable is returned if no rules or applicable policies can be found. Indeterminate indicates that an error occurred during

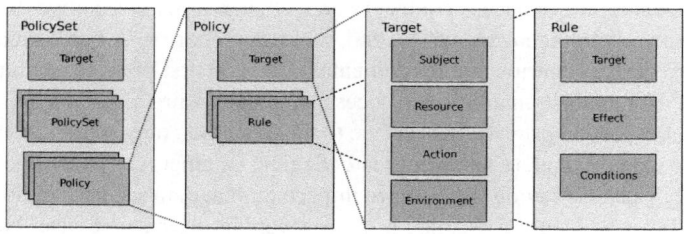

Fig. 1. XACML policy structure

the access decision. Obligations can be attached to the response and direct the Policy Enforcement Point (PEP), for example, to process an access in a designated way. However, XACML does not specify the communication protocol between PEP and PDP.

The XACML specification [5] defines additionally four combining algorithms to specify how policies of a policy set are combined during an access decision process.

2.1   Access Decisions

When connecting to a server, users need to authenticate themselves. Then they can use, for example, a file browser to browse a directory or to open a file. The client sends all actions as XACML requests to the PEP. The PEP resubmits the user's request to the Policy Decision Point. After receiving the request, the PDP starts to evaluate the top-level policy or the policy set. First, the target is checked whether the request matches the target specified in the policy. If a policy is evaluated to false, this policy is not applicable and a further evaluation of this policy is not necessary. Subsequently, the resources and actions specified in the target are evaluated as well. If the PDP evaluates the target to true the policies and rules in the next level below are evaluated. When the evaluation gets to a leaf of the XACML policy tree, the rule's conditions are executed. Every rule has an effect (permit or deny), which is sent back as decision if the condition is evaluated as true. Otherwise a non applicable is returned. The evaluation is performed with respect to the combining algorithm, which is specified for a policy or policy set and defines how the policies and rules need to be processed. During the evaluation, the PDP queries the attributes of the XACML request from the PIP, which collects subject, resource, environment, etc. On completion, the PIP sends the response to the PDP, which can then decide about the access. Finally, the PDP sends its access decision back to the PEP by using the XACML response language. The PEP executes the obligations bound to the policy and sends the final decision (permit / deny) to the user. If the system grants the access, the user is able, for example, to browse files on the file system or using a remote desktop solution.

3   Law-Awareness and Access Control

Remote access within a country, but especially cross-border access, implies that at least one set of legislation needs to be observed. In particular, financial institutions need to ensure both law-compliant access, which includes securing data against attacks, and serviceability. In general, laws define, among other things, conditions to satisfy and describe the handling of data and the access to data. When requesting access to data, a variety of context information can be used to support access decisions. In our approach, we mainly follow the definition of *Dey et al.* [3] for categories of context. In addition to their context types *identity*, *activity* (we use the term *task* as equivalent), *time*, and *location*, we extend the context by *legal constraints* and a *second identity*. Consequently, we describe our set of context information as: *Who* does *What* for *Whom*, *When*, *Where* and subject to *Which* legal constraints.

To improve readability, we elide the prefix *urn:oasis:names:tc:xacml:1.0:* in the urn-definitions of all policies in this document. For the same reason, we also elide *http://www.w3.org/2001/XMLSchema#* from the definition of the data type.

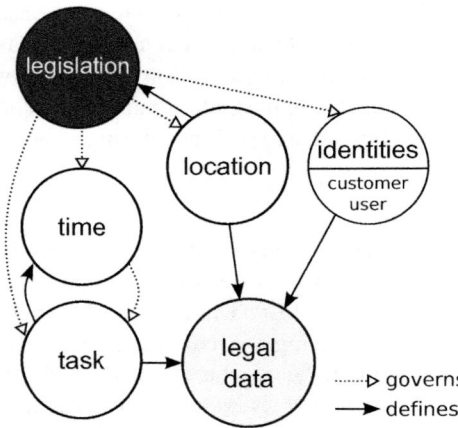

Fig. 2. Schema of a legislation government access control model

3.1 Modeling Legislation

Legislation defines a legal framework for daily business. Therefore, it governs every business action, every transaction and every data access. Figure 2 shows our used law model. Main component of the model is legislation of a region, country, or state. From the perspective of access control, legislation defines which data can be accessed legally with the given context. Laws govern the where (*location*), the who (*user*) and the whom (*customer*), the when (*time*) and the what (*task*). Contrariwise, the location defines, which legislation is valid and has to be observed. At the same time, legislation governs which location is permitted for a data access. Hence, location and legislation are mutually dependent. The identities of user and customer are also mutually dependent, as a user has various customers to advise and a customer may have various users as consultants. The relationship between both is defined in the law by a valid contract between both individuals, between an individual and a company/institution, or between two companies. Contracts including multiple parties are also possible. Time and task interact with each other. A task or an action that is legal is defined from the legislation. For this task, a specific point in time or a time range may be defined indirectly. Typically, the time range rather is specified by "as short as possible, as long as necessary" or during "working hours" than a specific time. Combining two or more sets of legislation will result in a set of legally accessible data for both sets of legislation with the given context. The various types of context information, which serve as basis for an law-aware access decision, remain the same. However, the result depends on the used sets of legislation.

3.2 Identities and Task

Identities (determined by *who* and *whom* in the context definition) are used in two ways: One identity identifies the user, e.g. all consultants of a company. This approach is well known for all access control systems. The other identity defines the customer who is the subject of the data.

A task describes what occurs in a specific situation (e.g. a customer advisory service). In our case, the notion task is a mandatory justification for mobile data access to sensitive data. However, a task is not necessarily required to access non-customer-related information or non-confidential data, such as product information. In our definition, a task is determined by an entry in the diary of a user that also may link the identities, data, location, and time.

3.3 Time

Time ranges represent a common access restriction. When a user requests access to specific information, the access control system checks whether the user is allowed to access this data at the current point in time. The context time is defined for customer-related data by a task. For access to non-customer-related data, the time range can be defined, for example, by a policy of the company to cover the itinerary working hours of a consultant. Due to the different time zones on a travel, the location context serves as input to calculate the correct local time. Sensitive data should be accessible for a limited time only to proactively minimize the risk of unauthorized access or data disclosure.

3.4 Location

Various approaches to context-aware access control systems [7,2] use a location as context information in decisions concerning access. Location describes the physical position of a mobile device. A position specifies not a single location point, but a location space. The method of determination defines the precision and the size of such location space. A position can be described by absolute values (GSM cell or GPS) or relative values (derived from an absolute position or from a proximity measurement).

In our approach, we distinguish between a *legislation location* and an *activity location*. The legislation location determines the validity area of a law – the country or region where a specific legislation needs to be observed. The activity location is the current location from where the user performs the access request. For the activity location, we differentiate between the *expected location* and the *current location*. An expected location is noted by the consultant in the diary. It specifies the location where the user will probably access the data, for example, at the location of the meeting with a customer. In our scenario, if a consultant has to travel abroad to accomplish a task, the supervisor gives an authorization for the travel in advance. This authorization confirms the expected location with a "second set of eyes". A current location is where a consultant is located during an access. This may be specified by an attribute of a subject. The OASIS organization designates the following attribute identifier for the current location of the subject: "urn:oasis:names:tc:xacml:3.0:ec-us:subject:location" [6].

We define a model called *zones$^+$* to categorize locations. Zones$^+$ is an XML location tree where a node can for example be a region (e.g. European Union), a country (e.g. the U.S.A, Japan or Germany) or a state (e.g. New York, Washington D.C.). The two children of such a node contain the areas separated into a restricted area where special law enforcement exist (e.g. in a customs duty area or a police station) and an unrestricted area, which contains all areas that are not defined as restricted. This bisection is used to support the insulation of sensitive data that should not be disclosed, for example, during a customs inspection where a consultant omitted to close the connection to a confidential resource at headquarters.

Policy 1. Definition of a legislation location within the target section

```
<Target>
  <Legislation>
  <LocationMatch MatchId="function:anyURI-equal">
    <AttributeValue DataType="string">Country A</AttributeValue>
    ...
    </LocationMatch>
  </Legislation>
</Target>
```

The definition of XACML itself provides no means of defining a legislative location. Thus, within the definition of the target an additional attribute extends the policy with an identifier that specifies the area where the policy is applicable (Policy 1). As the subject, resource, and action are checked, this attribute is included in the decision as to whether or not this policy is applicable to the current request. Therefore, if an access to data in country S is requested from another country D, both countries are checked. If the location defined in the policy equals either country S or D the policy becomes applicable unless subject, resource, and action do not match the request. Further on, a legislation tag can affiliate different countries. Provided that the countries concerned are listed as AttributeValues of a LocationMatch.

Remark that, in a standard XACML model the legislation-tag will be ignored and, therefore, the decision whether a policy applies to a request is based on subject, resource, and action only. A system, which does not evaluate the legislation tag, is rather over-restrictive than under-restrictive, as legislation-driven policies of different countries tend to constrain access. We also must point out, that instead of using the additional legislation tag to define a legislation location the property environment might be used. In our opinion, proceeding with this approach is more unambiguous than to merge the legislation location in an existing tag. However, this remains a subject to further discussions as the development of the XACML standard continues.

3.5   Legal Constraints

In the previous section we distinguished between *legislation location* and *activity location*. Legislation location means the validity area of a law, which we specify as a legal constraint. The *activity location*, which is the current position of the subject who requests the data, determines the legal constraints that have to be observed. A legislation location can be a *union* (e.g. European Union and the United States of America), a *country* (e.g. Germany, Japan, or Luxembourg), a *state* (e.g. Florida, California, or British Columbia) or an *organization* (e.g. Microsoft Corporation, Allianz SE), where the latter addresses organizational policies.

A single law can influence one or more of the other context information items, for example, if a law prohibits the use of strong encryption mechanisms, which another country presumes, a condition has to reflect this law. Additionally, laws can cause conditional constraints that relate to context information, but cannot be represented by one of the contexts described in this paper. Such a conditional constraint may be, for example, a signed customer agreement. In legal engineering two cases can occur. First, a policy is directly and unambiguously generated from the written natural form of an act. Secondly, a law cannot be transformed directly but has to be divided into several parts and has to be interpreted.

A Precedence-aware Policy-Combining Algorithm becomes necessary to handle the various levels of hierarchy that appear with laws of different controlling authorities. Depending on the controlling authority laws may overrule other laws, for example *national law* of a member state of the European Union overrules *European directives*.

Besides the *law* exists *precepts* (amongst others), which advise of a specific behavior or rule of action and is not mandatory. Existing XACML policy-combining algorithms do not take into account the priority of a law, which we call a precedence of a law. Algorithm 1 shows the pseudocode for a combining algorithm that respects the precedence of the sets of legislation. If the precedence of the first law (*law a*) is lower than the precedence of the second law (*law b*) the function is recalled with flipped variables, which does the implementation and, therefore, the cases to handle shorter (line 3). If both laws evaluate to indeterminate (Alg. 1, line 6) some error occurred and no decision can be made. Similarly, if laws have the same precedence but one evaluates to permit the other to deny, the total evaluation results in indeterminate (Alg. 1, line 10). This is,

---

Algorithm 1. Pseudocode of a Precedence-aware Policy-Combining Algorithm

```
1:  function Combine_{law−permit−override}(a,b)
2:     if (Precedence(b) > Precedence(a)) then
3:        return Combine_{law−permit−override}(b,a)      ▷ Re-call function with turned variables.
4:     end if
5:     if (a = indeterminate or b = indeterminate) then
6:        return indeterminate                           ▷ An error occurred and prevents a decision.
7:     end if
8:     if (Precedence(a) = Precedence(b) then
9:        if (a = deny and b = permit) or (a = permit and b = deny) then
10:          return indeterminate                        ▷ Conflicting decision.
11:       end if
12:    else if (a = permit) then
13:       return permit              ▷ Both legislation allow access or law a overrules law b.
14:    else if (a = not applicable) then
15:       if (b = permit) then
16:          return permit                               ▷ Only law b is applicable.
17:       else if (b = deny) then
18:          return deny                                 ▷ Law b denies access.
19:       else
20:          return not applicable                       ▷ Both laws are not applicable.
21:       end if
22:    else
23:       return deny                    ▷ For all remaining cases the access is denied
24:    end if
25: end function
```

however, a policy conflict that has to be resolved by an additional policy conflict handling. At the moment this has to be performed semi-automatic. Algorithm 1 respects the precedence, which means that as higher the number as "more" important is the law. In other words, if an organizational policy allows an access to data but a law with a higher precedence denies the same access, an access request has to be denied (Alg. 1, line 23). The combining algorithm denies by default (Alg.1, line 23).

It remains to mention that rules evaluating to "not applicable" neither permits nor denies.

This is different to "indeterminate" that indicates a problem that occurred during either the evaluation or the policy-combining process. The result indeterminate needs special attention. By default, indeterminate overrules a permit from another policy, which is required to keep data secure if a decision is not evaluated unambiguously.

Our proposed precedence-aware policy-combining algorithm is fully integrable into existing XACML systems to replace the default policy-combining algorithms.

4   Transformation

Transition of written legislation into a computer-useable form is an essential step preparing law-compliance of applications. As shown in Figure 3, taking legislation as input for a transformation is the first step. This transformation can either be performed automatically, semi-automatically or manually. In the latter case, lawyers and/or security officers read laws and interpret them. Mostly, they perform an interpretative translation and not a bijective transition of laws to policies. The documentation might refer to which policy reflects which law but is not necessarily mandatory. If a law changes the complete policy set needs to be revised in order to recover law-compliance. However, reassessing new or changed laws to existing policy sets is an error-prone and prolonged process if performed manually and includes more than one legislation, as it occurs for cross-border data access.

The output of the applying a transformation (see Fig. 3) is a single policy or set of policies. Such policies can be included into applications, application frameworks, or in any desired system that is able to interpret these policies. A verification process, which is denoted by (3) and (4) in Figure 3, shall cope with incompleteness or misinterpretations of the law-to-policy transformation process. Reporting missing legislative coverage Figure 3 (5) to the legislator to entail on future implementation of those missing laws completes the process of a law-to-policy transformation. Since laws are often written in a domain-specific, fuzzy, and stylistic-advanced notation steps (4) and (5) is

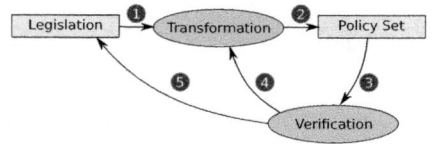

Fig. 3. Schema of a law to policy transition process

important for a transformation. Automatic transitions would require, however, a complete taxonomy of legal terms and their interrelations for an unambiguous interpretation.

Tanaka et al. presented in [12] a two-parted structure of law provisions of Japanese legislation, which consists of a *requisite part* (with subject and condition) and an *effectuation part* (with object, detail, and provision). This structure does also cover parts of the law sentences of the Luxembourgian law.

A single law can provide various types of information such as a: *a)* **condition**, which is a definition of exceptions or states for what a specific action is allowed or denied in a certain context. It defines what needs to be fulfilled to entitle an action; *b)* **default behavior** that defines the action if a decision cannot be made unambiguously; *c)* **detail**, which gives more information to specify a subject in more detail or to restrict effects, for example as annotation; *d)* **entitlement** that is a judgment-free statement of what right is granted; *e)* **general statement** specifying, for example, the purpose for which the law was defined; *f)* **link** as a reference to another law where, for example, a definition is refined or where exceptions may be specified; *g)* **nominal definition**, a definition of terms, which are used later within the document; *h)* **penalty**, a punishment, e.g. terms of imprisonment or payment of money, for breaking the specified law. Mostly, it defines a range from a minimum penalty to a maximum penalty.

To demonstrate the structure of laws, we use *Loi 02-08-2002* from Luxembourg, a law on the *Protection of Persons with regard to the Processing of Personal Data*. Figure 4 shows chapter IV, article 18 letter (1) – transfer of data to third countries of the Luxembourgian law for data protection and privacy. Article 18 describes the main principles.

As shown in Figure 4, letter (1) describes a resource (*data*), an action (*transfer to*) with an annotation of a location as detail (*a third country*), an effect (*may take place only*). The second part defines conditions for the effect (*only where*), which needs to comply to several provisions (*provides an adequate level of protection and complies with the provisions of this Law and its implementing regulations*). This annotation is currently performed by human experts. *A third country* attracts attention and on the first view all countries but Luxembourg are included. However, *Chapter 1, Art. 1:-Definitions, letter (m)* defines third countries as such that are not members of the European Union.

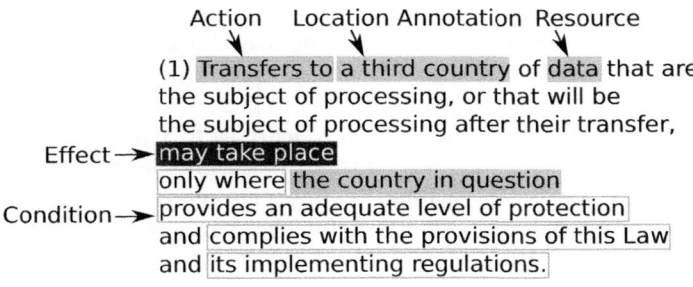

Fig. 4. Identifying transition properties in chapter IV, article 18, letter (1)

Policy 2. Target section of Loi 02-08-2002, chapter IV, article 18, letter 1

```
<Target>
  <!-- Applies to all Subjects -->
  <!-- Applies to all Resources -->
    <Actions>
      <Action>
        <ActionMatch
          MatchId="function:string-equal">
          <!-- Applies to the action transfer with a given destination -->
          <AttributeValue DataType="string">
            transfer
          </AttributeValue>
          <ActionAttributeDesignator AttributeId="action:action-id"
              DataType="string"/>
            <Apply FunctionId="function:not">
            <LocationMatch MatchId="function:string-equal">
      <AttributeValue DataType="string">
      destination
      </AttributeValue>
      <LocationMatch MatchId="function:destination-uri">
          <AttributeValue DataType="string">
          member of european union
      </AttributeValue>
      </LocationMatch>
        </LocationMatch>
      </Apply>
          </ActionMatch>
        </Action>
    </Actions>
    <!-- Applies to all Environments -->
  </Target>
```

Policy 2 shows an excerpt of the target section for chapter IV, article 18, letter (1) (see also Fig. 4). The policy is valid for any subject and any resource, which is designated by XACML through omitting the tags *<Subject>* and *<Resource>*. The policy uses, therefore, "transfer" as action in the target section including an additional location annotation that is expressed by a *<LocationMatch>* and a destination country that is not member of the European Union (*<Apply FunctionId="function:not">*).

The part "complies with the provisions of this Law" is quite straight as our approach includes by default the legislation of source country and destination country of a connection (Policy 3). However, the specification in XACML is more complicated as the AttributeValue is not a static value but dynamic. In practice, the destination of a connection remains statically at one location, in our case Luxembourg. The source country of a connection may change between requests and is, therefore, dynamic. The first part "provides an adequate level of protection" appears very fuzzy. The law does not specify how protective measures are defined and how to satisfy those requirements. However, Chapter IV-Section 23, *special security measures*, lists, for example, the protective measures of access control, usage control, transport control. To demonstrate how laws are we interpret and transform laws into policies we use, as example, *Loi 02-08-2002* from Luxembourg, a law on the *Protection of Persons with regard to the Processing of Personal Data*. Chapter IV, Section 5, letter (f) (Legitimacy of data processing) states that data processing is only permitted if the individual concerned has granted an agreement. Hence, it has to be checked whether a signed and valid agreement for this customer exist before deciding about access (Policy 4).

Policy 3. Condition to observe the legislation of source and destination of a connection

```
<Condition>
  <Apply FunctionId="setLegislation">
    <LegislationAttributeDesignator
      AttributeId="connectionSource" DataType="GEOLocation"/>
    <LegislationAttributeDesignator
      AttributeId="connectionDestination" DataType="GEOLocation"/>
  </Apply>
</Condition>
```

Policy 4. Condition for a signed customer agreement

```
<Condition FunctionId="function:not">
    <Apply FunctionId="urn:larbac:function:signedCustomerAgreement">
        <Apply FunctionId="function:boolean-equal">
            <AttributeValue DataType="string">true</AttributeValue>
        </Apply>
    </Apply>
</Condition>
```

This law is valid for any country, but needs only to be observed for customer-related data. Article 19. Derogations, letter (a) is written: "the data subject has given his consent to the proposed transfer", which results in the same condition as Article 5, letter (f) (Policy 4).

5 Conclusion and Future Work

In this paper, we addressed the problem of using the eXtensible Access Control Markup Language (XACML) for law-aware access control. We stressed the necessity to incorporate legislation into mobile cross-border access. To this end, we demonstrated how various types of context information are defined within an XACML policy, and described the different types of context and their interrelations. We introduced a new policy-combining algorithm that respects precedences of laws, which became necessary to facilitate the evaluation of transformed policies of different controlling authorities.

Currently, we are implementing a prototype of the law-aware access control system using our law-enriched XACML policies. We are also investigating semi-automatic methods to transform laws into XACML policies that support human experts during a law to policy transformation. For this, we build up an ontology consisting of legal terminologies and a mapping to XACML policies.

Acknowledgement

This work was supported by the Fonds National de la Recherche Luxembourg under grant number AFR 05/109.

References

1. Bertino, E., Bettini, C., Ferrari, E., Samarati, P.: A Temporal Access Control Mechanism for Database Systems. IEEE Transactions on Knowledge and Data Engineering 8(1), 67–80 (1996)

2. Damiani, M.L., Bertino, E., Catania, B., Perlasca, P.: GEO-RBAC: A Spatially Aware RBAC. ACM Trans. Inf. Syst. Secur. 10(1), 2 (2007)
3. Dey, A.K., Abowd, G.D.: Towards a Better Understanding of Context and Context-Awareness. In: Computer Human Intraction 2000 Workshop on the What, Who, Where (1999)
4. Katayama, T.: Legal Engineering - An Engineering Approach to Laws in e-Society Age. In: Proceedings of the 1st International Workshop on JURISIN (2007)
5. Moses, T.: eXtensible Access Control Markup Language TC v2.0 (XACML). In: Organization for the Advancement of Structured Information Standards (OASIS) (February 2005)
6. Organization for the Advancement of Structured Information Standards (OASIS). XACML 3.0 Export Compliance-US (EC-US) Profile Version 1.0 (September 2009)
7. Schilit, B., Adams, N., Want, R.: Context-Aware Computing Applications. In: IEEE Workshop on Mobile Computing Systems and Applications, Santa Cruz, CA, US (1994)
8. Serban, C., Chen, Y., Zhang, W., Minsky, N.: The Concept of Decentralized and Secure Electronic Marketplace. Electronic Commerce Research 8(1-2), 79–101 (2008)
9. Stieghahn, M., Engel, T.: Law-aware Access Control for International Financial Environments. In: MobiDE 2009: Proceedings of the Eighth ACM International Workshop on Data Engineering for Wireless and Mobile Access, pp. 33–40. ACM, New York (2009)
10. Stieghahn, M., Engel, T.: Using XACML for Law-aware Access Control. In: 3rd. International Workshop on Juris-informatics (JURISIN), pp. 118–129 (2009)
11. Strembeck, M., Neumann, G.: An Integrated Approach to Engineer and Enforce Context Constraints in RBAC Environments. ACM Trans. Inf. Syst. Secur. 7(3), 392–427 (2004)
12. Tanaka, K., Kawazoe, I., Narita, H.: Standard structure of legal provisions - for the legal knowledge processing by natural language (in Japanese). IPSJ Research Report on Natural Language Processing, 79–86 (1993)
13. Ungureanu, V., Minsky, N.H.: Establishing Business Rules for Inter-Enterprise Electronic Commerce. In: Herlihy, M.P. (ed.) DISC 2000. LNCS, vol. 1914, pp. 179–193. Springer, Heidelberg (2000)

Part II
Knowledge Collaboration in Software Development

3rd International Workshop on Supporting Knowledge Collaboration in Software Development (KCSD2009)

Masao Ohira and Yunwen Ye

[1] Nara Institute of Science and Technology, 8916-5 Takayama, Ikoma, Nara 630-0192, Japan
[2] SRA Key Technology Lab, 2-32-8 Minami-Ikebukuro, Toshima, Tokyo 171-8513, Japan
masao@is.naist.jp, ye@sra.co.jp

1 Introduction

The creation of modern software systems requires knowledge from a wide range of domains: application domains, computer hardware and operating systems, algorithms, programming languages, vast amount of component libraries, development environments, the history of the software system, and users. Because few software developers have all the required knowledge, the development of software has to rely on distributed cognition by reaching into a complex networked world of information and computer mediated collaboration. The success of software development, therefore, hinges on how various stakeholders are able to share and combine their knowledge through cooperation, collaboration and co-construction.

The overall goal of the series of the workshop seeks to gain an improved understanding on the theoretical, social, technological and practical issues related to all dimensions of knowledge collaboration in software development, and to explore opportunities for automated support, such as the timely acquisition of external knowledge and the facilitation of collaboration among developers.

KCSD2009 is the 3rd installment of the workshop. The first KCSD workshop (KCSD2005) took place on December 15, 2005, Taipei, as a part of the IEEE 12th Asia-Pacific Software Engineering Conference (APSEC 2005). The second KCSD workshop (KCSD2006) was collocated with the 21st IEEE/ACM International Conference on Automated Software Engineering (ASE2006), and took place on Sept 19, 2006, Tokyo.

2 Workshop Topics

KCSD2009 was a two-day workshop and took place on Nov.19-20, 2009 at Tokyo. It focuses on the transfer of knowledge among software developers and the collaborative creation of new knowledge that is needed for the development of software systems. The particular interests of KCSD2009 include:

- formation of shared understanding between users and developers during conceptualization, design, deployment and use of software systems;
- technical issues in accessing external knowledge resources and acquiring expertise from peer developers;
- social issues in facilitating knowledge transfer;

K. Nakakoji, Y. Murakami, and E. McCready (Eds.): JSAI-isAI, LNAI 6284, pp. 89–90, 2010.
© Springer-Verlag Berlin Heidelberg 2010

- socio-technical approaches to motivating participation in knowledge collaboration;
- utilization of social networks to connect developers for knowledge collaboration;
- understanding how knowledge is accumulated, transferred and shared among software developers;
- analyzing and understanding the unique features of knowledge collaboration specific to software development such as pair programming, inspection, maintenance and end-user development.

3 Workshop Organization

Workshop Co-Chaires:

Masao Ohira, Nara Institute of Science and Technology, Nara, Japan
Yunwen Ye, SRA Key Technology Lab, Tokyo, Japan

Program Committee:

Daniela Fogli (University of Brescia, Italy)
Mark Grechanik, (Accenture Technology Labs / University of Illinois, USA)
André van der Hoek (University of California, Irvine, USA)
Reid Holmes (University of Washington, USA)
Katsuro Inoue (Osaka University, Japan)
Yasutaka Kamei (NAIST, Japan)
Ken-ichi Matsumoto (NAIST, Japan)
Kumiyo Nakakoji (SRA Key Technology Lab / University of Tokyo, Japan)
Cleidson de Souza (Federal University of Para, Brazil)
Thomas Zimmermann (Microsoft Research, USA)

4 Workshop Outputs

The workshop featured two keynote talks. Professor André van der Hoek of University of California, Irvine presented on "Knowledge Collaboration in Distributed Software Development", and Dr. Shuichiro Yamamoto of NTT Data Corporation addressed on "Understanding Networked Collaboration".

KCSD2009 accepted 8 full papers and 3 position papers. All accepted papers were carefully reviewed by the program committee. After a second round of review, we selected 6 papers to be included in this post-workshop book.

On the Central Role of Mailing Lists in Open Source Projects: An Exploratory Study

Emad Shihab, Nicolas Bettenburg, Bram Adams, and Ahmed E. Hassan

Software Analysis and Intelligence Lab (SAIL)
Queen's University
Kingston, K7L 3N6, Canada
{nicbet,emads,bram,ahmed}@cs.queensu.ca

Abstract. Mailing lists provide a rich set of data that can be used to improve and enhance our understanding of software processes and practices. This information allows us to study development characteristics like team structure, activity, and social interaction. In this paper, we perform an exploratory study on the GNOME project and recover operational knowledge from mailing list discussions. Our findings indicate that mailing list activity is driven by a dominant group of participants, that it is greatly connected to development activity, yet influenced by external factors like market competition. Our results provide a broad picture of the central role played by mailing lists in open source projects.

1 Introduction

Most open source developers communicate through mailing lists. This style of communication makes mailing lists a rich source of information which researchers can use to understand software processes and improve development practices. Mailing lists have been used to infer social structure [4, 10], identify architectural changes [1], and most recently to study the code review process [3, 13, 17].

However, understanding the generality of the results derived from mailing lists requires that we first understand how mailing lists are used in practice and the impact of their usage patterns on the information in the lists. For example, previous studies (e.g. [4]) studied the social structure of developers using mailing lists, however, does this social structure change over time? How fast does the structure change?.

The central role played by mailing lists is depicted in Figure 1. Developers use mailing lists to discuss a variety of issues and project decisions [1, 9]. Many of these issues and decisions are related to and affect the source code. These issues are often driven by external factors such as the introduction of new features in competing products.

In this paper, we perform an exploratory study on the role played by mailing lists. Performing an exploratory study on mailing lists provides a holistic view of their role. This holistic view enhances the understanding of the findings of in-depth studies, unveils details which may not be apparent through in-depth studies and helps identify interesting directions for future research.

To perform our study, we use the mailing lists from 22 GNOME projects. The study centers around the following aspects, shown in Figure 1, in an open source project:

K. Nakakoji, Y. Murakami, and E. McCready (Eds.): JSAI-isAI, LNAI 6284, pp. 91–103, 2010.
© Springer-Verlag Berlin Heidelberg 2010

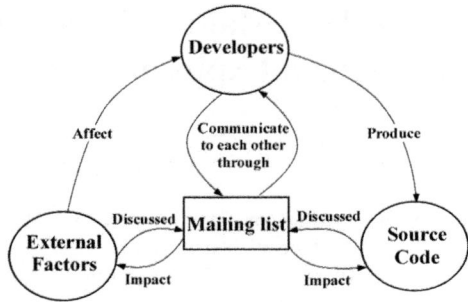

Fig. 1. The central role of Mailing lists in open source projects

- **Developers:** We characterized the communication style of mailing list's participants, i.e., the developers from the development mailing lists.
    *We found that a small number of developers play a central role in driving
    the mailing list activity. We also found that these developers remain stable
    throughout the lifetime of a project.*
- **Source code:** We explored the impact of mailing list activity on the source code
  activity, i.e., changes.
    *We found that there is a high correlation between mailing list activity and
    source code activity.*
- **External Factors:** We examined the effect of external factors, such as competing
  products on mailing list activity.
    *We found that competing products shape and drive many of the discussions
    on mailing lists.*

Overview of Paper. The rest of the paper is organized as follows: Section 2 discusses
the motivation for using the GNOME project as a case study and presents statistics
about the project. We present and analyze our findings in Section 3. The threats to validity are discussed in Section 4 and the related work is presented in Section 5. Section 6
concludes the paper.

2   GNOME as a Case Study

In this section, we detail the case study project used in our study. The GNOME project
is composed of approximately two million lines of code and has more than 500 different contributors from all over the world [8]. The GNOME project is composed of many
small projects that cover a wide range of applications, e.g., email client, text editor,
and file manager. The main source of communication for GNOME developers is the
developer mailing list for each project. These projects vary in size, age, user and developer base. We expect these differences in size, age, and domain to have an impact on
the mailing lists of these projects. Therefore, studying the mailing lists of the different
projects can lead to interesting and generalizable findings and open new directions for
future research.

Table 1. General overview of the GNOME mailing lists studied

Project name	Start date	Number of Messages	Participants	Age (months)	Threads	Application Domain
Deskbar Applet	Oct-05	1,098	106	39	340	Search interface
Ekiga	Aug-06	5,389	690	29	1,200	Teleconferencing
Eog	Mar-01	458	106	93	233	Image viewer
Epiphany	Dec-02	5,735	905	73	1,608	Web browser
Evince	Jan-05	1,358	415	48	566	Document viewer
Evolution	Jan-00	53,927	6,026	96	15,718	Email client
Games	Feb-03	1,590	190	71	531	Computer games
Gdm	Mar-00	2,578	675	105	1,040	Display manager
Gedit	Apr-00	2,237	530	104	919	Text editor
Multimedia	Oct-00	1,646	273	98	507	Multimedia library
Network	Aug-03	673	105	65	267	Network tools
Power Manager	Jan-06	1,059	199	36	305	Power management
Themes	Jan-98	1,310	221	132	447	Window manager
Utils	Oct-04	358	106	51	279	Utility applications
Control Center	Dec-99	1,478	168	97	311	Configuration
Libsoup	May-06	83	24	32	41	HTTP library
Metacity	Sep-05	262	48	40	59	Window manager
Nautilus	Apr-00	22,488	2,384	105	5,582	File manager
Orca	Jan-06	11,930	516	36	3,598	Screen reader
Screensaver	Oct-05	139	25	39	30	Screensaver
Seahorse	Jun-07	252	34	19	116	Encryption management
System tools	Nov-99	1,832	327	98	792	System admin tools

Table 1 presents a general overview of the mailing lists used for this study. The `Project name` column lists the name of the GNOME module. The `Start date`, `Number of Messages`, `Number of Participants`, `Age` and `Number of Threads` columns list the month and year of the first commit to the project's trunk (derived by examining the source control repository for the project), the total number of messages, the number of participants, the age, and the number of threads of the GNOME projects, respectively. In addition, the `Application Domain` column lists the application of the project. All calculations are based on the participation from the start date listed till the end of 2008, inclusive.

3 Results and Analysis

We now study the three aspects outlined in Figure 1 using the GNOME mailing list data. Subsections 3.1 and 3.2 cover the developers aspect, subsection 3.3 covers the source code aspect and subsection 3.4 covers the external factors aspect. We start each subsection by presenting our motivation to explore the aspect. We then describe the approach that we used to perform our exploration. Finally, we present our results and outline our main findings.

Since most of the GNOME mailing lists have low activity, we will often use the Evolution and Nautilus projects to more closely explore many of our findings since the two projects account for more than 65% of the total messages. We highlight the results that generalize for the rest of the 20 projects, where applicable.

3.1 Communication Style in Mailing Lists

Is mailing list activity mostly driven by a few participants (a dominant group) or is the participation evenly distributed? Does the dominant group engage in discussions with others or is it mostly involved in internal discussions?

Motivation. The Pareto principle (also known as the 80-20 rule), which states that the majority of the effects come from a minority of the causes, has applications in many fields. For instance, research shows that 20% of the code contains 80% of the bugs [7]. We hypothesize that there exist a few key participants (who we call the *dominant group*) in mailing lists, that are responsible for most of the messages posted on the mailing list. Most likely, they are members who are very knowledgeable about the project and use their knowledge to support newcomers and casual participants (who we call the *casual group*). It is important for us to investigate whether these experts exist on mailing lists for two reasons: 1) one can address his/her questions directly to such experts to receive a more accurate and speedy response and 2) the discussions of these experts can be used for future reference by others who are less knowledgeable about the project.

In addition, if such a group exists, we would like to know if they actively engage in discussions with others who are outside of the dominant group. If in fact they do engage with others then we can safely assume that newcomers and less experienced developers will benefit from these experts. If we determine otherwise, i.e. that the dominant group is a closed group, then newcomers and other participants may be better off reading previous discussions and learning from them rather than attempting to establish direct contact with the dominant group members.

Approach. We measured the number of messages contributed by the top 10% most active participants, who we call the dominant group. We found evidence that in fact there does exist a dominant group for each of the 22 GNOME mailing lists. The dominant group contributes a large amount of the messages posted.

Then, we examined the active discussion threads and classified these active threads into threads with:

- **Dominant group members only:** A high number of such threads implies that the dominant group is a closed group that does not engage with others.
- **Dominant and casual members together:** A high number of such threads is a good indicator of a stimulating mailing list where expert and casual participants actively engaging in discussions.
- **Casual group members only:** A high number of this type of discussion would indicate that the casual members are not integrated into the mailing list.

Results. In addition to finding out that there exists a dominant group in each mailing list, we quantified their contribution. We found that on average the dominant group accounts for approximately 60% of the messages. *This finding is consistent across all of the 22 GNOME projects.* We did not observe a consistent finding when we considered the top 20% of the participants (i.e. we did not find evidence of the Pareto principle).

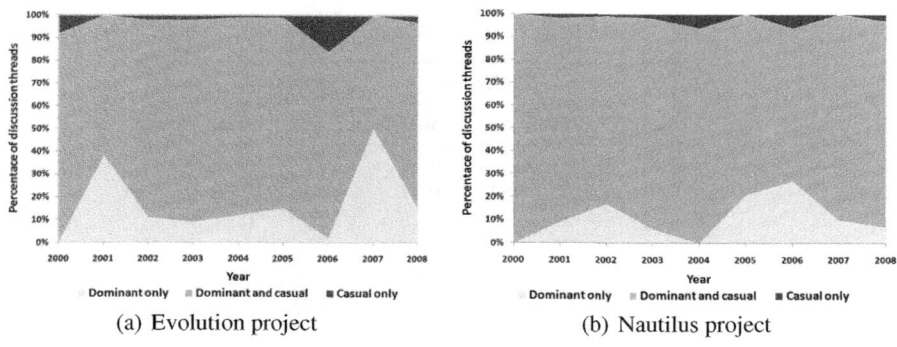

(a) Evolution project   (b) Nautilus project

Fig. 2. Distribution of discussion types

We plot the number of threads for the Evolution and Nautilus projects in Figures 2(a) and 2(b), respectively. In both projects, we found that the majority of the active discussions involve dominant and casual group members. On average, in 82% of the discussions dominant and casual group members were present. In 16% of the discussions, dominant group members were discussing exclusively and in the remaining 2% of the discussions the casual members discussing exclusively. We believe that it is a sign of a productive mailing list when the two groups actively engage in discussions, with the dominant group members most likely playing a supporting role for the casual group members.

However, in some cases a high percentage in discussions that involve dominant and casual members may not be desired. For example, some dominant group members may be overwhelmed by a high number of questions from casual members (since casual members may make unreasonable requests from more knowledgeable dominant group members). Whether a high number of discussions between casual and dominant group members is indicative of a productive mailing list depends on the product domain and the mailing list's members' knowledge.

Conclusion. 10% of mailing list participants (the dominant group) contribute 60% of the messages in a mailing list. The dominant group is very active and is engaging with outside-members, i.e. casual members.

3.2   Stability of Mailing List Participants

Do dominant group members change over time? If so, how much are they changing by? How is their stability compared to rest of the mailing list participants?

Motivation. As we have seen in the previous subsection, the dominant group plays an important role in the mailing list. They contribute the majority of messages posted and are involved in approximately 96% of active discussions. For this reason, it is quite important that dominant group members do not change frequently. We study the stability of the dominant group. In particular, we measure the variation in the dominant group over time. A relatively stable dominant group (i.e. one that does not change frequently)

Table 2. Cosine distance of dominant and casual groups of the Evolution and Nautilus projects

	Evolution		Nautilus	
Year	Dominant	Casual	Dominant	Casual
2000 - 01	0.68	0.11	0.73	0.20
2001 - 02	0.74	0.11	0.55	0.20
2002 - 03	0.63	0.16	0.40	0.21
2003 - 04	0.74	0.16	0.85	0.23
2004 - 05	0.84	0.16	0.76	0.24
2005 - 06	0.70	0.19	0.95	0.24
2006 - 07	0.35	0.17	0.88	0.19
2007 - 08	0.80	0.15	0.77	0.16
Average	**0.69**	**0.15**	**0.73**	**0.21**

is desirable because it means that dominant group members spend enough time in the project and achieve a higher level of expertise to better support casual group members.

Approach. To measure the stability of members in the dominant group, we performed two studies:

- **Dominant group change over time:** We measured the change between two consecutive years. This gives us a measure of how much a dominant group changes by from one year to the next.
- **Dominant group change compared to casual group change:** We measured the change of the casual group for two consecutive years and compared it to the change in the dominant group.

We used the Cosine Distance (CD) similarity metric to measure the similarity between the groups in two consecutive years. The CD metric outperforms other simple measures such as intersection or proportion which only measure the existence of a participant but not their level of contribution. The CD similarity is defined as:

$$CD(P, Q) = \frac{\sum_x P(X)Q(X)}{\sqrt{\sum_x P(X)^2}\sqrt{\sum_x Q(X)^2}}, \tag{1}$$

where $P(X)$ and $Q(X)$ represent the two input distributions to be compared. A value of 0 for the CD metric means that the group has changed drastically across two years with no members in common. A value of 1 for the CD metric indicates that the group is the exact same (i.e. is it a very stable group).

The Cosine Distance metric takes as input two participation distributions – one for each of the years under study. Each distribution has the contribution of each of the participants for that year. So when comparing the dominant group for the year 2000 and year 2001, the 2000 and 2001 participation distribution for the dominant group is used. One major challenge we faced when conducting this study was the use of multiple aliases by developers [4]. We used heuristics based on regular expressions to address this challenge as detailed in our previous work [2].

Results. The calculated CD values for the Evolution and Nautilus projects are shown in Table 2. It is observed that the dominant group is more stable than the casual group. On average, the dominant group is 3 times more stable than the casual group. *These two findings are observed across all of the 22 GNOME projects.* The same stability of social structures were also observed with the FLOSS projects [21]. This is a positive sign about the health of the dominant groups of many of these projects. Dominant group members, who are critically important to the mailing list of the project are stable enough to pass their knowledge to newcomers and casual group members.

Conclusion. The participants in the dominant group are very stable over time. On average, they are about 3 times as stable as casual participants.

3.3   Source Code Activity and Mailing List Activity

Can mailing list activity be used to infer information about source code activity (amount of work done on the source code)?

Motivation. Since mailing lists are the main source for developer communication [9], we expect that mailing lists contain useful information about the source code of a project. We want to explore if we can infer the types of source code changes and the level of activity done on the source code through the mailing list activity. Because developers often use the mailing list to discuss their source code changes and get assistance or feedback on these changes [13], we hypothesize that there will be high correlation between the mailing list activity and the code activity. Or in other words, the more work done on the source code, the more it will be discussed on the mailing list and vice-versa.

Approach. We mined the SVN source control repository and extracted the number of lines added, removed and modified per year for each project. We defined a Code Activity (CA) metric, defined as:

$$CA(Y) = A_Y + R_Y + M_Y, \tag{2}$$

where A_Y, R_Y and M_Y refers to the number of lines of source code added, removed and modified in year Y, respectively. We used this metric and measured the correlation between it and the mailing list activity, i.e., the number of messages per year. Furthermore, we examine the correlation between the number of messages and the type of the performed change (add, delete, modify).

Results. The number of messages per year and the Code Activity for the Evolution and Nautilus projects are plotted in Figures 3(a) and 3(b), respectively. It can be observed that there is a high correlation between the number of messages on the mailing list and the Code Activity metric. This finding shows that developers do rely heavily on the mailing list to discuss source code changes. As for the correlation between the level of mailing list activity and the type of change, we present the results in Table 3. We found that in the Evolution project, the highest correlation was between the number of messages and the lines of code added ($\rho = 0.83$). On the other hand, in the case of the Nautilus project, we found that the highest correlation is between the number of messages and the lines of code modified ($\rho = 0.85$). It seems that in the Evolution project, participants are discussing code additions more than they are discussing code removal

Table 3. Correlation between the number of messages per year and the type of source code change

	Type of change		
Project	Add	Remove	Modify
Evolution	**0.83**	0.60	0.61
Nautilus	0.32	0.53	**0.85**

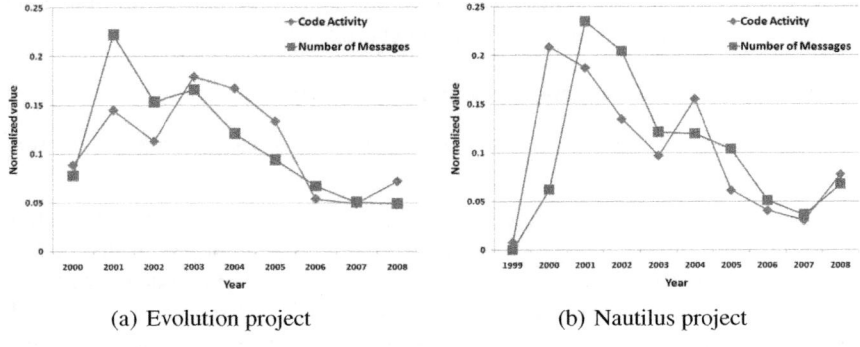

(a) Evolution project　　　　　　　　(b) Nautilus project

Fig. 3. Number of Messages and Code Activity

or modifications, while for the Nautilus project, code modifications are being discussed more than code additions and removals. We believe that further investigation is needed here to better understand the rationale for this discrepancy between both projects and whether it indicates different development and communication styles.

To verify, we measured the occurrence of terms that indicate code additions and code modifications in the mailing lists of the two projects. Since most commonly, code additions involve the introduction of new features, we classified the terms "new features" and "feature request" as indicators of code additions. Code modifications are usually carried out to fix bugs which are found during the testing phase and applied via patches. For this reason, we associate the terms "bug", "patch", "testing", and "maintain" to code modifications. We observed that in the Evolution mailing list, the terms associated with the addition of new features were mentioned in 57% more messages than on the Nautilus mailing list. On the other hand, the terms associated with code modifications were mentioned in 75% more messages in the Nautilus mailing list compared to the Evolution mailing list. The findings are consistent with our correlation results shown in Table 3.

Conclusion. Mailing list activity is closely related to source code activity. In addition, mailing list discussions are good indicators of the types of source code changes being carried out on the project.

3.4   Effect of External Factors on Mailing List Activity

Can we observe the effect of external factors on mailing list activity?

Motivation. One of the benefits of studying mailing lists is that they can provide us with knowledge about issues that indirectly affect a project, i.e., external factors. Market competition and management changes are examples of external factors. Such knowledge about external factors is often hard to uncover as it is not recorded in the source code or documentation. However, this knowledge is very important since it helps explain certain observed behaviors, such as an increase in bugs or the lack of interest in a project (and maybe its eventual death). We attempt to observe the effect of external factors on mailing list activity.

Approach. Due to space limitation, we perform the study of external factors on the Evolution project only. However, we note that our approach can be applied to any other project. We study the mailing list activity trend and perform two types of analysis: quantitative and qualitative analysis. In the quantitative analysis study, we treat the bodies of all email messages as a bag-of-words and compare the occurrence of the names of competing mail clients ("gmail", "outlook", and "thunderbird") to the occurrence of the terms: "evolution" and "evo" (a short hand form often used to refer to the evolution project). A rise in the number of times a term occurs indicates that it is being discussed more, hence it has a greater impact. In the qualitative study, we read through several email postings to better understand and clarify our quantitative findings.

Results.

Quantitative analysis: Looking at Figure 4(a), we observe that the activity on the Evolution mailing list is increasing from 2000 to 2001. This increase can be attributed to the creation of Ximian at the end of 1999, which was created to continue the development of the Evolution project [8]. This acquisition increased the attention and support for the Evolution project, hence the continuing increase in mailing list activity.

Then, from the year 2001 on, we observe a steady decline in mailing list activity (except for a small increase in activity in the year 2003). Market competition, along with organizational changes may have caused this decline. The results of the quantitative study (which measures the frequency of occurrence of terms in the message bodies per year) are shown in Figure 4(b). We observe a steady decrease in the use of the terms "evolution" and "evo", suggesting that the Evolution project is being discussed less frequently. At the same time, there is a steady increase in the number of times its market competitors "gmail", "outlook" and "thunderbird" are being mentioned.

Qualitative analysis: We read through several mailing list posting to better understand our aforementioned quantitative findings. The following quotations are excerpts from discussions that took place when a declining level of activity was observed:

"...Furthermore, I can't find where in the Tools menu to change this: the option is no longer present on any of the dialog boxes. Which is why I'm sending this with Thunderbird..."

"...Unless Ximian implements some features that aren't important to Ximian but are important to its users, evo will be relegated to "toy" status. I'm currently struggling to remain with my current distro of SuSE+Ximian in my business, but the lack of meaningful support in both components is forcing my hand to look around for another solution..."

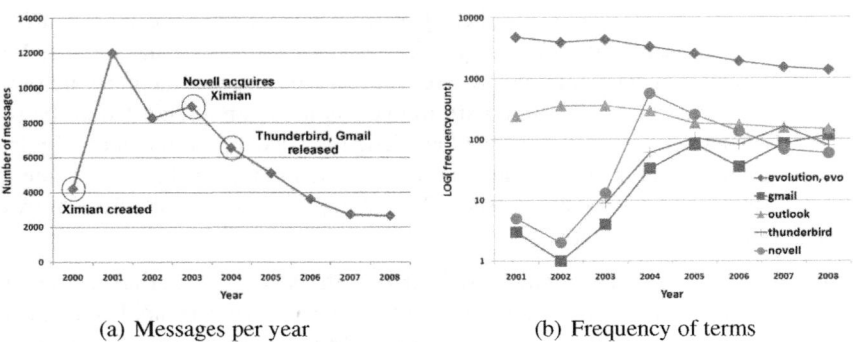

(a) Messages per year   (b) Frequency of terms

Fig. 4. Messages per year and Frequency of terms on the Evolution mailing list

We believe that these excerpts show that the Evolution mail client was and is losing market share due to competition from other competing mail clients, such as Thunderbird, with many of the postings pointing people to competing products.

As for the spike in activity on the Evolution mailing list in the year 2003, we believe this can be attributed to Novell's acquisition of Ximian in late 2003 [8]. We counted the occurrence of the term "novell" in the mailing list and found that the number of times the term "novell" was mentioned on the Evolution mailing list spiked from 13 in 2003 to 574 in 2004 (as depicted in Figure 4(b)). This spike is most likely due to hype surrounding Novell's acquisition, which quickly dies off in the coming years.

This study on external factors suggests that mailing lists can be leveraged to study the effect of external factors on a project. Furthermore, such information can be used to explain design decisions that happened in the past.

Conclusion. External factors affect mailing list activity.

4   Threats to Validity

In our stability analysis, we used the names of developers as identifiers. Although we used heuristics to resolve multiple aliases [2] (i.e. participants who use multiple email address and names), we were not able to deal with some rare cases. Additionally, in our study we assume that all mailing list participants are developers. This assumption is true for the vast majority of the cases (especially since we are considering developer mailing lists), but in some cases, it is possible that a participant on the developer mailing list is not engaged in any developmental effort.

In our studies on source code activity and external factors, we measure the frequency of key terms that we associate with specific topics (i.e. the term "maintain" with the topic maintenance). Although our list is not exhaustive and does not contain all the terms that may be associated with the respective topic, we believe that the terms used in our study are the most common and cover the majority of the terms that would be used to refer to the topic.

Finally, our findings may not generalize to all open source projects.

5   Related Work

Previous work used mailing lists to study the social structure of developers. Bird *et al.* [4] used mailing lists to study the social networks created by developers and non-developers. In their follow-on work [6], they extracted the sub-community structure from these social network and studied their evolution over time. Ogawa *et al.* [11] used Sankey diagrams to visualize evolving networks in mailing lists and concluded that social behavior can be related to events in a project's development.

In addition, several studies used mailing lists to study developer morale, work times and the code review process. Rigby and Hassan [14] performed a psychometric study on the Apache httpd mailing list to identify the personality types of open-source software developers and gain insight on the level of optimism in pre- and post release phases. Tsunoda *et al.* [16] used mailing lists to analyze developer work times and found that the ratio of committer messages sent during overtime periods is increasing every year. Weissgerber, Neu and Diehl [17] used mailing lists to study the likelihood of a patch getting accepted.

Furthermore, other studies used mailing lists to study developer coordination, motivation and knowledge sharing. Yamauchi *et al.* [18] studied the coordination mechanisms used by OSS developers to achieve smooth coordination. They found that spontaneous work coordinated afterward is effective, rational organizational culture helps achieve agreement among OSS members and communications media, such as CVS and mailing lists, moderately support spontaneous work. Lakhani and von Hippel [20] used mailing lists to study the motivating factors of OSS participants to perform mundane tasks. They found that direct learning benefits is one of the main motivators for these participants to conduct such tasks. Sowe *et al.* [19] studied knowledge sharing between developers in mailing lists. They found that developers share knowledge a lot.

Other work combined the information extracted from mailing lists with information from other repositories (e.g. the source code repository). Robles and Gonzalez-Barahona [15] used information from multiple historical archives to assist in accurately identifying actors. Baysal and Malton [1] used the similarity between mailing list and source code archives to identify architectural changes. Bird *et al.* [5] combined the use of mailing lists and the source code repository to study the time it takes for developers to be invited into the core group of a project.

Our work recognizes the central role played by mailing lists and, to the best of our knowledge, is the first to perform an exploratory study using a large number of mailing lists. The study on the communication style of participants and their stability is novel and complements previous work. For example, previous work on social network analysis, developer morale, work times and evolution could have treated dominant and casual group differently and put more emphasis on the dominant group findings. Doing so would enhance the impact of their findings and provide a better understanding of the phenomena being observed. The findings from our source code activity and external factors studies can assist researchers who use mailing lists in combination with source code repositories (e.g. [12, 1]) better understand the relationship between the two. Further, taking into account the effect of external factors may help explain some unexpected observations.

6   Conclusions

In this paper, the central role of mailing lists was studied through an exploratory study. The study centered around three aspects: developers, source code and external factors.

Our findings indicate that a small number of participants (dominant group) account for the majority of the messages posted on mailing lists. The dominant group is very active and engaging with others and its composition is very stable (3 times more stable than casual members). In addition, we found that mailing list activity is closely related to source code activity and mailing list discussions are good indicators of the types of source code changes being carried out on the project. Lastly, we showed that external factors affect mailing list activity.

References

1. Baysal, O., Malton, A.J.: Correlating social interactions to release history during software evolution. In: MSR 2007, p. 7 (2007)
2. Bettenburg, N., Shihab, E., Hassan, A.E.: An empirical study on the risks of using of off-the-shelf techniques to process mailing list data. In: ICSM 2009 (2009)
3. Bird, C., Gourley, A., Devanbu, P.: Detecting patch submission and acceptance in oss projects. In: MSR 2007 (2007)
4. Bird, C., Gourley, A., Devanbu, P., Gertz, M., Swaminathan, A.: Mining email social networks. In: MSR 2006, pp. 137–143 (2006)
5. Bird, C., Gourley, A., Devanbu, P., Swaminathan, A., Hsu, G.: Open borders? immigration in open source projects. In: MSR 2007, p. 6 (2007)
6. Bird, C., Pattison, D., D'Souza, R., Folkiv, V., Devanbu, P.: Latent Social Structure in Open Source Projects. In: Nyberg, K. (ed.) FSE 2008. LNCS, vol. 5086, pp. 24–35. Springer, Heidelberg (2008)
7. Boehm, B., Basili, V.R.: Software defect reduction top 10 list. Computer 34(1), 135–137 (2001)
8. German, D.M.: The gnome project: a case study of open source, global software development. Software Process: Improvement and Practice 8(4), 201–215 (2004)
9. German, D.M.: Using software trails to reconstruct the evolution of software: Research articles. J. Softw. Maint. Evol. 16(6), 367–384 (2004)
10. Hossain, L., Wu, A., Chung, K.K.S.: Actor centrality correlates to project based coordination. In: CSCW 2006, pp. 363–372 (2006)
11. Ogawa, M., Ma, K.-L., Bird, C., Devanbu, P., Gourley, A.: Visualizing social interaction in open source software projects. In: Asia-Pacific Symposium on Visualization, pp. 25–32 (2007)
12. Pattison, D., Bird, C., Devanbu, P.: Talk and work: a preliminary report. In: MSR 2008, pp. 113–116 (2008)
13. Rigby, P.C., German, D.M., Storey, M.-A.: Open source software peer review practices: A case study of the apache server. In: ICSE 2008, pp. 541–550 (2008)
14. Rigby, P.C., Hassan, A.E.: What Can OSS Mailing Lists Tell Us? A Preliminary Psychometric Text Analysis of the Apache Developer Mailing List. In: MSR 2007, p. 23 (2007)
15. Robles, G., Gonzalez-Barahona, J.M.: Developer identification methods for integrated data from various sources. SIGSOFT Softw. Eng. Notes 30(4), 1–5 (2005)
16. Tsunoda, M., Monden, A., Kakimoto, T., Kamei, Y., Matsumoto, K.-i.: Analyzing oss developers' working time using mailing lists archives. In: MSR 2006, pp. 181–182 (2006)

17. Weissgerber, P., Neu, D., Diehl, S.: Small patches get in! In: MSR 2008, pp. 67–76 (2008)
18. Yamauchi, Y., Yokozawa, M., Shinohara, T., Ishida, T.: Collaboration with lean media: how open-source software succeeds. In: CSCW 2000, pp. 329–338 (2000)
19. Sowe, S.K., Stamelos, I., Angelis, L.: Understanding knowledge sharing activities in free/open source software projects: An empirical study. J. Syst. Softw. 81(3), 431–446 (2008)
20. Lakhani, K.R., von Hippel, E., Lakhani, K.R.: How open source software works: Free user-to-user assistance. Research Policy 32, 923–943 (2003)
21. Howison, J., Inoue, K., Crowston, K.: Social dynamics of free and open source team communications. In: Second Intl. Conf. on Open Source Systems, June 2006, pp. 319–330 (2006)

A Proposal of TIE Model for Communication in Software Development Process

Masakazu Kanbe[1,2], Shuichiro Yamamoto[3,1], and Toshizumi Ohta[2]

[1] NTT DATA CORPORATION, 3-3-9 Toyosu Koutoku Tokyo, Japan
[2] The University of Electro-Communications, 1-5-1 Choufugaoka, Choufushi, Tokyo Japan
[3] Nagoya University, Furocho, Chikusaku, Nagoya, Japan
{kanbems,yamamotosui}@nttdata.co.jp, ohta@is.uec.ac.jp

Abstract. Communication is more important in software development fields. We proposed the intermediary knowledge model to analyze the enterprise communication by extending traditional knowledge creation model. In this article, we propose TIE models based on intermediary knowledge model. TIE model is the knowledge network model to explain the just in time documentation in the CMC (Computer Mediated Communication) tools like wiki. We analyzed the case of wiki based software development and showed the effectiveness and efficiency of the CMC tools in software development in certain conditions.

Keywords: Knowledge network, Software development, Communication.

1 Introduction

Software developments become more complex and many developers who have various backgrounds participant in its processes. Various communications occurred in the field of the software development. The communication style of software developments contains regular face to face meetings, ad hoc conversations in the local office, and acceptances of document by e-mail. Furthermore, CMC (Computer Mediated Communication) tools such as wiki, SNS, blogs and communication plug-ins of Integrated Development Environment support the developers' communication. In this article, we propose TIE model as knowledge network model for software development communication. TIE model has the aim to elaborate the knowledge transformation processes by network structure. TIE model is three layered network model. The three layers are tacit knowledge network, intermediary knowledge network and explicit knowledge network. We analyzed the case of wiki used software development process by TIE model. We also investigate the effectiveness and efficiency of wiki used software development process.

2 Related Works

In this chapter, we introduce the previous related works to explain our model.

K. Nakakoji, Y. Murakami, and E. McCready (Eds.): JSAI-isAI, LNAI 6284, pp. 104–115, 2010.
© Springer-Verlag Berlin Heidelberg 2010

2.1 Intermediary Knowledge Model

We proposed the intermediary knowledge model as knowledge sharing model in enterprises [1] [2]. Intermediary knowledge is the knowledge statement in which employees share the knowledge by the CMC tools. Intermediary knowledge model is one of the extended models of tacit and explicit knowledge concept [3].

The traditional knowledge creation model has tacit and explicit knowledge and four knowledge transformation modes; socialization, externalization, combination and internalization [4]. Intermediary knowledge model explains business problem solving without the knowledge spirals of the organizational process. Employees can share the knowledge in intermediary knowledge statement by using CMC tools. In intermediary knowledge model, employees can exchange the knowledge that can be shared in tacit knowledge with less cost compared to explicit knowledge.

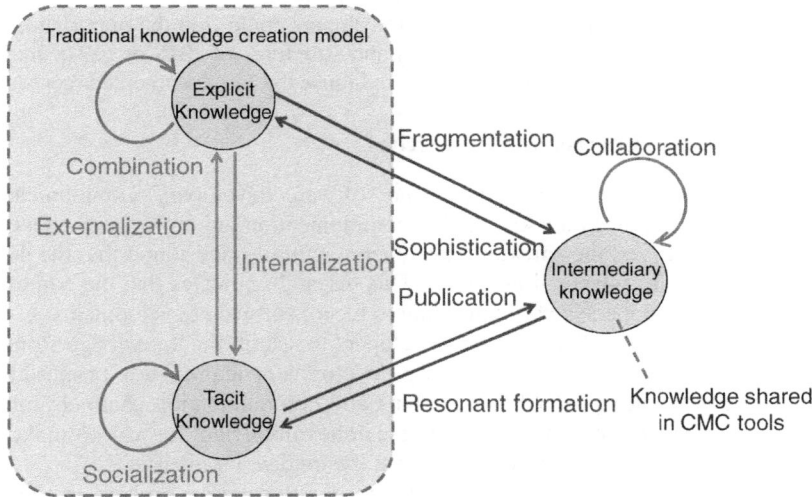

Fig. 1. Intermediary knowledge model

Fig. 1 shows the intermediary knowledge model. The dashed lined square in Fig. 1 indicates the traditional knowledge creation model. We add the intermediary knowledge and its knowledge transformation modes to the traditional model. This model indicates that the employees can rapidly develop the knowledge in the CMC tools by using the intermediary knowledge transformation modes. The modes consist of publication, fragmentation, collaboration, resonant formation, and sophistication. Publication means to publish individual experience or ideas. Fragmentation means to import the parts of explicit knowledge. Collaboration means to react with employees' problems or opinions. Resonant formation means to accept and understand the others' opinions. Sophistication means to develop explicit knowledge from intermediary knowledge.

According to the knowledge spiral condition of traditional knowledge creation model, if employees intend to use the knowledge formally and inter-organizationally, each employee in organization have to generate the explicit knowledge through the inner organizational knowledge spirals. Formally making explicit knowledge needs high cost and much labor through the inner organizational knowledge spirals. Intermediary knowledge transformation modes explain lower cost and labor in the knowledge exchange than the traditional knowledge transformation modes.

Also intermediary knowledge model explains the effectiveness of communication records. CMC tools create the more interaction occasions for employees than they do not use CMC tools. The employees have new communication in CMC tools. The communication in CMC tools are recorded as intermediary knowledge and employees reused the knowledge efficiently.

2.2 IBIS Model

IBIS [5] and gIBIS [6] [7] are the traditional software engineering method. One of the purposes of these IBIS methods is to fully record and structuralize the discussion processes and progresses in software developments. Recording and structuring all the discussion processes and progresses, software developers could find the important information in developments. Although the records and structures of IBIS or gIBIS may be useful, the costs or labors are too large to make and reuse the full documented records.

2.3 Recent Software Engineering Researches

Software engineering researches supports the software developers' communication. Software developments need the knowledge communication among software developers. Ko et al. [8] analyzed the software developers' activities and found that the developers used coworker as information source. This research indicates that the communication among the developers is very important in recent software developments.

Ye et al. [9] [10] helped the software developers to search the knowledge from the software libraries and members of software development team. They proposed the personalized search engine for API documents and communication channels for experts in software development team. Their researches implicate the way to make developers communicate each other for efficient software development.

Marczak et al. [11] indicated the importance of information brokers in requirement change management. As their research, the information brokers have important roles in the social network of software development team. The information brokers facilitate information flow to avoid misinterpretations of requirements. These researches indicate the importance of the developers' communication in software development.

3   TIE Model

3.1   Overview of TIE Model

We propose TIE model as CMC model for dynamic communication in software development process. TIE model has three layers consisted of Tacit Knowledge Network (TKN), Intermediary Knowledge Network (IKN) and Explicit Knowledge Network (EKN). Table 1 shows features of these three layers.

Table 1. Features of layers of TIE model

Knowledge Network	Network node	Media	Documentation	Examples of products
Tacit Knowledge Network	Human	Face to Face, Telephone, Video conference	No documentation	Discussions, Meetings
Intermediary Knowledge Network	CMC content	CMC tools (Wiki, SNS, blog, e-mail)	Just in time documentation	CMC logs
Explicit Knowledge Network	Document	Document management services	Full documentation	Requirements, specifications, codes, manuals, guidelines

TKN has roles to exchange the tacit knowledge. The network node of TKN is human. TKN is related to organization structures, roles of members, processes of decision making and so on. TKN is occurred in face to face meeting, telephone or video conference communication. It seems that TKN brings down no documentation for the software development. We assume TKN does not create any formal documents. The products of TKN are discussions and meetings. TKN does not always create the tangible products to be observed.

IKN has roles to exchange the intermediary knowledge. The network node of IKN is CMC content. IKN is related to CMC network in the software development team. These CMC contents grow up in CMC tools such as Wiki, SNS, blog and e-mail. IKN provides just in time documentation with the developers. If one needs to coordinate with others, one can use CMC tools to coordinate with others. And the coordination records are published for all the members of software developmental teams. These published coordination records are useful documents for software development. We call this process "Just in time documentation." Just in time documentation means that the necessary knowledge becomes documents when the developers communicate each other in CMC tools. The products of IKN are CMC logs.

EKN has roles to exchange the explicit knowledge. EKN is related to document network in the software development process. The network node of EKN is document. This document network grows up in document management services, which of functions are the document traceability, the historical management, the full text search and the document file sharing. EKN provides full documentation with developers. The products of EKN are documents, such as requirements, specifications, source codes, manuals and guidelines.

The network edges of TKN mean human communication in face to face meetings. The edges of IKN mean the concatenations of CMC contexts. The edges of EKN mean the concatenations and relation among documents. The edges between TKN and IKN mean the processes of intermediary knowledge provisions and acquisitions in CMC tools. The edges between IKN and EKN mean the processes quotations and documentation of explicit knowledge in CMC tools.

3.2 TIE Model for Software Development Communication

TKN do not create any formal document. We call this TKN statement "no documentation." EKN aims to create the documents elaborately. We call this EKN statement "full documentation." Traditional software developments use the TKN and EKN as the knowledge process. However, the knowledge processes in the traditional software developments has two problems. First problem is the loss of the important information of software developments. The knowledge processes in TKN are oral communication in discussions or meetings. Communication records of TKN are almost disappeared when the meetings or discussions ended. TKN contents are not always described in document and merely shared with all the members.

Second problem is the difficulty to record the all the important information of software developments. If all the events in software developments were documented fully at right time, each member could understand requirements, specification, and source codes perfectly. However, it is difficult to achieve full documentation because its cost is very high and its range is very ambiguous.

Fig. 2 shows TIE model we proposed. TIE model adds IKN to the traditional knowledge process in software developments. CMC tools support IKN. Balloons express the representative knowledge process of TIE model. The square balloon means the knowledge processes of traditional software development style. Round balloon means the knowledge processes of particular for TIE model.

The knowledge processes in IKN are open and agile communication on CMC tools. The CMC tools facilitate the communication of software development teams. IKN records the CMC logs. These CMC logs are not formal document, but very useful knowledge for software development. The development team members can read the knowledge processes each other in CMC tools. The knowledge processes in TIE model have correspondence relation with the knowledge transfer modes in the intermediary knowledge model in Fig. 1. Tacit knowledge, intermediary knowledge and explicit knowledge are corresponded with TKN, IKN and EKN respectively.

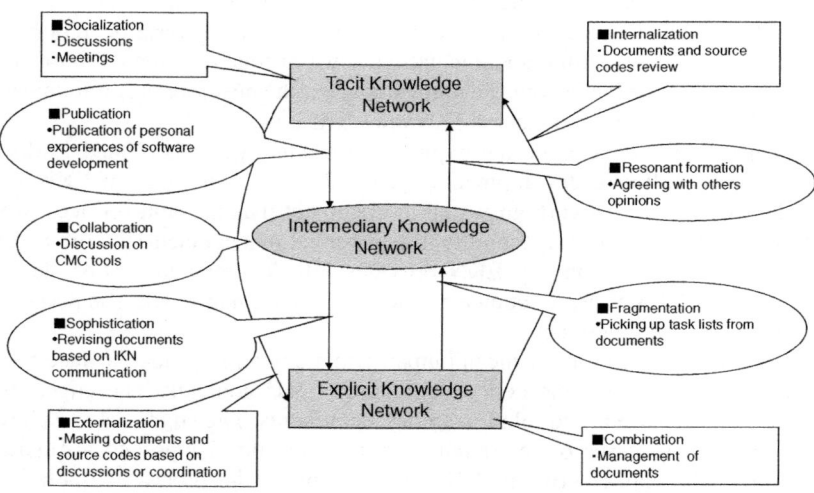

Fig. 2. TIE model and knowledge processes

4 Case Study of TIE Model for Software Development

To confirm the effectiveness and efficiency of TIE model, we analyzed the case of the wiki used software development. Wiki was used to facilitate the communication in software development team.

4.1 Aim of the Case Study

The case study aims to verify these assumptions form A1 to A3.

- A1: Software developers can make lists of necessary knowledge and gather them from members in CMC tools.
- A2: Software developers can supply and share the informal knowledge among members in CMC tools.
- A3: Software developers can supplement the knowledge sharing in face to face meeting in CMC tools.

These assumptions are set to confirm the effectiveness and efficiency of CMC tools in software development. We define that to verify these assumptions are to verify the effectiveness and efficiency of TIE model in software development communication because CMC tools support IKN and express the features of IKN.

4.2 Overview of Case

We selected the wiki used software development case. The number of software developers in this case was nine. Nine members belonged to two other Japanese companies. They cooperated to develop the software system development with unfamiliar devices. This software development had the processes; document production, program coding, and program test. These two companies office were in different location with no time zone difference. Although the members had the face to face meetings at once a week regularly, some communication mistakes occurred and caused negative effects for the software development. To deal with the communication mistakes, this software development team determined to use wiki to compliment with team communication. Appendix shows the outlines of the wiki communication. This wiki had 13 pages. There were 21 items in 13 wiki pages. We observed the knowledge processes of TIE model, 18 publications, 7 fragmentations, and 2 collaborations in wiki. The example of the publication was W3 in Appendix. In W3, a member imagined the tasks to be done for this process and published them. The example of fragmentation was W12 in Appendix. In W12, a member extracted documents lists from the development conference regulations in this organization. The example of collaboration was W8 in Appendix. In W8, members communicated about the test policy in wiki. We did not observe the knowledge process of resonant formation and sophistication. These two knowledge processes might occur outside wiki.

4.3 Verifying the Assumptions

To verify the assumptions, we picked up the evidences from the CMC in the wiki.

- Verifying A1: Software developers can make lists of necessary knowledge and gather them from members in CMC tools.

The wiki was used to make lists of necessary knowledge and gather them from members in software development. There were design documents lists in W1 and W2 of Appendix. These lists were fragmented intermediary knowledge from the explicit knowledge. A member considered which are necessary items from explicit knowledge for the software development team. This member extracted the document names from the document inventories. The document inventories are explicit knowledge. This member picked up and wrote the document names as intermediary knowledge in this wiki. Making lists of necessary knowledge in this wiki showed the edges between EKN nodes (documents) and IKN nodes (CMC contents) in TKN model. This member did not use e-mail to share the document lists, but used wiki.

Other members added the progress information of the document as intermediary knowledge. To add the progress information is publication of intermediary knowledge. All the member shared the dairy progress information of each document by wiki. Gathering necessary knowledge in this wiki showed the edges between TKN nodes (human) and IKN nodes (CMC contents) in TKN model.

The knowledge in the wiki is open to all the members and shared one target. If they shared this progress information by e-mail, they would not continue to refine the progress information because e-mail has the feature of cross in the post.

- Verifying A2: Software developers can supply and share the informal knowledge among members in CMC tools.

The wiki was used to supply and share the informal knowledge among members in software development. "Development know-how" in W15 of Appendix is one of the evidences to verify A2. This development know-how was supplied by a member who had a similar development experiences. This member wrote politely the knowledge to treat with specific devices. This knowledge was based on the member's own experiences and not formalized yet. In traditional software development, such know-how may be transferred by oral communication in TKN. On the other, this member had to write this know how as formal meeting document. In this case, the knowledge sharing in the wiki might avoid the rediscovery of this knowledge to treat specific devices. Supplying and sharing informal knowledge showed the edge between TKN node (human) and IKN node (CMC contents) in TKN model.

- Verifying A3: Software developers can supplement the knowledge sharing in face to face meeting in CMC tools.

The wiki was used to supplement the knowledge sharing in face to face meeting in software development. "Policy for test items (W7)" and "Comment for policy (W8)" of Appendix are the evidences to verify A3. In W7, a member published the policy for the test item for all the members. In W8, Another member in other location replied for the policy by the wiki. This type of knowledge sharing shows the edge between IKN node (CMC content) and IKN node (CMC content) in TIE model. In traditional software development, this communication between members might suspend until regular weekly meeting. In this case, using the wiki provided the appropriate communication occasions and eliminated the delay factor in software development.

5 Discussions

We analyzed the case in former section. In this section, we discuss on the effectiveness of CMC tools along with the case study. We also discuss on the limitation of our analyses for software development communication.

5.1 To Make Lists and Gather the Knowledge

We discuss the conditions that software developers made lists of necessary knowledge and gathered them from members in CMC tools. We assume two reasons why they wrote the necessary knowledge in the wiki.

First reason is that the contents to be shared should be open. The wiki provided the developers with the open communication environment consistently. If they did not use the wiki, they might communicate by face to face meeting, telephone or e-mail. This wiki is more open than these communication methods. The open feature of this wiki facilitates developers to write their knowledge. The open feature made casual communication and the developers published the progress information each other. In face to face meeting, powerful members may interfere in the remarks of other low powered members. Wiki may facilitate the remarks of low powered members. Wiki prepared the open IKN environment for the software development team. Although both e-mail and Wiki are CMC tool, we assume they might be different in openness.

Second reason is that the content in wiki is a single object. In the case, they added the progress information to the items of document lists. If they did not use the wiki, they might share the progress information of each document with e-mail. By e-mail, it is difficult to catch up with the progress information of all the members, because e-mail has the feature of cross in the post and makes multi objects to coordinate. This feature of e-mail made distribute their knowledge. It is not efficient that someone should gather the distributed knowledge in e-mail use. The members also need to read all the e-mail to comprehend the all the members' progresses. Because the shared contents feature is open and single object to edit, software developers record the important knowledge in CMC tools.

5.2 To Share Informal Knowledge

We discuss the conditions that software developers supplied and shared the informal knowledge among members in CMC tools. We assume that CMC usage was suitable for sharing developers' personal experiences about unfamiliar devices. The software development will advance smoothly with the developer's knowledge of unfamiliar devices. As these kinds of knowledge were not written in manuals, the developers cannot share the knowledge in EKN. CMC tools facilitate these kinds of knowledge to be share among the members. If this member wrote the knowledge in the wiki in 30 minutes and other eight members read the knowledge 5 minutes, the amount of the time is 70 minutes. If the rest of eight members acquired the knowledge by try and error in 120 minutes and the amount of the time is 960 minutes. Although this is an extreme example, knowledge sharing in the CMC tools may be very effective. We assume that sharing unfamiliar knowledge is effective usage of CMC tools. Wiki in this case created the edge between TKN node (human) and IKN node (CMC contents) in TKN model. There might be the feature of reciprocity in this team.

5.3   To Supplement the Face to Face Meeting

We discuss the conditions that software developers can supplement the knowledge sharing in face to face meeting in CMC tools. We suggested that two imaginary conditions of the software development communication; wiki style and traditional style. Wiki style has the regular weekly face to face meetings and wiki based communication. Traditional style has only the regular weekly face to face meetings. The relations between the amount of knowledge and time of each style are shown in imaginary chart Fig. 3. Fig. 3 expresses the only the amount of knowledge increased by developers' communication, does not express the amount of the software development works. When the amount of knowledge reaches at level K, the increases of knowledge may finished. Straight line in Fig. 3 shows the increase of knowledge in wiki style. The dashed line also shows the increase of knowledge in traditional style.

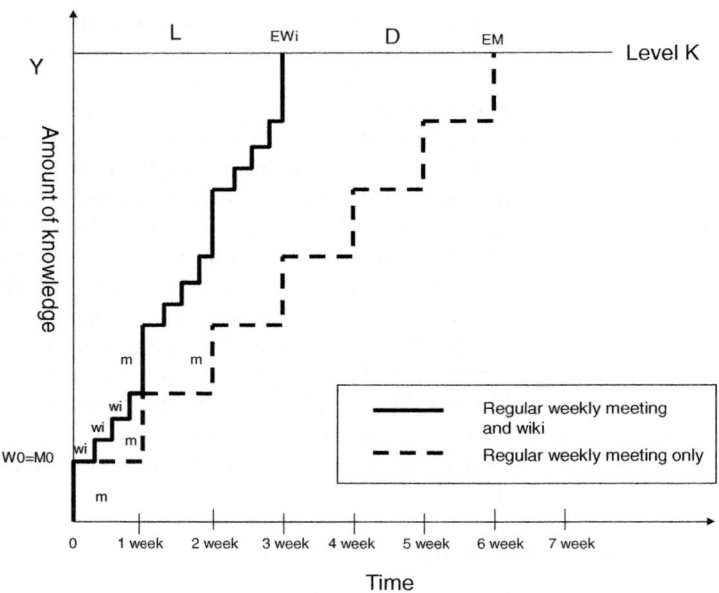

Fig. 3. The relation between time and amount of works

The amount of increase of knowledge in wiki style is Y(a,b).

$$Y(a,b) = a*wi + b*m \tag{1}$$

In formula (1), a is the number of communication of wiki, wi is the amount of knowledge increase per one wiki communication, b is the number of the face to face meetings and m is the amount of knowledge increase per one meeting.

The amount of increase of knowledge in traditional style is Y(c).

$$Y(c) = c*m \tag{2}$$

In formula (2), c is the number of the face to face meetings and m is the amount of knowledge increase per one meeting. To simplify explains, wi and m are fixed in this formula. However, wi and m are variable by the features of communication of wiki and meetings in the real case.

Both wiki and traditional styles may have the first meeting in zero point in Fig. 3. Both styles may gain the same amount of knowledge by first meeting. In wiki style, developers may increase their knowledge to communicate each other three times via wiki. In traditional style, developers may not try to increase the knowledge to communicate each other. Although both styles may increase the amount of knowledge by meeting, traditional style may gain the only half amount of the knowledge which wiki style may gain. As a result, wiki style will end the knowledge sharing at point EWi, its period L may be 3 weeks. Traditional style will end the knowledge sharing at point EM, its period L+D may be 6 weeks.

Fig. 3 is the suggestion that wiki supplemented the knowledge sharing of face to face meeting. Three interactions to gain the knowledge in wiki match one meeting communication to gain the knowledge. We suppose the wiki style may be more efficient knowledge communication environment than traditional style. However, we also found a condition to establish our estimation in Fig. 3. The condition is that the knowledge gained by wiki should be the same type by face to face meetings. We supposed that there might be two types of knowledge. One is the knowledge which only can be gained by CMC tools such as wiki. Another is the knowledge which only can be gained only by face to face meetings. We should distinguish the knowledge to investigate the knowledge communication conditions for CMC tools.

5.4 Limitations

In this article, we picked up the few positive effects of CMC tools. We should declare the conditions under which the CMC tools facilitate the software development more circumstantially. Also, we should analyze the negative factors of CMC tools such as false information in the wiki, information overload for developers. And we should investigate the relations among human, CMC content and document, which are the nodes of TIE model.

6 Summary

It is important to facilitate the knowledge communication among software developer. We proposed the intermediary knowledge model and TIE model for software development. Traditional software development researches focused on mainly human properties as experts and the quality of documentation. TIE model is software development network to express the CMC knowledge processes as network structure. We analyzed the case to show the effectiveness and efficiency of the wiki for software development. By analyzing the case, we discussed the conditions to facilitate the software development process by CMC tools. For further study, we should analyze more cases of software development by using our TIE model network structure. We should also investigate the relation between the communication works and the other important works such as making document or coding.

References

1. Yamamoto, S., Kanbe, M.: Knowledge Creation by Enterprise SNS. The International Journal of Knowledge, Culture and Change Management 8(1), 255–264 (2008)
2. Kanbe, M., Yamamoto, S.: An Analysis of Computer Mediated Communication Patterns. The International Journal of Knowledge, Culture and Change Management 9(3), 35–47 (2009)
3. Polanyi, M.: The Tacit Dimension. Routledge & Kegan Paul Ltd. (1966)
4. Nonaka, I., Takeuchi, H.: The knowledge creating company How Japanese Companies Create the Dynamics of Innovation, Oxford Univ. (1995)
5. Rittel, H., Kunz, W.: Issues as elements of information systems., Working paper# 131. Institute fur Grundlagen der Planung I.A.University of Stuttgart
6. Conklin, J., Begeman, M.L.: gIBIS: A Hypertext Tool for Exploratory Policy Discussion. ACM Transactions on Office Information Systems 4(6), 303–331 (1988)
7. Conklin, J., Selvin, A., Shum, S.B., Sierhuis, M.: Facilitated Hypertext for Collective Sensemaking: 15 Years on from gIBIS. In: Hypertext 2001 Conference (2001)
8. Ko, A.J., DeLine, R., Venoloa, G.: Information Needs in Collocated Software development teams. In: 29th International Conference on Software Engineering, ICSE 2007 (2007)
9. Ye, Y., Yamamoto, Y., Nakakoji, K.: Expanding the Knowing Capability of Software Developers through Knowledge Collaboration. International Journal of Technology, Policy and Management 8(1), 41–58 (2008)
10. Ye, Y., Yamamoto, Y., Nakakoji, K., Nishinaka, Y., Asada, M.: Searching the Library and Asking the Peers: Learning to Use Java APIs on Demand. In: Amaral, V., Veiga, L., et al. (eds.) Proceedings of 2007 International Conference on Principles and Practices of Programming in Java, pp. 41–50. ACM Press, Lisbon (2007)
11. Marczak, S., Damian, D., Stege, U., Schroter, A.: Information Brokers in Requirement-Dependency Social Networks. In: 16th IEEE International Requirement Engineering Conference, pp. 53–62 (2008)

Appendix: Contents and Intermediary Knowledge Processes in the Wiki

Names of pages	ID	Items	Contents	Knowledge transformation mode
Basic design	W1	- Basic design documents list	- Members added the progress for each items.	Fragmentation, Publication
Detail design	W2	- Detail design documents list	- Members added the progress for each items.	Fragmentation, Publication
	W3	- Tasks	- Tasks in this process	Publication
Make and unit test	W4	- FYI	- Discussion memo for the decision items	Publication
	W5	- Policy for test items	- Descriptions of policy for test items and conditions	Publication
	W6	- Estimation method of test density	- Estimation with number of test items and scales	Publication
System integration test	W7	- Policy for test items	- Descriptions of policy for test items and conditions	Publication
	W8	- Comment for policy	- Comment for policy of W7	Collaboration
	W9	- Estimation method of test density	- Estimation with number of test items and scales	Publication
	W10	- List of the test materials	- List of the materials; software and hardware	Publication
Run time test	W11	- Call for comments	- Message to make the run time test guideline	Publication
Development conference #1	W12	- Development conference #1 document list	- Document list for development conference #1	Fragmentation
Development conference #2	W13	- Development conference #2 document list	- Document list for development conference #2	Fragmentation
Graph	W14	- Graph description specification	- Graph specification of system	Publication
Know-how	W15	- Development know-how	- Know-how from the similar system experienced worker	Publication
Memo for project management	W16	- Items to be improved for project management	- Communication method for project management	Publication
Demonstration	W17	- Purpose	- Purpose of demonstration	Publication
	W18	- Scenario	- Description of use cases for office and factory	Publication
	W19	- Proposal of the demonstration	- Phases of demonstration and the To Do lists	Publication, Fragmentation, Collaboration
Name of documents	W20	- Chapters	- Chapter and correction comments for documents	Fragmentation, Publication
	W21	- Documents list	- Documents list and working memos	Fragmentation, Publication

Identifying the Concepts That Are Searchable with Keywords in Code Search Engines

Toshihiro Kamiya

Future University Hakodate
116-2 Kamedanakano-cho, Hakodate, Hokkaido, Japan 041-8655
`kamiya@fun.ac.jp`

Abstract. The (extended position) paper discusses the reason why keyword-based search engines may not be effective in code search, and shows an case study where which kind of concepts in source code can be effectively searched by keyword code search engines.

Keywords: Search-Driven Software Development, Code Reuse, Experiment.

1 Introduction

Many code search engines, such as Codase (www.codase.com), Codefetch (www.codefetch.com), Google code search (google.com/codesearch), JExamples (www.jexamples.com), Koders (www.koders.com), Krugle (www.krugle.org), and Merobase (www.merobase.com), have become available recently [1-4, 6, 7]. Most of them have Google-like interfaces through which a user can enter a set of keywords as a query to retrieve source code files that are related to the keywords in the query. Some code search engines also provide options that are specific to source code. For example, software developers can use options to specify the specific portions (such as comments, code, or functional definitions) in which the search keywords appear.

Such code search engines are important instruments to promote and support software reuse. However, their support for reuse may not be sufficient. When software developers consider reuse, they care about not only the functionality of the code, but also various characteristics such as performance ("Is the algorithm $O(N)$ or $O(N^2)$?"), usability ("Whether the API is easy to understand and use?"), and maintainability ("Can the code be easily customized to fit my code?"). This paper tries to evaluate the capabilities of keyword-based code search engines in terms of their support for searching reusable code based on multiple characteristics. The paper adopts what we call an oracle approach for the evaluation: it first identifies a classification schema that represents different dimensions of code characteristics, and then analyzes whether we are able to identify, for each dimension of characteristic, a set of intuitive keywords that can be used in a search query to retrieve effectively reusable code.

K. Nakakoji, Y. Murakami, and E. McCready (Eds.): JSAI-isAI, LNAI 6284, pp. 116–123, 2010.

2   Challenging Issues in Keyword-Based Code Search

Keyword-based code search systems are faced with two challenging issues:

(1)   an expensive process of selecting the right code from search results, and
(2)   indirect relations between keywords and source code concepts.

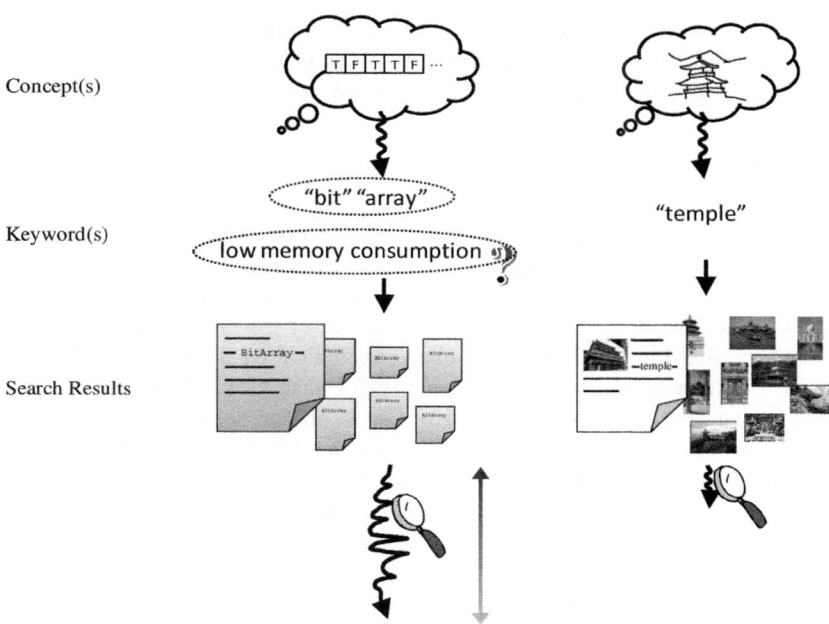

Fig. 1. Source code search vs. image/text search

Fig. 1 illustrates the two issues by contrasting the process of code search with that of text and image search. When a user searches in the web space with a keyword-based search engine, candidates of the search results are web pages (text, or images around text). The search engine performs matching between the keywords and the text of the candidates. What a user expects to receive is a set of web pages that directly include the keywords. For example, when a user types "temple" as a search keyword, he or she will receive web pages that have the word "temple" in text or embedded around the images in the web pages.

Moreover, it doesn't take much time for a user to evaluate whether such search results are what they need, or to select the best one from the results. When a user receives hundreds of images shown on a display, it is a matter of seconds to select some desired images from them. In contrast, it takes much longer time for a software developer to evaluate whether a source file in the search results are what he or she needs. In the case study described later, it took the author two person-days to read through 15 source files. When a search result includes a large number of source files, such tasks of evaluation and selection are very expensive.

In text or image search, when the search result includes many numbers of items, a user can reduce the size of the search result by adding extra keywords that are more specific to a concept to be searched. Such extra keywords should be similarly effective in code search. However, in the case of code search, except for the names of classes or methods, the semantics and qualities, such as comprehensibility and the order of algorithms used in the code, are often not easy to express and may become obvious only after reading through the whole text of the source file.

To deal with the above two issues, this paper introduces a case study where we try to explore whether non-functional concepts in source code are searchable with a keyword-based search engine, and what kind of keywords can be used for search.

3   The Oracle Approach

We describe the oracle approach using a case study of searching code in the following scenario. A developer is writing a Java program that needs an array of bits with low memory consumption, that is, a class of bit array that uses one bit in the memory for each element. The developer guesses, by analogy of the class Array of the standard Java library, that the name of such a class may be called BitArray. So the developer can use BitArray as the functionality query for searching. However, the developer also has other requirements such as performance, comprehensibility and maintainability, and the question is what kind of keywords that he or she should use to represent such requirements for the purpose of searching.

To evaluate the oracle approach of finding effective search keywords that capture the requirements of multiple characteristics, we will use a toy keyword based code search engine that is prepared for this evaluation. The oracle approach (Fig. 2) of finding highly discriminative search keywords works as follows. For each predefined "correct" answer set of desired code that we want to find, we create a series of keyword sets that are used as search queries. Each query will return a set of search results, the search results are then measured against the predefined correct answer set. Based on the measurement, we will find whether keywords with high discriminating power exist. More specifically, the approach consists of the following steps:

(1) Prepare a repository of source files, which are the candidates for code search.
(2) Examine each source file in the prepared repository and create a list of *concepts* that can be used to describe various features of source files, including basic functionality and other implementation details such as performance and potential usage pitfalls. This step needs to be performed by a subject expert.
(3) Extract a list of keywords from each source file to represent the source file.
(4) From the list of concepts, create a classification schema by putting each concept into different categories. This classification schema represents different dimensions of search *requirements*. A set of search requirements is created, and each search requirement contains a subset of the concepts, and in this case study, one search requirement contains one concept from each category.
(5) For each requirement,

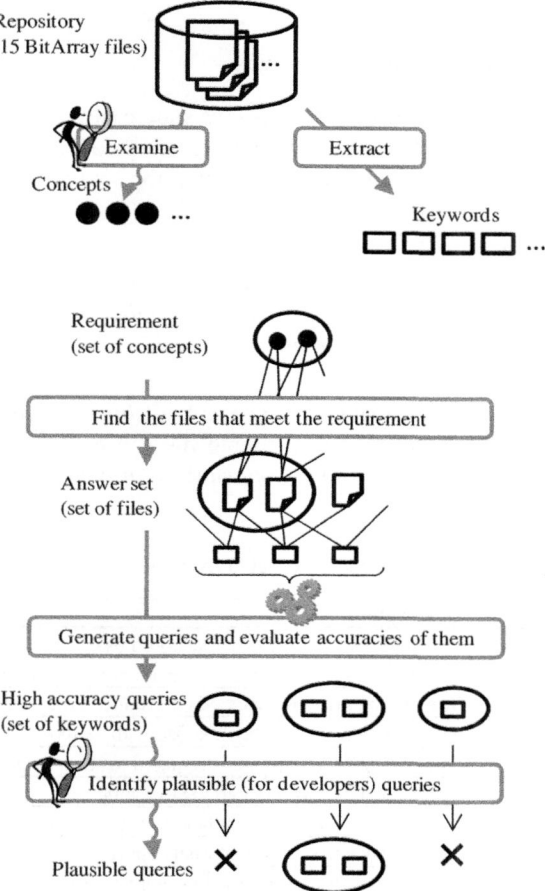

Fig. 2. Steps of the oracle approach

(5-1) Create an *answer set*, which is a set of source files from the repository that contain the concepts of the requirement.

(5-2) Determine what words can make a query that returns search results with high accuracy, namely, identifying the words that have the highest discriminative power in terms of search accuracy. To do this, we create search queries with arbitrarily selected keywords from the keyword lists, and then compare the search results of those queries with the predefined *answer set*.

(5-3) For each query that achieves high search accuracy, analyze whether it is plausible for a software developer to include such words in their search queries based on their search requirements. The search accuracy of a query is evaluated with precision and recall.

4 Case Study

4.1 Source Files, Concepts, and Keywords

The repository in the case study contains 15 Java source files of different implementations of the BitArray class that are found with existing code search engines. Table 1 shows the concepts that are identified by analyzing the source files. The concepts are classified into 4 categories:

> *basic operation*
> *implementation issue (scalability, optimization, pitfalls)*
> *extra operation (functions that can be implemented with combination of*
>   *basic operations)*
> *ease of development.*

In the *implementation issue* category, the concepts of *limited-size*, *unlimited-size* and *re-size* are mutually exclusive; that is, a source file can have only one concept from that category. *Non-pack* and *pack* in the *implementation issue* category are also mutually exclusive.

Table 1. Concepts extracted from the source files

Concept	Classi-	Description
basic	basic op.	Can store bits and retrieve the stored bits.
limited-size		The max count of bits are hard-corded in a source file (a constat).
unlimited-size		Size of a bit array is specified on instance creation.
re-size	impl.	Size of a bit array can be modified with methods.
non-pack	Issue	One byte or more stroage is required to store a bit.
pack		Less than a byte is required to stroe a bit.
mask		A workaround to prevent a time-consuming micro operation.
break-encap.		Has methods that return internal data (stored bits).
search		Has methods to search true bits in a bit array.
merge		Has methods to merge two bit arrays.
value		Has a predicate for equality/comparision between two bit arrays.
shift		Has methods to shift bits in a bit array.
logical	extra	Has methods to calculate "and" or "or" of two bit arrays.
range	op.	Has methods to obtain or modify bits within a range on a bit array.
XML		Convertible from/to XML strings.
booelan[]		Convertible from/to a "boolean[]" object.
byte[]		Convertible from/to a "byte[]" object.
file-io		Can write to/read from a file.
copy	e. o. d.	Has methods (or constructors) to duplicate a bit array.
tostring		Has a method of a "debug" print.

basic op. = basic operation,
break-encap. = break encapsulation,
extra op. = extra operation,
e. o. d. = ease of development,
impl. issue = implementation issue.

Keyword lists were generated from the 15 Java source files with a small script. Camel cased identifiers such as `getLength` are split into separate words (e.g. "get" and "length"). Tag names (e.g. `@author`) in JavaDoc comments were removed. Operators (such as <<=) were extracted as words too. The keywords extracted from source files were divided into equivalence classes: If keyword "a" appears in source files "f" and "g" only, and keyword "b" also appears in "f" and "g" only, then "a" and "b" are put into one equivalence class because they are equivalent in terms of their discriminative power. For each equivalence class of keywords, we need evaluate only one word from that class. In this case study, among the total 197 equivalence classes, 131 classes have only one word and the other 66 classes have an average 11.4 words, with the largest class having 236 words.

4.2 Evaluations

Table 2 shows queries with high accuracy (high precision and/or high recall) for each concept. We evaluated queries of one word, two words and three words. The queries shown in the table are queries of single word. If two or three word queries had higher precision and recall values than that of single-word queries for a particular concept, such queries are shown in the remark column in Table 2. Also, when the high discriminative queries in the second column are not intuitive ones, namely, those words seemed too difficult for a developer to guess from the concepts of the given requirement, the more intuitive queries were shown in the remark.

Table 2. High-accuracy queries for each concept

	Requirement (concept)	Max prec. * recall	Max prec.	Max recall	Remark
	basic	{ + } = 14/14 * 14/14	←	←	**{ & } = 13/14 * 13/14**
	limited-size	{ supplied }	←	←	**{ size } = 1/13 * 1/1**
	unlimited-size	**{ size } = 9/13 * 9/9**	{ || } = 5/5	{ size } = 9/9	{ limitations } = 2/4 * 2/9, { ++, size } = 9/12 * 9/9, { ++, copy, size } = 9/11 * 9*9
impl. Issue	re-size	{ initial } = 3/4 * 3/3	{ exceed } = 2/2	{ initial } = 3/3	**{ size } = 2/13 * 2/3**, { initial, size } = 2/3 * 2/3
impl. Issue	non-pack	{ name } = 2/2 * 2/2	←	←	{ size } = 2/13 * 2/2, { byte } = 0/9 * 0/2, { representing } = 2/3 * 2/2
impl. Issue	pack	{ copy } = 12/13 * 12/12	{ ~ } = **10/10**	{ copy } = 12/12	{ packed } = 0/4 * 0/2, { & } = 12/14 * 12/12. The equivalence class of "packed" includes "^".
impl. Issue	mask	{ prevent } = 2/2 * 2/2	←	←	**{ mask } = 2/8 * 2/2**, { fast, mask } = 2/3 * 2/2
impl. Issue	break-encap.	**{ mutable } = 1/1 * 1/1**	←	←	The equivalence class of "mutable" includes "corruption", "performance", and "sanity".
	search	{ serialized } = 3/3 * 3/3	←	←	{ first } = 3/5 * 3/3, { pos } = 3/5 * 3/3, **{ position } = 2/6* 2/3**, { <=, pos } = 3/3 * 3/3
	merge	**{ merge } = 2/2 * 2/2**	←	←	
	value	{ code } = 5/5 * 5/5	←	←	**{ equals } = 5/7 * 5/5**, { !, * } = 5/5 * 5/5
	shift	{ <<= } = 2/2 * 2/2	←	←	**{ shift } = 1/3 * 1/2**, { << } = 1/11 * 1/2, { >> } = 1/5 * 1/2, { >>> } = 1/5 * 1/2, { >>= } = { >>>= } = 1/1 * 1/2

Table 2. (*Continued*)

extra op.	logical	{ ^= } = 3/3 * 3/3	←	←	{ and } = 3/13 * 3/3, { or } = 2/5 * 2/3, { nor } = 2/2 * 2/3, { **and, or** } = 2/5 * **2/3**
	range	{ word } = 1/1 * 1/1	←	←	{ **range** } = **0/3** * **0/1**, { bounds } = 1/7 * 1/1
	XML	{ replace } = 1/1 * 1/1	←	←	{ **xml** } = **1/4** * **1/1**, { string, xml } = 1/1 * 1/1
	booelan[]	{ booleans } = 3/3 * 3/3	←	←	{ **boolean** } = **3/7** * **3/3**
	byte[]	{ gets } = 4/5 * 4/4	{ reserved } = 3/3	{ gets } = 4/4	{ **byte** } = **4/9** * **4/4**
	file-io	{ input } = 6/6 * 6/6	←	←	{ **file** } = **4/7** * **4/6**, { file, input } = 4/4 * 4/6
e. o. d.	copy	{ >= } = 8/10 * 8/8	{ >> } = 5/5	{ >= } = 8/8	{ copy } = 8/13 * 8/8, { **clone** } = **6/6** * **6/8**, { !=, >= } = 8/9 * 8/8, No three-word queryies outperformed.
	tostring	{ string } = 11/12 * 11/11	{ / } = 9/9	{ string } = 11/11	{ **to, string** } = **11/11** * **11/11**

The "{...}" are queries. The values at right side of "=" are precisions and recalls. Each of these values is denoted by a fraction, whose denominator and numerator are counts of source files, without canceling down (reduction). A bold-font query is the query that looks the most intuitive one for the given requirement. A left arrow "←" means the query in the cell is the same to one in the left cell.

Recall values in Table 2 are relatively high, and this is not surprising for this study because each concept and queries are both extracted from the source files. In other words, it is guaranteed that some source files contain words of that concept. If the set Conclusions.

The findings of the case study can be summarized as follows. Intuitive and high-precision queries are possible when (i) the name of the method that implements a concept is easy to guess from the coding conventions of Java, such as copy → clone(), value → equals(), and tostring → toString(), or (ii) some unique words (operators) are required to implement the concept, such as, pack → ~ and shift → <<=.

On the other hand, if the concepts are implemented without unique words, such as scale and conversion, it is difficult to find intuitive and high-precision queries. It is interesting to note that the former case is considered as searching concept from API (the exposed attribute of code) and the latter case is considered as implementation concept (the hidden attribute of entities).

5 Related Work

The CodeBroker [7] system is a code-search engine integrated with a source-code editor. It watches actions of a developer (user) in the editor and recognizes certain types of editing actions, such as typing a comment, as triggers for searching.

CodeBroker also incrementally builds discourse and user models about the knowledge of the user, and these models are used to provide search results that are tailored to the user: it will not present reusable components that the user has known already.

SparsJ [3] is another code search engine that provides a web interface like Google, in which a user types some keywords and receives a list of components as search results. It ranks search results using an algorithm named Component Rank: the most popular component appears in the top of the search results. This increases the chance for a user to find the components that are most likely to be reused.

As Marcus et al. pointed out in [5], most of the existing approaches for locating functions (that is, identify code fragments implementing a given domain concept) are built upon a model that represents a search query with some kind of internal representations with keywords.

6 Conclusion

This paper explored the possibility of finding effective keywords to search source code based on concepts that are important for code reuse but are difficult to express in words. As a case study, 20 concepts were extracted from 15 BitArray classes which have been developed independently by different developers (except for two). Queries (a set of keywords) were generated from texts of these source files, and their effectiveness are evaluated in terms of precision and recall values. The evaluation revealed that (1) if the concepts can be easily guessed from naming conventions of the programming language, effective keywords are easier to find; and (2) operators are also effective keywords for search components that implement the required concepts.

Acknowledgements. I am deeply grateful for Dr. Yunwen Ye for his comments, advice, and proofreading. This research is supported by JSPS Grant-in-Aid for Challenging Exploratory Research (No. 21650008).

References

1. Augusto, O., Lemos, L., Bajracharya, S., Ossher, J.: CodeGenie: a Tool for Test-Driven Source Code Search. In: Proceedings of the 22nd IEEE/ACM International Conference on Automated Software Engineering (ASE 2007), pp. 525–526 (2007)
2. Bajracharya, S., Ngo, T., Linstead, E., Dou, Y., Rigor, P., Baldi, P., Lopes, C.: Sourcerer: a Search Engine for Open Source Code Supporting Structure-Based Search. In: Proceedings of the 21th ACM SIGPLAN International Conference on Object-Oriented Programming, Systems, Languages, and Applications (OOPSLA 2006), pp. 681–682 (2006)
3. Inoue, K., Yokomori, R., Fujiwara, H., Yamamoto, T., Matsushita, M., Kusumoto, S.: Component Rank: Relative Significance Rank for Software Component Search. In: Proceedings of 25th IEEE International Conference on Software Engineering (ICSE 2003), pp. 14–24 (2003)
4. Mandelin, D., Xu, L., Bodik, R., Kimelman, D., Mining, J.: Helping to Navigate the API Jungle. In: Proceedings of the 2005 ACM SIGPLAN Conference on Programming Language Design and Implementation (PLDI 2005), pp. 48–61 (2005)
5. Marcus, A., Buchta, V., Petrenko, J., Sergeyev, A.: Static Techniques for Concept Location in Object-Oriented Code. In: Proceedings of 13th IEEE International Workshop on Program Comprehension (IWPC 2005), pp. 33–42 (2005)
6. Reiss, S.P.: Semantics-Based Code Search. In: Proceedings of the IEEE 31st International Conference on Software Engineering (ICSE 2009), pp. 243–253 (2009)
7. Ye, Y., Fischer, G.: Supporting Reuse by Delivering Task-Relevant and Personalized Information. In: Proceedings of the 24th IEEE International Conference on Software Engineering, pp. 513–523 (2002)

On the Use of Emerging Design as a Basis for Knowledge Collaboration

Tiago Proenca, Nilmax Teones Moura, and André van der Hoek

University of California, Irvine, Department of Informatics, Irvine CA 92697, USA
{tproenca,nmoura,andre}@ics.uci.edu

Abstract. Abstractions in software engineering have been used for guidance and understanding of software systems. Design in particular is a key abstraction in this regard. However, design is often a static representation that does not evolve with the code and therefore cannot help developers in collaborating after it becomes out-of-date. Our research group is exploring the use of Emerging Design, a dynamic abstraction, as the basis for knowledge collaboration through its implementation in a coordination portal called Lighthouse. This paper presents the state of the art of Lighthouse and discusses three knowledge collaboration problems that we are currently addressing.

1 Introduction

Collaboration is related to mutual sharing of knowledge [1] and has become an essential part of software development and indeed an important research field in software engineering. Today, most knowledge sharing is either informal or decoupled from the actual artifacts to which it pertains. For instance, in the Knowledge Depot [2], an email-based group memory tool, knowledge is stored in a separate repository that must be queried to find a particular piece of information. This not only creates a hurdle to accessing knowledge, but also leads to the update problem, i.e., in the presence of changes, one has to update two places: the artifacts themselves and the Knowledge Depot.

Our research group is exploring a different kind of solution, one where the knowledge is essentially attached to an abstraction that we are creating as part of a collaboration infrastructure. This abstraction is called Emerging Design [3] and is defined as the design representation of source code as it changes over time. With each code change, the Emerging Design is updated accordingly. Emerging Design satisfies the traditional roles of abstraction (guidance and understanding [3]) and includes support for new roles such as coordination, project management, and detection of design decay.

While the original focus of our use of Emerging Design was on detecting conflicts in code changes [3], we believe it is a particularly promising abstraction to address a broader class of collaboration issues. In this paper, we talk about three such collaboration problems and how we believe Emerging Design serves as good basis for exploring them.

K. Nakakoji, Y. Murakami, and E. McCready (Eds.): JSAI-isAI, LNAI 6284, pp. 124–134, 2010.
© Springer-Verlag Berlin Heidelberg 2010

The remainder of this paper is organized as follows. In Section 2, we review Emerging Design and its implementation in Lighthouse. Section 3 presents three knowledge problems in collaboration and how we believe they can be addressed by building upon Emerging Design. In Section 4, we summarize our ideas and discuss some challenges and future directions to improve our work.

2   Emerging Design

Since a design document illustrates the interactions among modules, it can help developers to gain an understanding of the high-level structure of the system and its interactions [3,4]. However, design is often a static representation that does not evolve (automatically) with the code. Therefore, as it becomes out-of-date, it loses value for developers who need to collaborate.

Our research group is exploring the use of Emerging Design as the basis for collaboration. Emerging Design is defined as the design representation of source code *as it changes over time*. It is a live document that stays up-to-date with all changes made to the system. It is annotated with information about the changes made, helping developers to be aware about how the code structure evolves, and with whom they may need to coordinate their actions in order to reduce and prevent conflicts.

We implemented this approach as an Eclipse plug-in called Lighthouse [5]. Lighthouse presents the Emerging Design view as a UML-like class diagram which is built dynamically as developers implement or make changes in the code. One particular characteristic of Lighthouse is that it does not require check-in of the changes made. Instead, it tracks workspaces, since the goal is help people collaborate and coordinate before sending the changes to the source code repository, so merge conflicts are avoided.

Figure 1 shows the Emerging Design basic representation. It shows the primary elements found in UML class diagrams, such as classes, fields, and relationships, as annotated with additional information. In particular, Lighthouse shows information about the evolution of the code. The *plus* symbol represents an addition of a class/method/field, *minus* represents a removal, and *triangle* represents a change. For instance, in the ATM class, the field *value* was added by Max. Another example is the field *balance_inquiry*. We can see that Theo and Bob changed that field, and finally Anna removed it. Notice that this history of changes is presented in a top-down manner, time-ordered with the most recent changes at the bottom.

The use of Lighthouse in a large software product naturally introduces scalability issues with respect to the visualization. This can harm a user's ability to spot a particular events of interest. As a first step to make the Emerging Design more scalable, we developed a variety of filters to improve the user's capability. With these filters, the information shown can be reduced by focusing on particular packages, developers, modifications or some combination of them. As a result, crowded visualizations that clearly indicate a problem can be examined for what that problem exactly is.

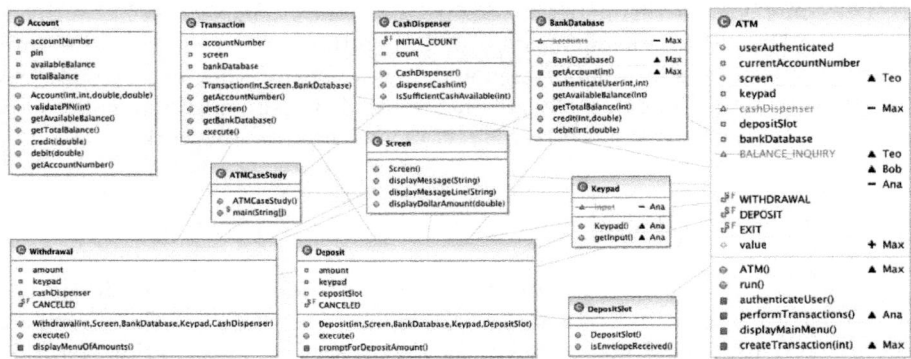

Fig. 1. Emerging Design basic representation

Figure 2 shows an Emerging Design representation with several classes, where each class has numerous events representing the activities of four developers (Max, Bob, Ana, Jim), who are all coding a particular part of the project. The following picture (Figure 3) illustrates how Lighthouse allows users to turn on the filter by developer. In this specific example, the user has chosen to show Bob's code changes. At this manner the user could be aware of any event that is happening in Bob's workspace.

The second filter uses the concept of Java packages. We believe that this feature is very pertinent, because most of the time, while developing, users interact

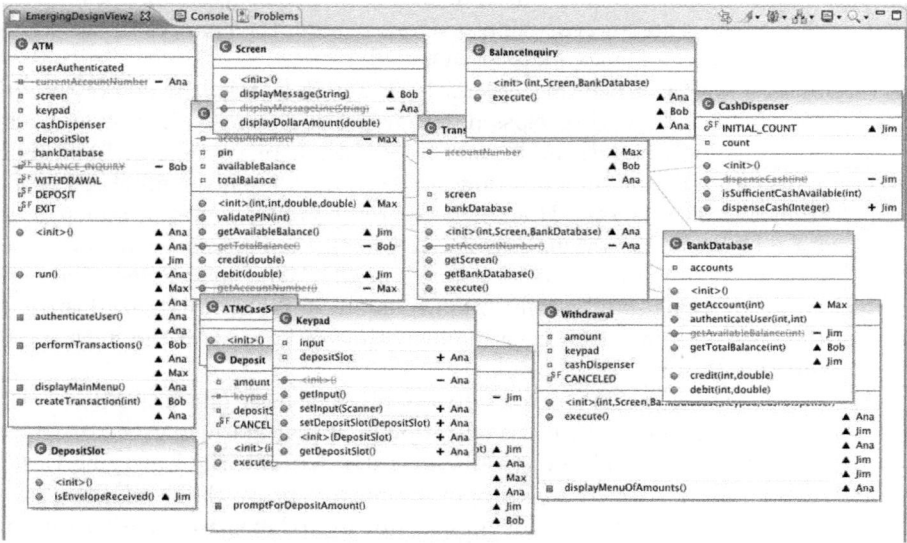

Fig. 2. Emerging Design basic representation with four developers (Max, Bob, Ana, Jim)

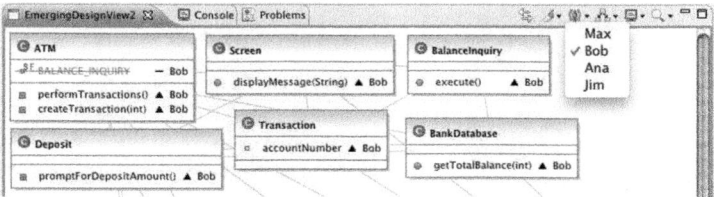

Fig. 3. Emerging Design basic representation with filtering. Just modifications in Bob's workspace are shown.

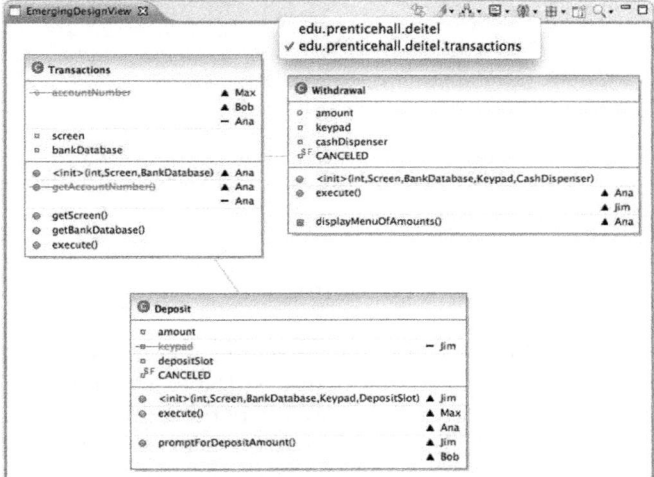

Fig. 4. Emerging Design basic representation with filtering. Just modifications in the selected package are shown.

with only a few package of the whole project. By using this filter, users could pay more attention to specific packages that are related with the task in hand, as can be seen in Figure 4.

The third and last filter only shows the classes that have any modification. So, instead of showing all the classes as in Figure 1, the tool decrease the numbers of nodes from the visualization by hiding the classes that are not being modified for any member, as shown in Figure 5.

To date, Lighthouse is a collaboration portal focused on detecting conflicts. It uses the Emerging Design to show who is making the changes where, and by looking at that, enabling developers to find where their changes may be conflicting with somebody else's. In this paper, we take this work a step further. We outline how we believe the concept of Emerging Design is not only useful for detecting conflicts, but also as a basis for knowledge collaboration. In the next section we talk about three particular knowledge collaboration problems and how Emerging Design can be used as a basis for exploring them.

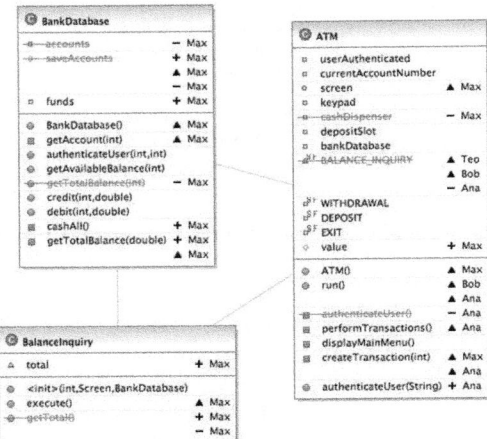

Fig. 5. Emerging Design basic representation with filtering. Just classes with modifications are shown.

3   Three Knowledge Problems

Knowledge collaboration manifests itself in many different forms and may revolve around many issues. In this section, we discuss the following three problems: (1) How to support developers in determining where the implementation is deviating from the original design; (2) How to support finding the right expert related to a given design; and (3) How to support identification of those parts of the program with less than ideal quality.

3.1   Design Decay

It is well known that software changes, and that such changes involve modifications to the original design that may lead to design decay [6]. Prior to the implementation phase, some conceptual design diagrams are usually constructed to guide developers and help them understand the project's high-level picture. The reasons *why* a particular design decays generally are not available, and therefore could be said to represent a knowledge collaboration problem: at some future point in time, other developers must understand why a certain piece of code is like it is, and much rationale resides behind the code changes from the original design to the current state.

In order to illustrate this problem, consider the following scenario: Ana is a developer in a large team and has been assigned to a task that involves making changes in a part of the system with which she is unfamiliar. The previous person that developed that specific piece of code is on vacation and is not available for questions. However, Ana remembers that the project has some documentation, including a detailed UML design diagram that was made before the system's

implementation. Ana finds out that this document includes some notes on rationale for some structural decisions and uses it to find the information she needs to
complete the task. Ready to work, Ana realizes that the design does not match
with the source code. Some elements have changed, others elements are missing, and no rationale was provided to understand why this has happened and
whether or not the changes happened accidentally or intentionally. Ana is left
to study the source code in detail to try and understand how to accomplish her
task, an unfortunate situation [3].

We can address this issue by marking the Emerging Design, so it shows devi-
ation, and providing facilities for developers to provide contextual information
pertaining to the changes they make. Imagine a developer restructuring a certain piece of code in a certain way that is counter-intuitive. By leaving a note,
directly visible on the diagram (Figure 6) they now can motivate their change.
Other developers can respond either in the affirmative or by expressing concerns
and such. A discussion can ensue, for which it is crucial to note that the discussion takes place directly in the context of the artifacts and as the changes are
happening. Design decay can be avoided this way, and design evolution becomes
under joint ownership of the developers.

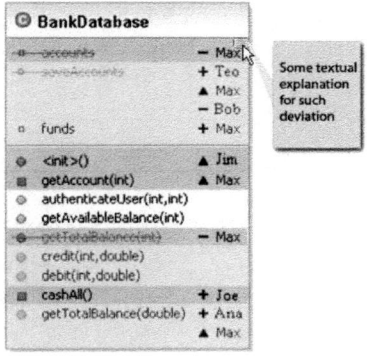

Fig. 6. Design Decay Representation

In Figure 6, the green overlays are used for elements that are present in
both the conceptual and emerging design, i.e., the ones that were implemented
according to the original design. Red overlays are used for items that are in the
emerging but not in the conceptual design, meaning that the implementation
diverges from the original design. Elements left in white are the ones that are
in the conceptual but not in the emerging design. These elements have not been
implemented yet.

The Emerging Design provides a natural basis for addressing design decay
because it already tracks design evolution. By now using this basis with simple but powerful extensions, the Emerging Design provides instant knowledge
collaboration, both implicitly because it makes visible the design as it evolves

and explicitly because its evolution can be gauged, questioned, discussed, and resolved as needed.

We also note that this can take place both among individual developers at the level of individual or small sets of changes, and by team leaders and architects based upon views of the code as a whole.

3.2   Expertise

The time taken to find an expert is one of the major reasons that co-located work tends to take less time than similar development work split across sites [7]. Quickly finding the right expert related to a given design and/or implementation issue is critical to the success of any software development project. There is a clear knowledge collaboration problem when one needs to understand how some class/method works, why it is as it is, and how it may need to evolve. For instance, in the previous section, because of the absence of the expert, Ana found herself in a situation where she had to study the source code in detail, which implies more time in order to accomplish her task. Often, an expert can provide useful assistance in this regard.

We again explore how the basis of Emerging Design can be leveraged to address this problem. Particularly, we envision exploring the use of a visualization to allow users to browse through the Lighthouse diagram in order to find the proper expert. Since Lighthouse already provides the basis for who made which

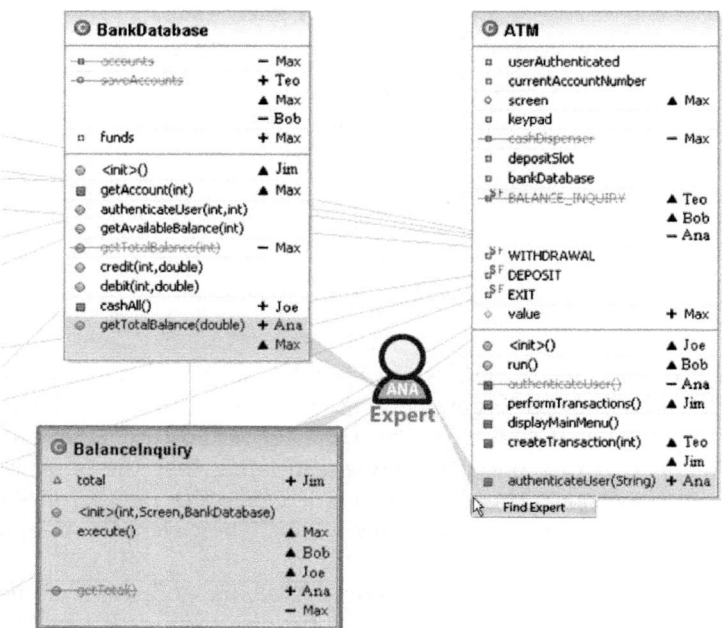

Fig. 7. Expertise Representation

changes, now we can actually build various overlays that make it possible, for instance, to click on one of the authors of a particular method and have the other pieces that they changed highlighted.

In another form, we note that it is often difficult to find someone with broader knowledge pertaining to multiple artifacts and methods. We plan to develop a feature that allows the user to select a group of methods and classes in order to find the expert related to that set of artifacts, as shown in Figure 7.

The advantage here is that, while most expertise systems are limited to work at the level of artifacts, our approach can provide more fine-grained as well as a broader range of answers.

3.3   Code Quality

Software quality metrics can drive software process improvement [8]. Explicit attention to characteristics of software quality can lead to significant savings in software life-cycle costs [9]. Some information that could be useful in this regard is the overall quality of each class, which if available would enable the identification of the most problematic or complex parts of a project. This kind of information is not usually accessible, representing the third knowledge collaboration problem that we address in this paper.

The use of Emerging Design in this situation would help developers and managers to quickly spot code that is growing without proper quality. We envision a software quality visualization that will show individual factors, such as *number of developers*, *number of recent bugs*, *how well the class/method was tested*, and *number of changes/code volatility* at the bottom of each class. We also take these

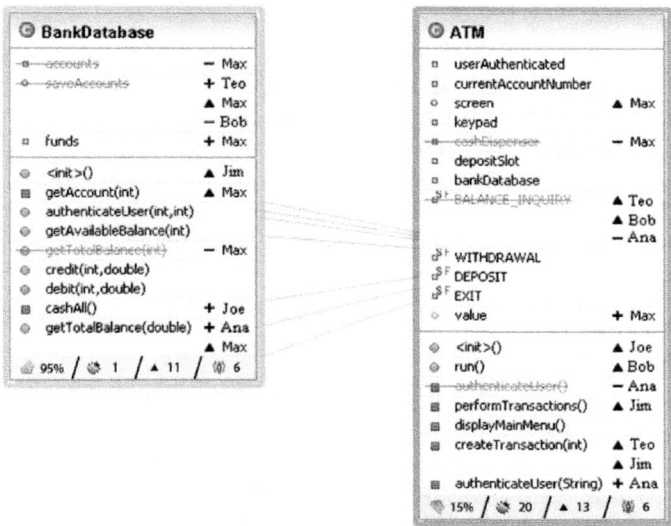

Fig. 8. Code Quality Representation

individual factors in consideration to provide an overall quality measure, and we represent this high level awareness information by using colored border, in which green means good quality and red means bad quality (Figure 8).

We extend the capability of Emerging Design to deal with software quality issues. This approach can help understanding which classes/methods are producing higher quality code. In this way, managers would be able to identify areas that need attention, and also tell what parts of the project are in need of more tests and what parts have enough coverage already.

4   Related Work

Several tools have been created to help people collaborate and to enhance individuals' awareness. The War Room Command Console [10] shows in a public display the current state of a system across workspaces in real-time. The visualization shows the ongoing changes made by developers in thumbprints, a graphical representation of the source code, displayed in a topographic layout. This work, like Lighthouse, uses a program-centered approach to show how changes made by developers are related with the artifacts and how the system is evolving. Its display, however, is in a central location and not on a per-developer basis. Furthermore, the information that it shown is compacted, and does not allow easy access to details.

Palantír [11] provides real-time awareness of changes made by developers and estimates the impact of how severe these changes are. Palantír, like Lighthouse, does not require developers to check-in the changes made and presents a view with information of all developers' workspaces. However, Palantír differs from Lighthouse since it uses a low-level abstraction that focuses on files, while Lighthouse uses the concept of Emerging Design.

FastDash [12] and CollabVS [13] both use a collaboration-centered approach to display the artifacts' interaction among developers. Unlike Lighthouse, this approach uses real-time awareness of developers' activities instead of focusing on program artifacts. The visualization shows people and the activities they currently undertake, e.g., who has which file open or who is editing which file. This approach has the drawback of not providing a spatial awareness of artifacts and it does not provide a historical view of changes made.

5   Summary

In this paper we recapped Emerging Design and presented our vision of the potential role it can play in knowledge collaboration. We described Lighthouse briefly and addressed three knowledge collaboration problems: *design decay* by providing developers with the rationale resides behind the code changes from the original design to the current state; *expertise* by finding the proper expertise for a particular group of methods and/or classes; and *code quality* by providing developers the identification of parts of the program with less than ideal quality.

The benefit we can see is that the knowledge is directly anchored to the artifacts to which it pertains and is thereby easily accessible and intuitive since it fits with the task that a developer is currently working on. Presently, we are engaged in providing this support and we will perform various explorations and evaluations as we build our extensions to Lighthouse. A particular question is whether Emerging Design is useful to support other knowledge collaboration problems as well. Another question is how it can support multiple problems in parallel, as some of our solutions use similar techniques and thus cannot be used at the same time.

Acknowledgments

Effort partially funded by the National Science Foundation under grant number 0920777.

References

1. Rus, I., Lindvall, M.: Knowledge management in software engineering. IEEE Software 19(3), 26–38 (2002)
2. Kantor, M., Zimmermann, B., Redmiles, D.: From group memory to project awareness through use of the knowledge depot. In: CSS 1997: California Software Symposium (1997)
3. Van der Westhuizen, C., Chen, P.H., van der Hoek, A.: Emerging design: New roles and uses for abstraction. In: ROA 2006: Proceedings of the 2006 International Workshop on Role of Abstraction in Software Engineering, pp. 23–28. ACM, New York (2006)
4. Parnas, D.L., Clements, P.C.: A rational design process: How and why to fake it. IEEE Transaction on Software Engineering 12(2), 251–257 (1986)
5. da Silva, I.A., Chen, P.H., Van der Westhuizen, C., Ripley, R.M., van der Hoek, A.: Lighthouse: Coordination through emerging design. In: Eclipse 2006: Proceedings of the 2006 OOPSLA Workshop on Eclipse Technology Exchange, pp. 11–15. ACM, New York (2006)
6. Eick, S.G., Graves, T.L., Karr, A.F., Marron, J., Mockus, A.: Does code decay? assessing the evidence from change management data. IEEE Transactions on Software Engineering 27(1), 1–12 (2001)
7. Herbsleb, J.D., Mockus, A., Finholt, T.A., Grinter, R.E.: An empirical study of global software development: distance and speed. In: ICSE 2001: Proceedings of the 23rd International Conference on Software Engineering, Washington, DC, USA, pp. 81–90. IEEE Computer Society, Los Alamitos (2001)
8. Livingston, J., Prosise, K., Altizer, R.: Process improvement matrix: A tool for measuring progress toward better quality. In: Proceedings of 5th International Conference on Software Quality (1995)
9. Boehm, B.W., Brown, J.R., Lipow, M.: Quantitative evaluation of software quality. In: ICSE 1976: Proceedings of the 2nd International Conference on Software Engineering, pp. 592–605. IEEE Computer Society Press, Los Alamitos (1976)
10. O'Reilly, C., Bustard, D., Morrow, P.: The war room command console: shared visualizations for inclusive team coordination. In: SoftVis 2005: Proceedings of the 2005 ACM symposium on Software visualization, pp. 57–65. ACM, New York (2005)

11. Sarma, A., Noroozi, Z., van der Hoek, A.: Palantír: raising awareness among configuration management workspaces. In: ICSE 2003: Proceedings of the 25th International Conference on Software Engineering, Washington, DC, USA, pp. 444–454. IEEE Computer Society, Los Alamitos (2003)
12. Biehl, J.T., Czerwinski, M., Smith, G., Robertson, G.G.: Fastdash: a visual dashboard for fostering awareness in software teams. In: CHI 2007: Proceedings of the SIGCHI conference on Human factors in computing systems, pp. 1313–1322. ACM, New York (2007)
13. Hegde, R., Dewan, P.: Connecting programming environments to support ad-hoc collaboration. In: 23rd IEEE/ACM International Conference on Automated Software Engineering, ASE 2008, pp. 178–187. ACM Press, New York (2008)

A Time-Lag Analysis for Improving Communication among OSS Developers

Masao Ohira, Kiwako Koyama, Akinori Ihara,
Shinsuke Matsumoto, Yasutaka Kamei, and Ken-ichi Matsumoto

Graduate School of Information Science, Nara Institute of Science and Technology,
8916-5, Takayama, Ikoma, Nara, Japan
{masao,kiwako-k,akinori-i,shinsuke-m,yasuta-k,matumoto}@is.naist.jp

Abstract. In the open source software (OSS) development environment, a communication time-lag among developers is more likely to happen due to time differences among locations of developers and differences of working hours for OSS development. A means for effective communication among OSS developers has been increasingly demanded in recent years, since an OSS product and its users requires a prompt response to issues such as defects and security vulnerabilities. In this paper, we propose an analysis method for observing the time-lag of communication among developers in an OSS project and then facilitating the communication.

Keywords: time-lag analysis, OSS, distributed development.

1 Introduction

Open source software (OSS) such as Linux and Apache is generally developed by globally distributed developers. Unlike commercial software development in a company, OSS development does not necessarily request developers to engage in development at a designated time and location. OSS developers may voluntarily decide whether they continue to dedicate themselves to OSS development or not.

In this OSS development environment, a time-lag occurs in communication among developers more than a little, because of differences of time zones among geographically-distributed developers with a variety of lifestyles. For instance, according to the geographical distribution of registered users at SourceForge which was reported by Robles and Gonzalez-Barahona [1], the top three regions by the number of registered developers at SourceForge are North America, West Europe, and China. Since the time-lag among those regions is at least more than five hours, it would not be easy to discuss among developers in real-time. Furthermore, even if developers reside in the same time zone, it is not still guaranteed that developers can communicate each other in real time, because each developer has no constraint on working hours.

The goal of our research is to construct a support mechanism for effective communication among geographically-distributed OSS developers. As a first step toward achieving the goal, in this paper we present an analysis method for helping

K. Nakakoji, Y. Murakami, and E. McCready (Eds.): JSAI-isAI, LNAI 6284, pp. 135–146, 2010.

OSS developers comprehend the whole picture of the communication time-lag occurred in a OSS project. The analysis method targets a mailing list archive as a data source, and consists of three kinds of analyses as follows;

1. analysis of a geographical distribution and activity time of OSS developers
2. analysis of a distribution of time required for information exchanges among OSS developers in different locations, and
3. analysis of appropriate timing for sending messages.

From a case study with Python project data, this paper explores the usefulness of the analysis method.

2 Analysis Method

This section describes data extraction, conversion and classification which are necessary in advance of performing our analysis.

2.1 Preparation

(1) Data extraction and conversion. The target data source for our analysis is archives of mailing lists which are used by OSS developers to exchange information. The reason we select mailing list archives as the target data for our analysis is because mailing lists are widely used in OSS projects. We consider that data of mailing list archives allows us to reveal the whole picture of the existence of the time-lag in many OSS projects.

In order to apply the analysis method to the target data, firstly we need to extract information of **posted date and time**, and **posted locations** from mailing list archives (i.e., from e-mail headers). In what follows, "posted date and time" means local date and time of a message's sender, and "posted locations" is presented as a time-lag between Coordinated Universal Time (UTC) and local time. For instance, "**UTC+9**" means the location of Japan because the standard time of Japan is nine hours prior to UTC.

Fig. 1 shows the procedure of data extraction and conversion. When a developer posts a message to a mailing list, the message is delivered to subscribed developers of the mailing list. Replying to the post, the other developers can discuss the message. Using such the post-reply relationship (i.e., thread structure) in a mailing list, we extract[1] information on posted/replied date and time, and locations (time zones) from mailing list archives. For instance, from a thread structure illustrated in Fig. 1 (a), we extract information of posted and replied messages as the table in Fig. 1 (b). Then we convert the information into post-reply relationships as the table in Fig. 1 (c) and calculate a time-lag from a difference between posted and replied date and time. Note that we suppose that message B replied to message A can be a posted message for message C.

---

[1] We do not collect data from posted messages with no replies.

(a) a thread in a mailing list

post A: _____
  └ reply B : Re:_____
    └ reply C : Re:Re:_____
  └ reply D : Re:_____

(b) a list of information of posted/replied messages

ID	posted date and time	location (time zone)	Coordinated Universal Time (UTC)
message A	2009/04/29 10:33:53	UTC + 9	2009/04/29 01:33:53
message B	2009/04/28 22:40:04	UTC − 4	2009/04/29 02:40:04
message C	2009/04/29 09:12:30	UTC + 3	2009/04/29 06:12:30
message D	2009/04/29 02:26:59	UTC − 10	2009/04/29 12:26:59

(c) a relationship among posted and replied messages

info. of posted messages			info. of replied messages			time lag (hours)
ID	posted date and time	location	ID	replied date	location	
message A	2009/04/29 10:33:53	UTC + 9	message B	2009/04/28 22:40:04	UTC − 4	1.10
message B	2009/04/28 22:40:04	UTC − 4	message C	2009/04/29 09:12:30	UTC + 3	3.54
message A	2009/04/29 10:33:53	UTC + 9	message D	2009/04/29 02:26:59	UTC − 10	10.88

Fig. 1. Data extraction and conversion

(2) Classification of data. Several factors such as differences of time zones (i.e., countries and/or regions) and differences of developers' working hours may have an influence on time-lags between posted time and replied time. For instance, communication among developers living in different time zones might be prolonged due to differences of lifestyles (e.g., dinner time or sleeping time). Developers in the same time zone also might be difficult to communicate each other in real time, because each developer has no constraint on working hours.

In order to distinguish between the time-lag due to time zone differences and the time-lag due to lifestyle differences, the collected data described above is classified into data **within** and **over** the time-lag of 24 hours. Many of replied messages within 24 hours after a post would be affected by differences of time zones, while replied messages over 24 hours after a post would be generated by differences of developers' lifestyles and/or difficulty of the content of a posted message, rather than geographical differences among developers. For these reasons, our analysis method targets the data of posted and replied messages within the 24 hours time-lag.

2.2   Procedure

(1) Geographical distribution and activity time of OSS developers. In order to understand the existence of the communication time-lag in an OSS project, the analysis method firstly identifies a geographical distribution of developers of the project, counting the number of replied messages by each location (UTC−11∼UTC+12). The analysis method also identifies a distribution of the number of replied messages by local time in each location in order to understand working hours of developers by each location, since developers' working hour can

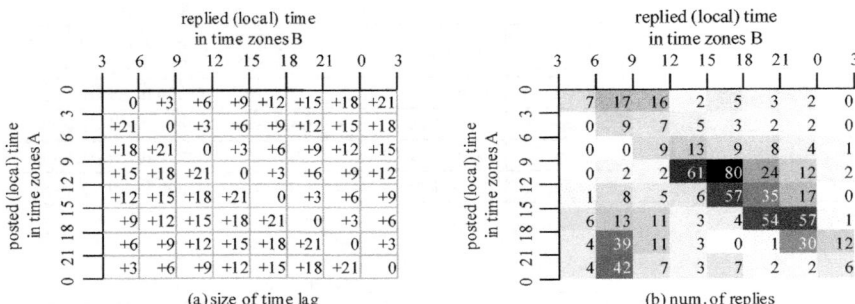

Fig. 2. Distribution of posted and replied time

differ even in the same location. By this means, we can identify active or inactive locations and working hours of OSS developers.

(2) Distribution of time required for information exchanges. In order to understand the communication time-lag due to the geographical (time zone) differences, the analysis method calculates distributions of time required for information exchanges among OSS developers in **different** locations and the **same** locations respectively. This helps us more clearly distinguish between the time-lag by the geographical differences and the time-lag by the differences of developers' lifestyles.

(3) Appropriate timing for sending messages. In order to identify the appropriate timing for communication which resolves communication time-lags as much as possible, the analysis method calculates the number of replied messages by each hour, using **posted (local) time** and **replied (local) time**. A numerical number in Fig. 2 (a) shows the size of a time-lag (hours) between time zones A and B. Fig. 2 (b) shows the number of pairs of posted messages from time zone A and replied messages from time zones B. For instance, suppose that one developer in A post a message between 9 and 12, and other developer in B replies a message between 15 and 18. In this case, the time-lag is +3 hours and the number of post/reply pairs is 80.

Time zones A and B are fixed after selecting target locations for analysis. Time zones B in Fig. 2 is arranged as replied messages within an hour correspond to posted messages on the diagonal. In Fig. 2, the size of time-lag and the number of posted/replied messages are counted by three hours, but the length may be changed depends on analysis needs. Furthermore, the all cells in Fig. 2 (b) are gray-scaled according to the number of posted/replied pairs of messages, to grasp a big picture of time slots with a large or small number of replied messages.

Using Fig. 2 (a) and (b), it is possible to identify a time slot with a large or small time-lag. For instance, we can see that messages posted between 21 and 0 in time zones A (the bottom row in Fig. 2) tend to be replied after 6 hours. That is, to post messages from 21 to 0 would not be the appropriate timing for less time-lag communication.

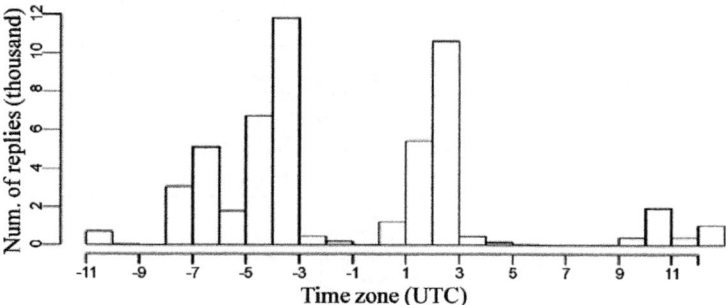

Fig. 3. Distribution of the number of replied messages by time zones

3   Case Study

This section describes a case study with a mailing list for developers in the Python project. Through the case study, we would like to confirm whether the analysis method can help us understand the existence of the time-lag in communication among OSS developers.

3.1   Python Project

Python[2] is an object oriented script language developed by OSS. It is very popular in Europe and the United States as well as Perl. Because it supports various platforms and provides rich documentations and libraries, it is used in a broad range of domains (e.g., Web programming, GUI-based applications, CAD, 3D modeling, formula manipulation, and so forth).

3.2   Target Data

We selected the mailing list archive called Python-Dev[3] which is for discussing development of Python such as new features, release and maintenance. We use the Python-Dev mailing list archive from April 1999 to April 2009, which have 89,301 messages. Excluding posted messages with no replies and messages with no information on posted/replied time and locations, posted and replied messages were 56,707. 51,830 of 56,707 messages were sent within 24 hours.

3.3   Analysis Results

(1) Geographical distribution and activity time. Fig. 3 shows a distribution of the number of replied messages by time zones. The X-axis and Y-axis respectively mean time zones and the number of replied messages. It indicates

---

[2] Python Programming Language, http://www.python.org/
[3] Python core developers ML, http://mail.python.org/mailman/listinfo/python-dev/

Table 1. Target locations for the case study of Python

region	time zone	location
North and South American continent	UTC−8∼ UTC−4	United States, Canada, West of Brazil, Chile, Bolivia, Mexico, etc.
European and African continent	UTC+0∼ UTC+3	Europe, Africa, Moscow, Iran, Saudi Arabia, etc.

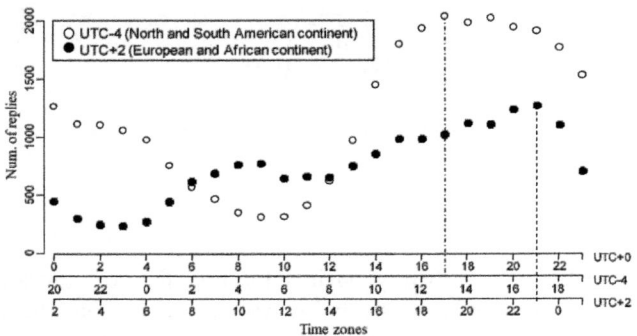

Fig. 4. Distribution of the number of replied messages by time slots (white circles: North and South American continent, black circles: European and African continent)

that a large number of messages are replied by developers from UTC−4 (East of the United States) and UTC+2 (central Europe) in the Python project. This result is not surprising at all, because Python is mainly used and developed by European and American developers. It would be natural that developers living in these locations actively communicated.

Many of countries in the locations of UTC−4 and UTC+2 is utilizing daylight-saving time. And countries around the countries in UTC−4 and UTC+2 also have many messages. So, we selected two regions around UTC−4 (the North and South American continent: UTC−8∼UTC−4) and UTC+2 (the European and African continent: UTC+0∼UTC+3) as the analysis target in this paper. Table 1 shows major countries included in these regions.

Fig. 4 shows transitions of replied messages by hour in the two regions which are determined from Fig.3. The X-axes shows time in the three time zones (UTC+0, UTC−4, UTC+2) and the Y-axis is the number of replied messages. It indicates that the maximum and minimum number of replied messages from the North and South American continent are attained respectively at 13 and 5 in the local time (UTC−4). Python developers in the North and South American continent seem to mainly communicate during daytime hours. In contrast, Python developers in the European and African continent actively communicate during nighttime hours, because the number of replied messages from the European and African continent is peaked at 23 in the local time (UTC+2). In this

Table 2. Statistics of time-lags by region (A: North and South American continent, E: European and African continent)

posted region → replied region	the number of replies	maximum (hours)	median (hours)	minimum (hours)
A → A	18,901	11.55	1.24	0.00
A → E	6,942	16.34	2.07	0.00
E → E	9,426	14.69	1.59	0.00
E → A	7,215	13.91	1.80	0.00

way, analyzing activity time of OSS developers by using the number of replied messages helps us understand the existence of the difference of working hours by region.

Although Fig. 4 provides an overview on the difference of working hours of OSS developers by region, however, it does not tell us anything about time-lags. In fact, developers in the both regions actively communicate each other from 12 to 23 in UTC+0. Communication time-lags might not exist in the regions. In contrast, developers in either one region or the other region does not actively communicate from 12 to 23 in UTC+0. Communication time-lags between developers living different locations might exist in this time period.

(2) Distribution of time required for information exchanges. Table 2 shows time spent to reply messages to the same and different time zones, the number of replied messages, and time-lags (maximum/median/minimum). A pair of a post from location X and a reply from location Y is represented as "X → Y".

The median hours of the time-lag among the same time zone was 1.24 hours for A → A and 1.59 hours for E → E. The median hours of the time-lag between the different time zones was 2.07 hours for A → E and 1.80 hours for E → A. Developers in the same time zone can expect to have a reply within 90 minutes, and developers between different time zones also can expect to have a reply within about 2 hours. Since the actual difference of the time-lag between the target regions is nearly 6 hours, we can consider that communication time-lags in the Python project are relatively small.

(3) Appropriate timing for sending messages. Fig. 5 (a), (b), (c) and (d) are distributions of the number of replied messages between two regions. For the simplicity, only gray-scaled figures without the number of replied messages are shown in Fig. 5.

We can see that the zero time-lag (i.e., dark gray cells near the diagonal line) is expected from 10 to 17 in Fig. 5 (a), from 9 to 17 in posted local time and from 15 to 23 in replied local time in Fig. 5 (b), from 16 to 23 in Fig. 5 (c), and from 16 to 23 in posted local time and from 10 to 17 in replied local time in Fig. 5 (d). For these time periods, developers would timely communicate each other.

In contrast, reply time seems to be delayed from 18 to 23 in posted local time in Fig. 5 (b) and from 7 to 13 in replied local time in Fig. 5 (d), because there

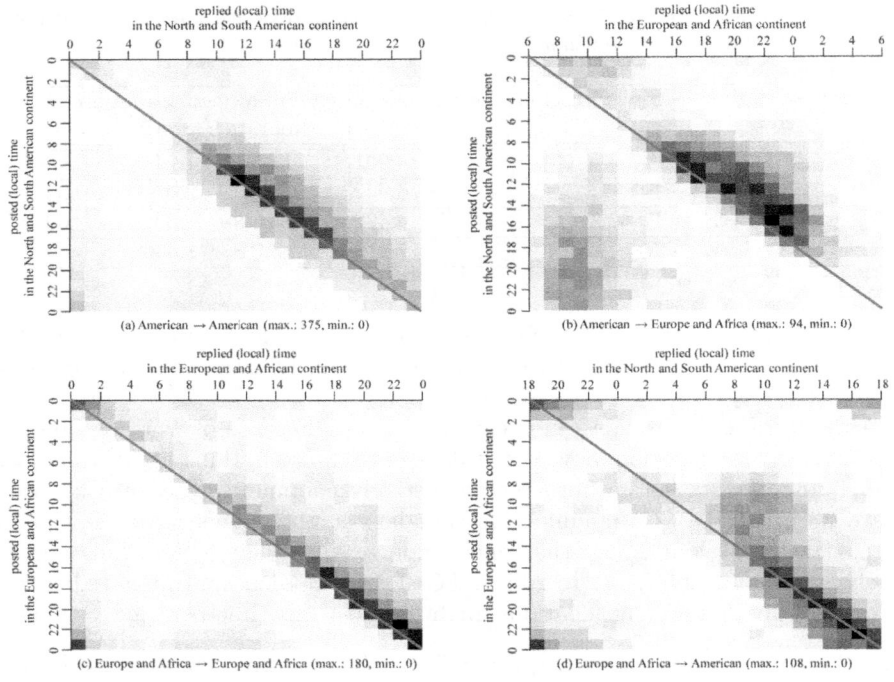

Fig. 5. Distributions of posted/replied local time between two regions

are darker cells a short distance away from the diagonal line. These two posted local time periods correspond to the time period from midnight to early morning i0 to 6) in replied locations, which means that developers in replied locations was sleeping at the posted time.

From the result of Fig. 5, in order to receive a quick reply, it would be desirable to post a message from 10 to 17 in the North and South American continent, and from 16 to 23 in the European and African continent. On the contrary, it is not appropriate timing to post a message from 18 to 23 in the North and South American continent, and from 7 to 13 in the European and African continent, since the time-lag is likely to occur.

In this way, our analysis method helps OSS developers know the appropriate timing so that they can resolve a time-lag of information exchange in an OSS project as much as possible.

4   Discussions

Opposite to what we expected before our case study, we have confirmed in Table 2 that the influence of the time-lag due to the time zone difference was relatively small in the Python project. One reason of this result might be that active time of Python developers is partly overlapping in the two regions. Although there are about 6 hours time-zone difference between the two regions, the active time in the North and South American continent was different from that in the European and African continent as shown in Fig. 4. Therefore, active hours of Python developers in the two regions might overlap by coincidence from 10 to 17 in the North and South American continent (from 16 to 23 in the European and African continent). Another reason may be that the number of Python developers subscribed to the "Python-Dev" mailing list is sufficiently-large to quickly respond to a posted message at any time.

Our analysis method is not only useful in knowing the appropriate timing for communication among geographically-distributed OSS developers, but also useful in changing communication media used in an project. For instance, when a project replaces mailing lists with IRCs (Internet Relay Chat) as communication media, developers would be required to more precisely understand the appropriate timing for communication to resolve the time-lag. In that case, our method would help developers know the better timing for real-time communication.

OSS developers are not necessary to be geographically-distributed, but they may be at the same region or location. Though our analysis method mainly aims to understand the communication time-lag arising from time-zone differences, it can be used for the time-lag due to lifestyle differences of OSS developers in the same region or location. OSS developers have no constraint on their working hours and they can freely engage in OSS development. At the same region, some developers can work in the morning and other developers can develop OSS at midnight. Depending on the differences of lifestyles of developers, time-lags could happen even if they live close to each other. In this situation, our method can provide an insight on the differences of active time in the same region and help developers understand the appropriate timing for sending messages.

The analysis method also can be used for distributed development in a company. Working hours in a company are fixed to some extent, but it is not necessarily that a developer in one site can communicate with other developers in another site at a particular time. In the prior study [2], time zone differences are visualized to understand and exploit overlapping hours in a distributed environment. Our method can not only visualize the time zone differences, but also allows developers to understand the easiness of communication at a particular time period, using the number of replied messages (i.e., density of working activity at a particular time period).

In this paper, we introduce the time-lag analysis method toward improving the communication efficiency of geographically-distributed OSS developers. The analysis method targets mailing list archive data as communication logs to reveal the existence of communication time-lags. Although IRC communication is often used in OSS projects and they can be our analysis target, communication using IRC do not work when developers one wishes to talk are off-line. So, IRC communication logs are not likely to well-capture communication time-lags.

In this paper, we have conducted a case study of the Python project, using the "Python-Dev" mailing list archive. Python-Dev consists of mailing list archive data for about 10 years. So, it might be too large to show communication time-lags among Python developers at the fine-grained level. Actually, we have observed that communication time-lags in the Python project were relatively small. We suspect that this results from the size population of developers (subscribers) of Python-Dev. In Python-Dev, a posted message must be read by a number of developers in the world and so it might be easy to have replies. In order to emphasize the existence of time-lags and its issues, in the near future, we need to analyze more specific situations such as the level of communication among module owners, reviewers and patch contributors.

5   Related Work

The issues on the communication time-lag or delay in OSS development have been intensively studied in relation to bug modification processes with bug tracking systems [3,4,5,6,7,8,9,10,11,12,13]. For instance, Wang et al. [12] proposed several metrics to measure the evolution of open source software. The metrics include the number of bugs in software, the number of modified bugs and so on. As a result of the case study using the Ubuntu project (one of Linux-based operating system distributors), the study found that about 20% of all the reported bugs were actually resolved and over ten thousand bugs were not assigned to developers. These findings indicate that it takes a long time to resolve all bugs reported into bug tracking systems and that it also takes a long time to start modifying bugs. The study, however, did not reveal the amount of time or communication time-lags to resolve bugs.

Mockus et al. [10] and Herraiz et al. [6] have reported studies on the mean time to resolve bugs in open source software development. Mockus et al. have conducted two case studies of the Apache and Mozilla projects to reveal success factors of open source software development. In the case studies, they analyzed the mean time to resolve bugs because rapid modifications of software bugs are generally demanded by users. As a result of the analysis, they have found that the mean time to resolve bugs were short if bugs existed in modules regarding to kernel and protocol, and existed in modules with widely-used functions. They also found that 50% of bugs with the priority P1 and P3 were resolved within 30 days, 50% of bugs with P2 were resolved within 80 days, and 50% of bugs with P4 and P5 were resolved within 1000 days. While [10,6] mainly focused on precise understandings of bug modification processes in open source software development, we are interested in the influence of communication time-lags among developers on the bug modification process.

The issues on differences of time-zone and/or geographical distance in distributed development rather have been discussed in terms of the context of corporate (proprietary) software development [14,15,16,17,18]. For instance, Harbsleb et al.[16] have compared single-site development with multi-sites development and then revealed that development in the distributed environment introduced the delay of development speed. In contrast, Bird et al. [19] analyzed the development of Windows Vista by comparing distributed teams with collocated teams from the aspect of the post-release failures of components. They have found a slight difference in failures, but the difference has been less significant. Nguyen et al.[20] also reported the similar phenomena in the Eclipse Jazz project. Although the lessons learned from these studies on distributed software development provides us a lot of useful insights, they are partly applicable to geographically-distributed OSS development due to the differences of lifestyles of developers even in the same region or location. In this paper, we tried to tackle this unique feature of time-lags in OSS development.

6   Conclusion and Future Work

In this paper, we proposed a method for analyzing a communication time-lag among OSS developers. As a result of our case study of the Python developers' mailing list archive, we could confirm that our analysis method helps geographically-distributed OSS developers understand: (1) active time of developers are different from regions, (2) communication time-lags in the Python project is relatively small, and (3) there exists the appropriate timing for resolving communication time-lags as much as possible. In this paper, our analysis method targets communication time-lags in the two regions with the time zone difference. In the future, we need to analyze regions and/or locations without time zone differences in order to better understand the influence of lifestyle differences of developers on communication time-lags. As described before, we still need to analyze more specific situations of time-lags at the fine-grained level.

Acknowledgment

This research is being conducted as a part of the Next Generation IT Program and Grant-in-aid for Young Scientists (B)–20700028, 21–8995C20–9220 by the Ministry of Education, Culture, Sports, Science and Technology, Japan.

References

1. Robles, G., Gonzalez-Barahona, J.M.: Geographic location of developers at source-forge. In: The 2006 international workshop on Mining Software Repositories (MSR 2006), pp. 144–150 (2006)
2. Laredo, J.A., Ranjan, R.: Continuous improvement through iterative development in a multi-geography. In: The 2008 IEEE International Conference on Global Software Engineering (ICGSE 2008), pp. 232–236 (2008)
3. Bettenburg, N., Just, S., Schröter, A., Weiss, C., Premraj, R., Zimmermann, T.: What makes a good bug report? In: The 16th ACM SIGSOFT International Symposium on Foundations of Software Engineering (SIGSOFT 2008/FSE-16), pp. 308–318 (2008)
4. Colazo, J.A.: Following the sun: Exploring productivity in temporally dispersed teams. In: The Fourteenth Americas Conference on Information Systems (AMCIS 2008). Paper no. 240 (2008)

5. Godfrey, M.W., Tu, Q.: Evolution in open source software: A case study. In: 16th IEEE International Conference on Software Maintenance (ICSM 2000), pp. 131–142 (2000)
6. Herraiz, I., German, D.M., Gonzalez-Barahona, J.M., Robles, G.: Towards a simplification of the bug report form in eclipse. In: The 2008 international Working Conference on Mining Software Repositories (MSR 2008), pp. 145–148 (2008)
7. Ihara, A., Ohira, M., Matsumoto, K.: An analysis method for improving a bug modification process in open source software development. In: The joint international and annual ERCIM workshops on Principles of Software Evolution and Software Evolution Workshops (IWPSE-Evol 2009), pp. 135–144 (2009)
8. Kim, S., Pan, K., Whitehead, E.J.: Memories of bug fixes. In: The 14th ACM SIG-SOFT international symposium on Foundations of software engineering (SIGSOFT 2006/FSE-14), pp. 35–45 (2006)
9. Kim, S., Zimmermann, T., Whitehead, E.J.: Automatic identification of bug-introducing changes. In: The 21st IEEE/ACM International Conference on Automated Software Engineering (ASE 2006), pp. 81–90 (2006)
10. Mockus, A., Fielding, R.T., Herbsleb, J.D.: Two case studies of open source software development: Apache and mozilla. ACM Transactions on Software Engineering and Methodology (TOSEM) 11(3), 309–346 (2002)
11. Śliwersk, J., Zimmermann, T., Zeller, A.: When do changes induce fixes? In: The 2005 International Workshop on Mining Software Repositories (MSR 2005), pp. 1–5 (2005)
12. Wang, Y., Guo, D., Shi, H.: Measuring the evolution of open source software systems with their communities. ACM SIGSOFT Software Engineering Notes 32(6), Article No.7 (2007)
13. Yilmaz, C., Williams, C.: An automated model-based debugging approach. In: The twenty-second IEEE/ACM international conference on Automated Software Engineering (ASE 2007), pp. 174–183 (2007)
14. Carmel, E.: Global software teams: collaborating across borders and time zones. Prentice Hall PTR, Upper Saddle River (1999)
15. Karolak, D.W.: Global Software Development: Managing Virtual Teams and Environments. Wiley-IEEE Computer Society Press, Los Alamitos (1999)
16. Herbsleb, J.D., Mockus, A., Finholt, T.A., Grinter, R.E.: An empirical study of global software development: distance and speed. In: The 23rd International Conference on Software Engineering (ICSE 2001), pp. 81–90 (2001)
17. Milewski, A.E., Tremaine, M., Egan, R., Zhang, S., Kobler, F., O'Sullivan, P.: Guidelines for effective bridging in global software engineering. In: The 2008 IEEE International Conference on Global Software Engineering (ICGSE 2008), pp. 23–32 (2008)
18. Sangwan, R., Bass, M., Mullick, N., Paulish, D.J., Kazmeier, J.: Global Software Development Handbook. Auerbach Series on Applied Software Engineering Series. Auerbach Publications, Boston (2006)
19. Bird, C., Nagappan, N., Devanbu, P., Gall, H., Murphy, B.: Does distributed development affect software quality? an empirical case study of windows vista. In: The 2009 IEEE 31st International Conference on Software Engineering (ICSE 2009), pp. 518–528 (2009)
20. Nguyen, T., Wolf, T., Damian, D.: Global software development and delay: Does distance still matter? In: The 2008 IEEE International Conference on Global Software Engineering (ICGSE 2008), pp. 45–54 (2008)

Comparison of Coordination Communication and Expertise Communication in Software Development: Motives, Characteristics, and Needs

Kumiyo Nakakoji[1], Yunwen Ye[1], and Yasuhiro Yamamoto[2]

[1] Key Technology Laboratory, SRA Inc., Japan
[2] Precision and Intelligence Laboratories, Tokyo Institute of Technology, Japan
{kumiyo,ye}@sra.co.jp, yxy@acm.org

Abstract. The research question we pursue is how to go beyond existing communication media to nurture communication in software development. Nurturing communication in software development is not about increasing the amount of communication but about increasing the quality of the communication experience in the context of software development. Existing studies have shown that different motives and needs are inherent when developers communicate with one another. Identifying *coordination communication* (*c-comm* for short) and *expertise communication* (*e-comm*) as two distinct types of communication, we characterize the difference between the two and discuss important factors to take into account in designing mechanisms to support each type of communication.

Keywords: nurturing communication in software development, knowledge collaboration, continuous coordination, unified interface for communication, coordination communication, c-comm, expertise communication, e-comm, design considerations.

1 Introduction

Communication has been regarded as an important element in software development. Increasingly more studies argue that socio-technical aspects of software development need to be seriously taken into account in supporting software development. The underlying premise is that peer developers are important knowledge resources in the same way as other artifacts, such as source code, comments, design documents, release notes, and bug reports, and that obtaining knowledge and information from peers is quintessential in software development. Communication should not be regarded as something to eschew, but instead as something to be nurtured [10].

The media currently used in such communication demonstrate a variety of means, including face-to-face, telephone, personal email, mailing-list, Wiki, Internet Relay Chat (IRC), video conferencing, or digital and physical artifacts (e.g., comments inserted in source code or post-it notes pasted on a printed document). Awareness mechanisms may also be regarded as a form of communication media in the sense that one can obtain information about what other members of the projects are doing. As communication media vary, styles of communication in software development

K. Nakakoji, Y. Murakami, and E. McCready (Eds.): JSAI-isAI, LNAI 6284, pp. 147–155, 2010.

range from indirect to direct, from asynchronous to synchronous, and from intentional to unintentional. It might be one to one, one to a designated some, or one to unknown numbers of many.

Most of communication media that software developers currently use have been built for general purposes (with few exceptions such as Wiki). The goal of our research is to design innovative communication media to nurture communication for software developers. Nurturing communication in software development is not about increasing the amount of communication but about increasing the quality of the communication experience in the context of software development. The primary task of a software developer is to develop software, and not to communicate. Communicative activities should be seamlessly integrated within the context of software development activities. Communication is a means, not a goal.

In order to address the research question of how to go beyond existing communication media to nurture communication in software development, we need to better understand *why* software developer communicate with each other. By looking into the motives of communicative activities of software developers, we have identified two distinctive types of needs in such communications: *coordination communication* and *expertise communication* [10].

In coordination communication, or *c-comm* for short, a developer tries to coordinate his or her task with dependent peers in order to avoid and/or to solve emerging or potential conflicts. In expertise communication, or *e-comm* for short, a developer seeks information to solve his or her task at hand and asks peers for help. Note that by expertise communication, we do *not* mean that a certain group of developers who have general expertise thereby transfer their knowledge to novice developers through communication. In contrast, our view is that expertise is always defined in terms of some context, for instance, in terms of a particular method, a particular class, a particular release, or a particular bug report at a particular point in time; and that expertise is not something definable without context. In this view, each developer has his or her own expertise in some aspects of the system and the project. Expertise communication, therefore, may take place among all of the peer developers in every direction [16].

Developers currently do not distinguish the two types of communication, which are driven by their "information needs" and are carried out through common communication channels. Coworkers were the most frequent source of information for software developers, and the two types of information most frequently sought by software developers from their coworkers were "What have my coworkers been doing?" and "In what situations does this failure occur?" [7]. The former information is sought primarily for the purpose of coordinating the work, and the latter is for the purpose of getting some knowledge about the source code. Data on three well-known open source projects have shown that text-based communication (mailing lists and chat systems) is the developers' primary source of acquiring both general knowledge about other developers (who has the necessary expertise) and specific awareness (who is working on their relevant parts of the system—to coordinate their tasks) [4].

It seems that developers often mix the two types of communication within a single discourse session without paying any attention to distinguishing the two. For instance, developer John first asks his colleague Mary over the cubicle wall whether she knows why class C calls a method X instead of Y; then Mary answers that it is because Y is designed to be thrown away, and that, by the way she has just been working on X and checked-in the changes, so he had better check the latest version of X if he is working on C. Thus, while the initial question posed by John is e-comm (i.e., he wanted to ask Mary to give him the answer as to why C calls X instead of Y), the subsequent conversation provided by Mary turns out to be c-comm (i.e., C that John is working on depends on X that Mary is working on).

Why does it matter then to distinguish the two types of communication if developers do not distinguish them? It matters because when it comes to design computational mechanisms for supporting communication in software development, each type of communication demands different types of design concerns.

In this paper, we first describes what fundamental differences exist between the two types of communication in software development. We then explain how different aspects need to be considered in designing computational support mechanisms. We conclude with a list of research issues to be considered in developing such support.

2 Expertise Communication and Coordination Communication

A few features distinguish e-comm (expertise communication) from c-comm (coordination communication).

We first illustrate c-comm. Suppose developer X initiates communication with developer Y, which turns out to be c-comm. The purpose of the c-comm is to coordinate tasks to resolve emerging conflicts or to avoid possible future conflicts among the tasks in which X and Y are engaged. Developers X and Y are called "socially dependent" [2] in the sense that they have to coordinate their tasks through social interactions when it becomes necessary to resolve the perceived conflicts. X and Y together form an "impact network" [3]. Coordination communication is a part of impact management, which is "the work performed by software developers to minimize the impact of one's effort on others and at the same time, the impact of others into one's own effort" [3]. X may need to further involve those developers who are part of the impact network.

In contrast, suppose developer A initiates communication with developer B, which turns out to be e-comm. The purpose of this e-comm is for A to get some information about A's task at hand; A is asking B to help A by providing some information for A's particular task. As noted earlier, e-comm refers to the activities to seek information that is essential to accomplish A's software development activities, not for the purpose of learning, but for the purpose of performing A's job. If A does not get satisfying information from B, A might need to ask other peers the same question.

Thus, while the relation between X and Y in the c-comm is reciprocal, that of A and B in the e-comm is not. In c-comm, there is a symmetric or reciprocal relation between those who initiate communication and those who are asked to communicate, with roughly equal interests and benefits. In e-comm, in contrast, there is an asymmetric and unidirectional relation between the one who asks a question and the one who is asked to help. The benefit would primarily for the communication initiator, and the cost (i.e., the additional effort) is primarily paid by those who are asked to participate in the communication; that is, the cost of paying attention to the information request; of stopping their own ongoing development task; of composing an answer for the information-seeking developer, including collecting relevant information when necessary; and of going back to the original task [15].

The role and value of the resulting communicative actions would also differ between the two types of communication. When developers communicate with one another, their conversations as well as produced artifacts (mail message contents or white board drawings, for instance) can be stored (if appropriate media is used). Such recorded communication can be useful if generated through e-comm. Email exchange about a particular design of a class, for example, would serve as a valuable auxiliary document for the class because another developer might find it useful to read when using the class at a later time.

Archived communication generated through c-comm might be useful to inform other developers within the same impact network for the time being. However, the impact network constantly changes over time, and such information communicated over a particular class may soon become obsolete. Moreover, c-comm without its temporal context could be quite harmful when misused. A collection of the coordination communication about a particular object over a long period of time may serve as the object's development log but it would not be more than the existing developmental records captured within current development environments.

Table 1 summarizes the differences between c-comm and e-comm.

Table 1. Comparing Coordination Communication (c-comm) and Expertise Communication (e-comm)

	Coordination Communication (c-comm)	Expertise Communication (e-comm)
purpose	to coordinate work	to get information
needs	conflict avoidance, conflict resolution	problem solving
cost & benefit	reciprocal between a communication initiator and the other communication participants	asymmetric between a communication initiator and the other communication participants
expanding participants	when others are part of the impact network	when the initiator could not get satisfactory information
recorded communication	useful for the time being until the impact network changes	becomes valuable for later use

The next section compares the different aspects of concern in designing mechanisms for supporting each of the two types of communication.

3   Different Needs for Supporting the Two Types of Communications

A thing is available at the bidding of the user--or could be--whereas a person formally becomes a skill resource only when he consents to do so, and he can also restrict time, place, and method as he chooses [6].

In talking about depending on other people, such as teachers, as knowledge re-sources, Illich argued that their willingness to participate is essential in regarding them as information resources. Using peers as potentially relevant information resources is likely to increase the cognitive load for both those who initiate communication and those who are asked to participate in the communication. Unlike a Help Desk, where it is the job of those who are asked to answer [1], peer developers are there not to commu-nicate but to perform their own development tasks in a time-critical fashion. They might be willing to communicate if they had more time and less stressful situations; otherwise, they might not be willing unless they see an immediate need to communicate.

Therefore, the asymmetric nature of the beneficiary and benefactors in e-comm demands critical attention in designing communication support mechanisms. For an information-seeking developer, involving more participants in the communication means having more potential information resources, implying a better chance of ob-taining the necessary information but at the cost of information overload; thus, high-quality ranking and triaging mechanisms would become essential. For those who are asked to participate in the communication and provide information, however, re-sponding to the request becomes yet another task [15].

On the one hand, when the relation between the communication initiator and the rest of the communication participants is symmetrical and reciprocal, those who are asked to participate in the communication would feel an equal importance of engag-ing in the communication and would therefore participate. On the other hand, when the relation is asymmetrical, where the initiator would be a beneficiary and the other participants would be benefactors, mechanisms to persuade people to participate in the communication are necessary.

Although there had been no explicit distinctions of the two types of communica-tions in software development, existing research currently demonstrates different emphases on supporting each aspect of communication with regard to key concepts, tools, and the primary functionality. Both approaches stress the importance of taking socio-technical aspects into account, but in different contexts. Table 2 illustrates the two distinctive approaches.

Table 2. Different Present Research Emphases on the Two Types of Communication

	Coordination Communication (c-comm)	**Expertise Communication (e-comm)**
key concepts	continuous coordination [11] impact management [3]	developer as knowledge resources [9] communication channel [15]
primary functionality	awareness visualization	finding expertise choosing experts socially-aware communication channel
tools	Palantir [12] Ariadne [2]	Expert Finder [13] Expertise Browser [8] STeP_IN_Java [15]
socio-technical aspects	social interaction needs are inferred from the technical (structural) dependencies of the tasks [5]	communication participants are selected based on their technical experiences on sought information and previous social relations with an information seeker [15]

Supporting c-comm has been studied primarily in such research areas as coordinating programmers and programming tasks. Supporting e-comm has been studied primarily in such research areas as knowledge sharing and expert finding.

Although they do not explicitly use the term "coordination communication," Redmiles et al. [11] present the continuous coordination paradigm for supporting coordination activities in software development. The paradigm contains the following four principles: (1) to have multiple perspectives on activities and information; (2) to have nonobtrusive integration through synchronous messages or through the representation of links between different sites and artifacts; (3) to combine socio-technical factors by considering relations between artifacts and authorship so that distributed developers can infer important context information; and (4) to integrate formal configuration management and informal change notification via the use of visualizations embedded in integrated software development environments [11].

This paradigm stresses the importance of integrating coordination activities within the programming environment, and of making developers aware of the need for communication and simultaneously minimizing the distraction of software developers by using formal configuration management mechanisms and informal visual notification and awareness techniques. Redmiles et al. (2007) focus on socio-technical factors in the sense that peer-to-peer coordination communication needs are inferred by analyzing structural (technical) dependencies of the system components on which developers are working because they have to coordinate their tasks through social interactions when the resolution of perceived conflicts becomes necessary [3], [14].

Nakakoji et al. [10] present nine design guidelines for expertise communication support mechanisms. The guidelines state that expertise communication support mechanisms should be integrated with other development activities, be personalized and contextualized for the information-seeking developer, be minimized when other types of information artifacts are available, take into account the balance between the cost and benefit of an information-seeking developer and group productivity, consider social and organizational relationships when selecting developers for communication, minimize the interruption when approaching those who are selected for communication, provide ways to make it easier for developers to ask for help; provide ways to make it easier for developers to answer or not to answer the information request, and be socially aware.

The guidelines presented by Nakakoji et al. [10] stress the importance of finding communication participants who not only have necessary information, but are also willing to provide the sought information in an appropriate way in a timely manner. The guidelines also pay attention to the cost to those who are asked to engage in expertise communication, and argue for the use of socially aware communication channels. They focus on the socio-technical aspect in the sense that finding potential communication participants takes into account not only technical skills of developers but also their social relationship with the information-seeking developer.

Each approachs take socio-technical aspects into account differently. Research on c-comm focuses on *socio-technical congruence*, where the structural similarities between an organizational structure and software structure are primarily studied. Research on e-comm focuses on a *socio-technical space*, where social relations among developers are considered in finding communication partners who would be willing to engage in the communication.

Such differences of the two types of communication necessitate fundamental differences in designing communication support mechanisms, specifically,

- how to select participants for the communication,
- what timing to use to start communication,
- how to invite people to participate in the communication,
- which communication channel to use
- how to use the resulting communicative session (i.e., communication archives).

Table 3 lists factors that are common and those that are distinctive to the two types of communication in software development.

Table 3. Comparison of Design Factors

	Coordination Communication	**Expertise Communication**
in relation to the development environment	integrate with the development environment	
disturbance	minimize	
when communication needs are identified	conflicts are detected or possible conflicts are detected	a developer is in need of information about the task at hand
trade-off of not communicating	potential risks of rework caused by conflicts that might arise by not coordinating	potential risks of slowing work when appropriate information is not provided to the information-seeking developer
alternative means to reduce communication	to visualize the status of the potential conflicts so that by glancing at the visualized information a developer may not need to engage in explicit communication	to guide the information-seeking developer to relevant artifacts such as source code and documents so that a developer may not need to engage in explicit communication
the use of the object on which a developer is working	by looking at what objects a developer presently works on in order to infer the impact network	by looking at what objects a developer previously worked on in order to infer the technical expertise of the developer
the use of who is initiating the communication	by using the communication initiator's impact network in selecting communication participants	by using the communication initiator's social relations in selecting communication participants
helping one in initiating communication	mechanisms to switch to an explicit communication mode with the peers in the impact network when urgent communication needs are detected	mechanisms to ask without worrying about bothering peers
helping those who are asked to participate in the communication	mechanisms to judge how urgent and important the conflict is	mechanisms to minimize feeling guilty for not responding to the request
awareness of communication channel	impact-aware so that developers can easily judge and communicate how much impact the emerging conflict might have and how to avoid and solve the conflict.	socially aware so that developers use the right channel instead of the channel that is easier to use (whom to ask, through which media)

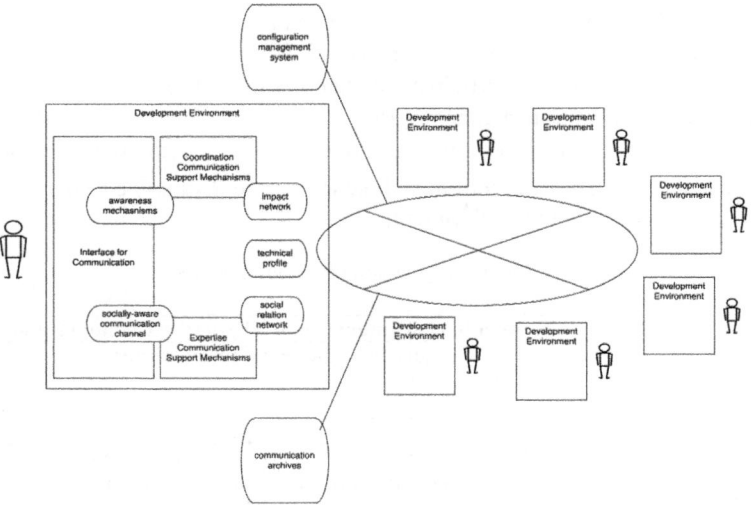

Fig. 1. An Architecture of Communication Support Mechanisms that Takes into Account Two Types of Communication

Figure 1 illustrates how communication support mechanisms should be built in support of software developers. On the one hand, there should be a unified interactive framework with communication for the software developer that is integrated within a development environment. Developers should not need to explicitly choose which communication type in which they would like to be engaging. Communication with peer developers should be supported as another type of information usage during software development, and needs to be integrated with a program- and document-authoring and browsing environment. On the other hand, how the communication is designed and structured needs to be tuned to each of the two types of communication. What is needed is to take the above differences seriously into account and design the communication support mechanisms accordingly.

4   Concluding Remarks

Nurturing communication in software development is not about increasing the amount of communication but about increasing the quality of the communication experience in the context of software development. Although having been recognized merely as communicative acts, different motives and needs are inherent when developers communicate with one another. Different computational mechanisms are necessary to realize successful communication. This paper presents our initial attempt to list different aspects necessary to take into account in designing mechanisms to support coordination communication and expertise communication. As opposed to general communication needs, there are either coordination communication needs or expertise communication needs. A real challenge would be to design a developer-centered unified interactive framework that seamlessly integrates the two.

References

1. Ackerman, M.S., Malone, T.W.: Answer Garden: a tool for growing organizational memory. In: Proceedings of the ACM Conference on Office Information Systems, Cambridge, MA, pp. 31–39 (1990)
2. [de Souza et al. 2007] de Souza, C.R.B., Quirk, S., Trainer, E., Redmiles, D.: Supporting collaborative software development through the visualization of socio-technical dependencies. In: Proceedings of GROUP 2007, pp. 147–156 (2007)
3. [de Souza & Redmiles 2008] de Souza, C.R.B., Redmiles, D.: An empirical study of software developers management of dependencies and changes. In: Proceedings of International Conference on Software Engineering (ICSE 2008), pp. 241–250 (2008)
4. [Gutwin et al. 2004] Gutwin, C., Penner, R., Schneider, K.: Group awareness in distributed software development. In: Proceedings of the ACM Conference on Computer Supported Cooperative Work (CSCW 2004), pp. 72–81 (2004)
5. [Herbsleb & Grinter 1999] Herbsleb, J., Grinter, R.E.: Splitting the organization and integrating the code: Conway's Law revisited. In: Proceedings of International Conference on Software Engineering (ICSE 1999), pp. 85–95 (1999)
6. [Illich 1971] Illich, I.: Deschooling Society. Harper and Row, New York (1971)
7. [Ko et al. 2007] Ko, A.J., DeLine, R., Venolia, G.: Information needs in collocated software development teams. In: Proceedings of International Conference on Software Engineering (ICSE 2007), pp. 344–353 (2007)
8. [Mockus & Herbsleb 2002] Mockus, A., Herbsleb, J.: Expertise Browser: a quantitative approach to identifying expertise. In: Proceedings of International Conference on Software Engineering (ICSE 2002), pp. 503–512 (2002)
9. [Nakakoji 2006] Nakakoji, K.: Supporting software development as collective creative knowledge work. In: Proceedings of International Workshop on Knowledge Collaboration in Software Development (KCSD 2006), Tokyo, pp. 1–8 (2006)
10. [Nakakoji et al. 2010] Nakakoji, K., Ye, Y., Yamamoto, Y.: Supporting expertise communication in developer-centered collaborative software development environments. In: Finkelstein, A., van der Hoek, A., Mistrik, I., Whitehead, J. (eds.) Collaborative Software Engineering, January 2010, ch. 11. Springer, Heidelberg (2010)
11. [Redmiles et al. 2007] Redmiles, D., van der Hoek, A., Al-Ani, B., Hildenbrand, T., Quirk, S., Sarma, A., Filho, R.S.S., de Souza, C., Trainer, E.: Continuous coordination: a new paradigm to support globally distributed software development projects. Wirtschaftsinformatik J. 49, S28–S38 (2007)
12. [Sarma et al. 2003] Sarma, A., Noroozi, Z., van der Hoek, A.: Palantir: raising awareness among configuration management workspaces. In: Proceedings of International Conference on Software Engineering (ICSE 2003), pp. 444–454 (2003)
13. [Vivacqua & Lieberman 2000] Vivacqua, A., Lieberman, H.: Agents to assist in finding help. In: Proceedings of Human Factors in Computing Systems (CHI 2000), pp. 65–72 (2000)
14. [Wagstrom & Herbsleb 2006] Wagstrom, P., Herbsleb, J.: Dependency forecasting. Communications of ACM 49(10), 55–56 (2006)
15. [Ye et al. 2007] Ye, Y., Yamamoto, Y., Nakakoji, K.: A socio-technical framework for supporting programmers. In: Proceedings of ESEC/FSE 2007, pp. 351–360 (2007)
16. [Ye et al. 2008] Ye, Y., Yamamoto, Y., Nakakoji, K.: Expanding the knowing capability of software developers through knowledge collaboration. International Journal of Technology, Policy and Management (IJTPM), Special Issue on Human Aspects of Information Technology Development 8(1), 41–58 (2008)

Part III
Logic and Engineering of Natural Language Semantics

6th International Workshop on Logic and Engineering of Natural Language Semantics (LENLS 6)

Daisuke Bekki

Ochanomizu University, 2-1-1 Ohtsuka, Bunkyo-ku, Tokyo 112-8610, Japan
bekki@is.ocha.ac.jp

The annual international workshop LENLS (Logic and Engineering of Natural Language Semantics) covers topics in formal linguistics and related fields, such as theoretical computer science, mathematical logic and formal philosophy, and includes the following:

✠ Dynamic syntax/semantics/pragmatics of natural language
✠ Categorical/topological/coalgebraic approaches to natural language syntax/ semantics/pragmatics
✠ Logic and its relation to natural language and linguistic reasoning (especially dynamic logics)
✠ Type-theoretic approaches to natural language
✠ Formal philosophy of language
✠ Formal pragmatics (especially game- and utility-theoretic approaches)
✠ Substructural expansion of Lambek Lambda Calculi
✠ Many-valued/Fuzzy and other non-classical logics and natural language

Formal linguistics is intrinsically an interdisciplinary field. Indeed, most of the formalisms which have been adopted in formal linguistics have their origin in the field of theoretical computer science, and in programming semantics especially. To take a few examples:

Lambda calculus originally appeared in Church (1941) and was subsequently incorporated into Montague Grammar (Montague (1973)).
Dynamic logics has its root in the programming semantics of the mid 1970s such as in the work of Harel (1979), which inspired the seminal paper by Groenendijk and Stokhof (1991).
Continuations first appeared in Plotkin (1975) and later in Danvy and Filinski (1990) before being extended to the field of formal semantics by de Groote (2001) and Barker (2001).

The average time for these notions to jump the boundary from the originating field into another was around 23 years. During such time they matured, became widely accepted in their respective fields as important notions, were introduced in many lectures, before eventually becoming known to formal linguists.

Besides programming semantics, other fields of theoretical computer science have influenced formal linguistics. Game theory, Bayesian networks and stochastic processes are among the many examples, and have helped to shape the discipline over time and remain influential in formal linguistics today.

K. Nakakoji, Y. Murakami, and E. McCready (Eds.): JSAI-isAI, LNAI 6284, pp. 159–160, 2010.

On the other hand, only a very small number of theories emerging from formal linguistics have influenced theoretical computer science, to say nothing of to mathematical logic. Algebraic automata theory, which originates in automata theory, is one of the few exceptions. Lambek calculus is another exception, if we take the view that it comes from categorial grammar, which, together with linear and fuzzy logics, gives its own insights into substructural/algebraic logics.

This means that formal linguistics has been, rightly or wrongly, almost a 'net-importer of formal theory'. Whether we accept this situation and want to accelerate the cycle of importing new formal theories, or whether we regard it as a problem and try to develop linguistic-oriented formal theories, it seems that one of our prime tasks is to promote and maintain people-to-people and theory-to-theory exchanges between formal linguistics and adjacent fields.

The aim of LENLS, in the spirit of its founder Norihiro Ogata, is to revitalize such exchanges, rediscover the connections between the formal disciplines (theoretical computer science, logic and mathematics) and empirical disciplines (syntax, semantics, pragmatics and philosophy), and revive the interdisciplinary nature of formal linguistics to what it once was.

In LENLS6, we focused particularly on work related to the interplay between logic, philosophy of language and formal semantics and pragmatics. This postproceedings volume contains 8 papers, selected from the 20 presentations given at the LENLS 6 workshop.

LENLS6 was held as part of JSAI International Symposia on AI (JSAI-isAI 2009). I hope that LENLS will continue its international development, alongside the other workshops of JSAI-isAI, within this new scheme.

References

Barker, C.: Introducing Continuations. In: Hastings, R., Jackson, B., Zvolenszky, Z. (eds.) The Proceedings of SALT 11. CLC Publications, Ithaca (2001)

Church, A.: The Calculi of Lambda Conversion. Annals of Mathematical Studies, vol. 6, ii+77. Princeton University Press, Princeton (1941)

Danvy, O., Filinski, A.: Abstracting Control. In: The Proceedings of LFP 1990, the 1990 ACM Conference on Lisp and Functional Programming, pp. 151–160 (1990)

de Groote, P.: Type raising, continuations, and classical logic. In: van Rooij, R., Stokhof, M. (eds.) The Proceedings of the 13th Amsterdam Colloquium, Institute for Logic, Language and Computation, Universiteit van Amsterdam, pp. 97–101 (2001)

Groenendijk, J., Stokhof, M.: Dynamic Predicate Logic. Linguistics and Philosophy 14, 39–100 (1991)

Harel, D. (ed.): First-Order Dynamic Logic. LNCS, vol. 68. Springer, Heidelberg (1979)

Montague, R.: The Proper Treatment of Quantification in Ordinary English. In: Hintikka, J., Moravcsic, J., Suppes, P. (eds.) Approaches to Natural Language, pp. 221–242. Reidel, Dordrecht (1973)

Plotkin, G.D.: Call-by-Name, Call-by Value and the Lambda Calculus. Theoretical Computer Science 1, 125–159 (1975)

Representing Covert Movements
by Delimited Continuations

Daisuke Bekki and Kenichi Asai

Department of Information Science, Faculty of Science,
Ochanomizu University,
Ootsuka 2-1-1, Bunkyo-ku, Tokyo, Japan

1 Background

1.1 Motivation: Covert Movements and Delimited Continuations

In phenomena which have been claimed to require "covert movements" in generative terms, a relevant lexical item seems to require a means to somehow refer to the meaning of its *surroundings* in order for the meaning of the whole sentence to be properly computed. This has motivated generative/transformational grammars to adopt a movement of the relevant item to the position where its scope contains surroundings that influence its meaning, while it remains as an issue to be solved for categorial/Lambek-style grammars, namely, grammars without movements.

Analyses that represent "covert movements" by means of *delimited continuations* (Danvy and Filinski (1990)) have recently attracted much attention in the field of natural language semantics (de Groote (2001), Shan (2002), Barker (2002), Barker (2004), Shan and Barker (2006), Barker and Shan (2006), Shan (2007), Otake (2008)). Delimited continuation is a notion which originates in the theory of functional programming languages; it enables each subterm to refer to "the rest of the computation", i.e. its surroundings.

In this paper, we propose a new method to define delimited continuations in terms of an *internal monad*, which is described by *Meta-Lambda Calculus* (cf. Bekki (2009)), which thereby enables us to analyse linguistic phenomena such as focus and inverse scope. In the following sections, we will review the notion of continuations both in programming and natural languages, then point out some problems that the preceding continuation theory of natural language semantics, especially in what is called *Continuized Semantics*, advocated in Barker (2002) and Barker (2004), and demonstrate how our approach successfully overcomes those problems. In the appendix, we will briefly present the language and theory of Meta-Lambda Calculus used throughout the paper.

1.2 Continuations and CPS Transformation

The concept of a Continuation (Stratchey and Wadsworth (1974)) was devised originally to capture the semantics of control operators such as jumps, which otherwise do not fit into a purely functional view of programming languages.

K. Nakakoji, Y. Murakami, and E. McCready (Eds.): JSAI-isAI, LNAI 6284, pp. 161–180, 2010.

For any subterm $N : \alpha$ of a lambda term $M : O$, a continuation for N is a function $k : \alpha \to O$, which returns $M : O$ when $N : \alpha$ is passed as its argument. Intuitively, this function corresponds to the *rest* of the computation for N in M.

It is often the case that the whole term M is not specified and only its type, which is called the **answer type**, is implicitly given. Let us use the following notation, using T as a type operator:

$$T\alpha \overset{def}{\equiv} (\alpha \to O) \to O$$

where O is an answer type implicitly assumed.

Example 1. In the term $1 + (a * 2) : int$ (the answer type is int), the continuation of a subterm $a : int$ is a function $\lambda\kappa.(1 + (\kappa * 2)) : (int \to int) \to int$.

Example 2. In the proposition $love(john, mary) : t$ (the answer type is t), the continuation of a subterm $mary : e$ is a function $\lambda\kappa.love(john, \kappa) : (e \to t) \to t$.

In order for each subterm to refer to its continuation, terms are transformed using the continuation-passing style (CPS).

Definition 3 (The transformation rules for call-by-value CPS). *(Slightly adapted from the definition originally proposed in Plotkin (1975).)*

$$[\![x]\!] \overset{def}{\equiv} \lambda\kappa.\kappa(\overline{x})$$

$$[\![fM]\!] \overset{def}{\equiv} \lambda\kappa.[\![M]\!](\lambda a.\kappa(\overline{f}a))$$

$$[\![\lambda x.M]\!] \overset{def}{\equiv} \lambda\kappa.\kappa(\lambda\overline{x}.[\![M]\!])$$

$$[\![MN]\!] \overset{def}{\equiv} \lambda\kappa.[\![M]\!](\lambda f.[\![N]\!](\lambda a.(fa)\kappa))$$

$\overline{x}$ is just the variable x but assigned a different type: if the type of x is α, then the type of $\overline{x}$ is α^*, where the operator * is recursively defined as follows:

Definition 4 (* operator).

$$b^* \mapsto b$$

$$(\alpha \to \beta)^* \mapsto (\alpha^* \to T(\beta^*))$$

By CPS transformation, a lambda term of type α is uniformly transformed into one of type $T(\alpha^*)$. It should be noted that a linguist who is familiar with the continuation analysis of Barker (2002) and Barker (2004) must note that the CPS of value types and functional types are of different type.

Example 5.

$$e^* = e$$

$$(e \to t)^* = (e \to Tt)$$

$$(e \to e \to t)^* = (e \to T(e \to Tt))$$

Not all terms in CPS have their counterparts in terms of the direct style. Such terms include a term in which there is reference to its continuation. This is the power resulting from the CPS transformation and an example of such an extended term is the shift/reset operator (Danvy and Filinski (1990)), which is a static way of capturing these operators.

Definition 6 (Shift/Reset operators)

$$[\![shift\ f.M]\!] \stackrel{def}{\equiv} \lambda\kappa.([\![M]\!](\lambda x.x)[\lambda a.\lambda\kappa'.\kappa'(\kappa a)/\overline{f}])$$

$$[\![reset(M)]\!] \stackrel{def}{\equiv} \lambda\kappa.\kappa([\![M]\!](\lambda x.x))$$

1.3 Barker's Continuized Semantics for Natural Language

Barker (2001) and de Groote (2001) first pointed out the similarity between Montague-style type raising and CPS transformation. Barker (2002) and Barker (2004) regarded the type raising operation as an instance of translation into CPS, and this is the strategy that succeeding analyses (Shan and Barker (2006), Barker and Shan (2006) have adopted. Let us call this approach *Continuized Semantics* after Barker (2002). Barker's CPS transformation rules are defined as follows.

Definition 7 (The transformation rules for call-by-value CPS). *(Slightly adapted from the definition proposed in Barker (2002) and Barker (2004).)*

$$[\![x]\!] \stackrel{def}{\equiv} \lambda\kappa.\kappa(x)$$

$$[\![f]\!] \stackrel{def}{\equiv} \lambda\kappa.\kappa(f)$$

$$[\![MN]\!] \stackrel{def}{\equiv} \lambda\kappa.[\![M]\!](\lambda f.[\![N]\!](\lambda a.\kappa(fa)))\ \textit{(Call-by-value)}$$

$$[\![MN]\!] \stackrel{def}{\equiv} \lambda\kappa.[\![M]\!](\lambda f.[\![N]\!](\lambda a.\kappa(fa)))\ \textit{(Call-by-name)}$$

While Plotkin's original definition in Definition 3 transforms a lambda term of type α to a lambda term of type $T(\alpha^*)$, Barker's rules transform a lambda term of type α to a lambda term of type $T(\alpha)$. Taking the complexity of the star operator (Definition 4) into account, Barker's transformation rules are substantially simpler, and contain substantial flaws as well. We will come back to this point in Section 2.2.

In Barker (2004), quantifiers such as *everyone* and *someone* are assigned their semantic representations in CPS style in the lexicon as shown below.

$$everyone \vdash \lambda\kappa.\forall x.\kappa x : (e \to t) \to t$$

$$someone \vdash \lambda\kappa.\exists x.\kappa x : (e \to t) \to t$$

With the presence of the type raising operation (indicated by (T)), the type mismatch problem can be resolved as in the following derivation.

(1)

$$
\cfrac{
  \cfrac{\text{somebody}}{\cfrac{\lambda P.\exists x(Px)}{:(e\to t)\to t}}
  \quad
  \cfrac{
    \cfrac{\cfrac{\text{loves}}{\cfrac{\lambda y.\lambda x.love(x,y)}{:e\to e\to t}}\ (T)\quad \cfrac{\cfrac{\text{everybody}}{\cfrac{\lambda P.\forall y(Py)}{:(e\to t)\to t}}}{\cfrac{\lambda P.\lambda x.\forall y(Pyx)}{:(e\to e\to t)\to e\to t}}}{\cfrac{\lambda x.\forall y(love(x,y))}{:e\to t}}
  }{\ (<)}
}{\cfrac{\exists x(\forall y(love(x,y)))}{:t}}\ (>)
$$

On the other hand, Continuized Semantics achieves the same effect in the following way:

(2)

$$
\cfrac{
  \cfrac{\text{somebody}}{\cfrac{\lambda\kappa.\forall x(\kappa x)}{:T(e)}}
  \quad
  \cfrac{
    \cfrac{\cfrac{\text{loves}}{\cfrac{\lambda\kappa.\kappa(\lambda y.\lambda x.love(x,y))}{:T(e\to e\to t)}}\quad \cfrac{\text{everybody}}{\cfrac{\lambda\kappa.\exists y(\kappa y)}{:T(e)}}}{\cfrac{\lambda\kappa.\forall y(\kappa(\lambda x.love(x,y)))}{:T(e\to t)}}
  }{\ (>CBN)}
}{\cfrac{\lambda\kappa.\exists x(\forall y(\kappa(love(x,y))))}{:T(t)}}\ (<CBN)
$$

Barker claims that this method uniformly yields an NP, be it a proper name or quantifier, a representation of type e without recourse to type raising. Moreover, a boolean type approach to coordination (Partee and Rooth (1983), Winter (2001), among many others) can be integrated into this system as an instance of CPS.

2   Problems of Continuized Semantics

In this section, we will show that Continuized Semantics implies at least four problems that need to be addressed; the first three are theoretical, and the last one is empirical.

2.1   fcontrol and run

Barker (2004) mentioned the interaction between the adverbial *only* and focus within its scope as an instance of applying delimited continuation to natural language semantics, which can be defined by using run and fcontrol operators (Sitaram and Felleisen (1990), Sitaram (1993)) as follows (slightly adapted).

$$
[\![only\ P]\!] \overset{def}{\equiv} \mathrm{run}(P)(\lambda x.\lambda\kappa.\lambda y.(\kappa xy \wedge \forall z.(\kappa zy \to x=z)))
$$
$$
[\![[M]_F]\!] \overset{def}{\equiv} \mathrm{fcontrol}([\![M]\!])
$$

It is clear that the behaviour of Sitaram's `fcontrol` and `run` fits in with this phenomenon. However, this analysis is defective in the sense that Barker has not given any definition of those operators in terms of his CPS.

CPS transformation is regarded as a static simulation of control operators and Sitaram's `fcontrol/run` operators (or, equivalently, Felleisen (1988)'s `control/prompt` operators) are defined only dynamically, and it is still controversial whether they can be statically represented in terms of CPS.

Recently, Dybvig et al. (2007) has given a *small step semantics* for `control/prompt` operators, and mentions its relation to CPS (in section 2.3.4); nevertheless that relation is still far from obvious at present. Fairly speaking, Continuized Semantics has left much work to be done, including the definition of `fcontrol/run` operators, in order to argue that it has good prospects for the analysis of adverbial *only* and focus.

2.2   Lambda Abstractions

As we mentioned in Section 1.3, the simplification of the transformation rules in Continuized Semantics comes at a cost; there is no rule for transforming a lambda abstraction term. According to Barker's rule, a lambda term $\lambda x.M$ of the type $\alpha \to \beta$ is transformed into one of type $T(\alpha \to \beta)$, namely, $((\alpha \to \beta) \to O) \to O$. Then it seems natural to assume that this term has the form $\lambda \kappa.N$, where κ is a variable of type $(\alpha \to \beta) \to O$ and N is some lambda term of type O, and it also seems natural to assume that N is constructed only from a combination of the following components (where $[\![M]\!]$ contains x as a free variable).

$$\kappa : (\alpha \to \beta) \to O$$
$$x : \alpha$$
$$[\![M]\!] : (\beta \to O) \to O$$

But this does not seem to be achievable[1]. The main difficulty lies in the impossibility of obtaining a term of type β from $[\![M]\!]$.[2]

Barker defends the fact that his theory lacks a transformation rule for a lambda abstraction term by claiming that it might be possible to eliminate any lambda abstraction terms from logical forms of natural language. For example, Barker's mechanism can deal with the type mismatch problem of quantifiers without recourse to quantifier raising (QR), and Shan and Barker (2006) mentions a treatment of relative clauses without operator movements.

However, in natural language semantics, lambda abstractions are more abundant than his optimism allows for. We may even say that formal semantics

---

[1] For example, the term $\lambda \kappa.[\![M]\!](\lambda v : \beta.\kappa(\lambda x : \alpha.v))$ is of type $T(\alpha \to \beta)$. But this term contains x as a free variable and the inner variable x is vacuous, which obviously does not serve as a translation of $\lambda x.M$. This argument is related to our realization of the transformation rule of a lambda abstraction term via meta-lambda calculus. See Section 4.2 for details.

[2] This is only possible in the special case in which the answer type is t; then $\bigwedge [\![M]\!]$ is of type β if β is a complete lattice.

without lambda abstraction is bound hand and foot: How can we formalize comparatives without lambda abstractions? How can we extend our system with event variables and possible worlds without lambda abstractions?

2.3   The Type of Determiners

Another theoretical problem of Continuized Semantics concerns the type that it assigns to determiners. According to the standard theory of generalized quantifiers (in direct style), a determiner (e.g. "every") has a type $(e \to t) \to (e \to t) \to t$, and a common noun (e.g. "man") has a type $e \to t$.

$$\text{every} \vdash \lambda Q.\lambda P.\forall x(Px \to Qx) : (e \to t) \to (e \to t) \to t$$
$$\text{man} \vdash man : e \to t$$

But this view, as it stands, does not fit into Continuized Semantics for the following reasons:

1. In the first place, the semantic representation of "every" is of lambda abstraction form, therefore it cannot be transformed into CPS.
2. Suppose that the representation of "every" is somehow transformed into CPS; it should have type $T((e \to t) \to (e \to t) \to t)$. Suppose further that the representation of "man" is transformed with CPS into one whose type is $T(e \to t)$. Then the composition of these two yields the representation of "every man" whose type is supposed to be $T((e \to t) \to t)$. However, this is not consistent with one of the selling points of Continuized Semantics; namely, that the representation of "everyone" is of type $T(e) \equiv (e \to t) \to t$ and already in CPS form. This means that "every man" and "every" have different types!

In order to get around this problem, Barker (2002) assumes that the representation of a determiner such as "every" is of type $T((e \to t) \to e)$, namely, $(((e \to t) \to e) \to t) \to t$, instead of $T((e \to t) \to (e \to t) \to t)$.

The first question that naturally comes to mind is whether one can define a representation of generalized quantifiers having this type, since this intuitively means that we define the representation of "every" as having type $(e \to t) \to e$: in other words, a function that takes a set like "man" and returns an object of type e.

This seems impossible within the compass of direct semantics. But Continuized Semantics gets round this problem by employing choice functions of type $(e \to t) \to e$ (cf. Kratzer (1998)). Determiners are defined in the appendix of Barker (2002) as follows, in which the variable f is a choice function that plays the role of choosing one entity from a given set, and what is universally quantified here is this f.

$$\text{every} \vdash \lambda d.\forall f(d(f)) : T((e \to t) \to e)$$
$$\text{some} \vdash \lambda d.\exists f(d(f)) : T((e \to t) \to e)$$

The derivation goes as follows. The resulting representation states that for any of those choice functions, $\kappa(f(man))$ is true.

(3)

$$(>CBV)\dfrac{\dfrac{\text{every}}{\lambda d.\forall f(d(f)) \;:\; T((e \to t) \to e)} \quad \dfrac{\text{man}}{\lambda\kappa.\kappa(man) \;:\; T(e \to t)}}{\lambda\kappa.\forall f(\kappa(f(man))) \;:\; T(e)}$$

Generalized quantifiers like "most" are also definable in terms of a choice function.[3]

$$\text{most} \vdash \lambda d.\exists C \in \mathbf{MOST}(\forall f \in C \; (d(f))) : T((e \to t) \to e)$$

$$\mathbf{MOST} \overset{def}{\equiv} \{\, C \mid \forall P(|P| < 2 * |\,\{x \mid \exists f \in C \; (x = f(P))\}\,|)\,\}$$

However, there are determiners, such as *only* and *even*, which cannot be given a representation of this type. Let us take "only" for example. Putting aside the pre-suppositional content of "only", the following representation is plausible for "only" in direct style.

$$\text{only} \vdash \lambda z.\lambda P.(Pz \wedge \forall x(Px \to x = z)) : e \to (e \to t) \to t$$

This means that "only" must be represented as a type $T(e \to e)$ in Continuized Semantics, because the derivation must go as follows.

(4)

$$(>CBV)\dfrac{\dfrac{\text{only}}{T(e \to e) \equiv ((e \to e) \to t) \to t} \quad \dfrac{\text{John}}{T(e) \equiv (e \to t) \to t}}{T(e) \equiv (e \to t) \to t}$$

Now it seems impossible to define a representation of "only" as having type $T(e \to e)$, considering its propositional content.[4] The same argument applies to "even". Moreover, there are other kinds of determiners which lead to the same sort of difficulty, such as "same" or "different", among many others.

Therefore, the central view of Continuized Semantics that type raising can be reduced to CPS transformations, although appealing at a glance, cannot be maintained as it stands if we consider a wider range of phenomena in natural language semantics.

---

[3] Barker (2002) and Barker (2004) did not mention anything about downward entailing quantifiers, but they are also definable in terms of negated forms of upward entailing quantifiers as follows:

$$\text{few} \vdash \lambda d.\neg\exists C \in \mathbf{MOST}(\forall f \in C \; (d(f))) : T((e \to t) \to e)$$

[4] Barker (2004) includes a lexical entry for adverbial "only" which is defined via the **run** operator. But this does not remedy the above problem; in the sentence "only John runs", "John" is focused and marked with the $[\;\;]_F$ operator, but since the continuation that "John" receives is empty, "John" cannot refer to the representation of the verb phrase.

2.4   Inverse Scope

Barker (2002) suggests that the scope ambiguity between linear and inverse scope readings (in sentences with more than two quantifiers) can be reduced to a choice between two evaluation strategies: *call-by-value* and *call-by-name* in the interpretation of functional applications.

For example, the derivation of the sentence "Everybody loves somebody" in (2) uses only the call-by-name strategy for the interpretation of functional applications, which yields the linear scope reading. On the other hand, by using only the call-by-value strategy, the inverse scope reading yields the interpretation shown by the following derivation.

(5)

$$
\cfrac{
  \cfrac{\text{somebody}}{
    \begin{array}{c}\lambda\kappa.\exists x(\kappa x)\\ : T(e)\end{array}
  }
  \quad
  \cfrac{
    \cfrac{
      \cfrac{\text{loves}}{\begin{array}{c}\lambda\kappa.\kappa(\lambda y.\lambda x.love(x,y))\\ : T(e \to e \to t)\end{array}}
      \quad
      \cfrac{\text{everybody}}{\begin{array}{c}\lambda\kappa.\forall y(\kappa y)\\ : T(e)\end{array}}
    }{\begin{array}{c}\lambda\kappa.\forall y(\kappa(\lambda x.love(x,y)))\\ : T(e \to t)\end{array}}\ (>CBV)
  }{}
}{\begin{array}{c}\lambda\kappa.\forall y(\exists x(\kappa(love(x,y))))\\ : T(t)\end{array}}\ (<CBV)
$$

It is true that its conception and repercussions (we mean, if natural language has any phenomena which can be explained by assuming that there are two different strategies, and more importantly, we have the chance to *choose* one of them, for each evaluation of a functional application) are worth pursuing, as Shan and Barker (2006) does.

However, we believe that any analysis of scope ambiguity has to be verified by sentences which contain more than three quantifiers. This is not a flaw only of Continuation Semantics, but rather is an inherited vice in the field of formal semantics, which has unquestioningly assumed that there exists a reading for every combination of quantifiers contained in a given sentence; namely, $n!$ different readings for a sentence with n quantifiers, based only on the fact that a sentence which contains only two quantifiers has two different readings: linear and inverse scope readings. But the situation becomes different when $n > 3$.

For example, more deliberate studies of scope ambiguity, such as Hayashishita (2003) among others, point out that a sentence with three quantifiers, Q_1, Q_2 and Q_3 in their linear order, hardly shows the $Q_3 > Q_2 > Q_1$ reading (which we may call a "reversed reading"), such as the *every > most > some* reading in the following sentence.

(6)   Some teachers introduce most students to every company.

The analysis of Barker (2002) and Barker (2004) wrongly generates the reversed reading for the above sentence, if every functional application is evaluated by the call-by-value strategy.

On the other hand, the *intermediate*-inverse scope readings such as *most > some > every* or *every > some > most* seem to exist, but these are not derivable in the analysis of Barker (2002) nor Barker (2004), because there are only two choices in the scope relation betwen *most* and *every*, namely *most* > *every* or *every* > *most*, and the scope of the subject is either higher than both of them, or lower than both of them, yielding only the following four combinations (the fourth of which is the reversed reading).

$$some > most > every$$
$$most > every > some$$
$$some > every > most$$
$$every > most > some$$

Thus the analysis of Continuized Semantics does not properly predict the empirical combination of linear/intermediate-inverse/inverse scope readings. This will be solved in our analysis, in which scope ambiguity is not due to the choice of evaluation strategy, but rather to the use of the *inverse scope operator*; See Section 4.4.

2.5   Summary of the Problems

To summarize, Continuized Semantics has the following set of problems:

- Lack of definitions for `fcontrol` and `run` in terms of CPS
- Empirically wrong predictions for inverse scope readings
- No transformation rule for lambda abstraction construction
- Existence of undefinable determiners

We believe that these problems are particular to the formulation of Continuized Semantics and not problems for the use of continuations in general for natural language semantics.

The central idea of Continuized Semantics is to regard CPS transformation as a generalization of type raising operations and thus it enables us to do without type raising operations. In other words, Continuized Semantics represents the whole semantic system in CPS, in order to obviate the need for type raising.

This view is the root cause of the problem that we discussed in Section 2.3, but we think that it is burdened by a more conceptual problem; can we really consider Continuized Semantics as doing without type raising operations? As the conclusion of Barker (2002) stated, CPS transformation can also be seen as a kind of raising. If this is so, then it can be regarded as raising everything by CPS transformation!

Moreover, every semantic representation in Continuized Semantics is described in CPS. This opposes the original concept of CPS transformation in Danvy and Filinski (1990) , in which the advantage of CPS transformation is that we can write representations in *direct style* while making use of control operators such as `shift`/`reset`, whose interpretations are defined in terms of CPS; nevertheless they are allowed to appear within other representations in direct style.

The achievement of Barker (2002) and Barker (2004) was to introduce the notion of (delimited) continuations to the field of natural language semantics and indicate that continuations may have interactions with many linguistic phenomena in an important way. At the same time, its motivation to reduce type raising operations has been misunderstood to be the central merit of continuations, which leads to a situation where the real descriptive power of continuations in linguistics is hard to grasp.

Unlike Continuized Semantics, our strategy is to write semantic representations in direct style according to the line of Danvy and Filinski (1990) . Then we will define control operators in terms of CPS transformations, allowing them to appear among direct style representations. This strategy places an emphasis on bringing out the original expressive power of delimited continuations.

In the next section, we will present our proposal in the following way.

1. We define an *internal monad* for delimited continuations in term of the *meta-lambda calculus* (Bekki (2009)). This enables delimited continuations to co-exist with other monadic analyses in Bekki (2009).
2. We define CPS transformation (of simply-typed lambda terms) by means of internal monads. This transformation, unlike Barker's, is defined generally enough to treat any lambda terms, including lambda abstraction constructions.
3. We will demonstrate how the resulting analysis solves the problems we have pointed out in this section: the problem of "only" and focus, the transformation of lambda abstraction terms, the problem of types of determiners, and the empirical problem of inverse scope readings.

3   Proposal: Delimited Continuations via Meta-Lambda Calculus

3.1   Transformation Rules by Continuation Monad

To begin, we define a general translation rule in terms of *internal monads* (Bekki (2009)p.202) which is robust enough to deal with any term defined in typed lambda calculus.

Definition 8 (Internal monad for deliminted continuations). *The internal monad for delimited continuations is a triple* $\langle T, \eta, \mu \rangle$*, each of which is defined using the following meta-lambda terms.*

$$T = \zeta f. \zeta X. \zeta \kappa. (X \lhd (\zeta v. \kappa \lhd (f \lhd v)))$$
$$\eta = \zeta X. \zeta \kappa. (\kappa \lhd X)$$
$$\mu = \zeta X. \zeta \kappa. (X \lhd (\zeta v. v \lhd \kappa))$$

Definition 9 gives a set of transformation rules for simply-typed lambda terms, which are parametrized by internal monads.

Definition 9 (Transformation with Internal Monad (call-by-value))

$$[\![x]\!]_T = \eta \lhd x$$
$$[\![c]\!]_T = \eta \lhd c$$
$$[\![\lambda x.M]\!]_T = (T \lhd (\zeta X.\lambda x.X)) \lhd [\![M]\!]_T$$
$$[\![MN]\!]_T = \mu \lhd (T \lhd (\zeta X.((T \lhd (\zeta Y.XY)) \lhd [\![N]\!]_T) \lhd [\![M]\!]_T)$$

Definition 10 (Translation for continuation monad). *Given an internal monad for continuations, the translation rules for lambda terms with delimited continuations are accordingly defined as follows.*

$$[\![x]\!]_c = \zeta \kappa.(\kappa \lhd x)$$
$$[\![c]\!]_c = \zeta \kappa.(\kappa \lhd c)$$
$$[\![\lambda x.M]\!]_c = \zeta \kappa.([\![M]\!]_c \lhd (\zeta v.\kappa \lhd (\lambda x.v)))$$
$$[\![MN]\!]_c = \zeta \kappa.([\![M]\!]_c \lhd (\zeta m.[\![N]\!]_c \lhd (\zeta n.\kappa \lhd (mn))))$$

3.2 Control Operators

In the light of this setting, shift/reset operators are defined in the following way.

Definition 11 (Control operators).

$$[\![shift\ \kappa.M]\!] = \zeta \kappa.([\![M]\!] \lhd (\zeta x.x))$$
$$[\![reset(M)]\!] = \zeta \kappa.(\kappa \lhd ([\![M]\!] \lhd (\zeta x.x)))$$

The definition above looks too simple compared with that of Danvy and Filinski (1990) but has the same effects, as indicated by the following computations.

$$[\![1 + reset(10 + shift f.(f(f(100))))]\!] = \zeta \kappa.(\kappa \lhd 121)$$
$$[\![1 + reset(10 + shift f.(100))]\!] = \zeta \kappa.(\kappa \lhd 101)$$
$$[\![1 + reset(10 + shift f.(f(100) + f(1000)))]\!] = \zeta \kappa.(\kappa \lhd 1121)$$

4 Solutions

4.1 "Focus Movement" as shift/reset

Rooth (1992) discussed the truth conditions of the two sentences (7a) and (7b) which differ only in the location of the focus. As a description of the situation where Mary introduced Bill and Tom to Sue, with no other introductions, (7a) is false and (7b) is true.

(7)   a.   Mary only introduced $[Bill]_F$ to Sue.

b.   Mary only introduced Bill to $[Sue]_F$.

In order to account for this, Wagner (2006), among others, adopts an operation called "focus movement", which is an instance of covert movements.

We can, however, directly compute the semantic representation without covert movements under the mechanism of delimited continuations with `shift`/`reset` operators. This is achieved by the following definition, where focus is interpreted by means of the `shift` operator, and the adverbial "only" is the `reset` operator.

Definition 12 (Focus operator). *For any meta-lambda term* $\phi : e$ *and* $\psi :$ $(e \to t) \to e \to t$,

$$[\phi]_F \overset{def}{\equiv} shift\ f.\lambda x.\forall z((f \lhd z) \lhd x \to z = \phi) : (e \to e \to t) \to e \to t$$

$$only(\phi) \overset{def}{\equiv} reset(\phi) : (e \to t) \to e \to t$$

The semantic representations for (7a) and (7b) are respectively calculated as follows:

$$[\![((only\ ((introduce\ [b]_F)\ s)\ m)]\!]_c$$
$$= \zeta\kappa.[\![only\ ((introduce\ [b]_F)\ s)]\!]_c(\zeta x.\kappa(x(m)))$$
$$= \zeta\kappa.\kappa([\![(introduce\ [b]_F)\ s]\!]_c(\zeta x.x)(m))$$
$$= \zeta\kappa.\kappa([\![[b]_F]\!]_c(\zeta w.introduce(w)(s)))(m))$$
$$= \zeta\kappa.\kappa([\![b]\!]_c(\lambda y.\forall z(introduce(z)(s)(m) \to z = y))$$
$$= \zeta\kappa.\kappa(\forall z(introduce(z)(s)(m) \to z = b))$$

$$[\![((only\ ((introduce\ b)\ [s]_F)\ m)]\!]_c$$
$$= \zeta\kappa.[\![only\ ((introduce\ b)\ [s]_F)]\!]_c(\zeta x.\kappa(x(m)))$$
$$= \zeta\kappa.\kappa([\![(introduce\ b)\ [s]_F]\!]_c(\zeta x.x)(m))$$
$$= \zeta\kappa.\kappa([\![[s]_F]\!]_c(\zeta w.introduce(b)(w)))(m))$$
$$= \zeta\kappa.\kappa([\![s]\!]_c(\zeta y.\forall z(introduce(b)(z)(m) \to z = y))$$
$$= \zeta\kappa.\kappa(\forall z(introduce(b)(z)(m) \to z = s)))$$

4.2   Transforming Lambda Abstractions

The transformation rules in Definition 9 properly transform a lambda term of lambda abstraction form in spite of the difficulty we mentioned in Section 2.2. The rules in Definition 9 transform a lambda term of the form $\lambda x.[\![M]\!]$ into the following form:

$$\zeta\kappa.([\![M]\!]_c \lhd (\zeta v.\kappa \lhd (\lambda x.v)))$$

Notice this form resembles the one we mentioned in the footnote in Section 2.2, which is as follows.

$$\lambda\kappa.[\![M]\!](\lambda v.\kappa(\lambda x.v))$$

Recall that this form does not serve our purpose, because 1) the variable x remains free in $[\![M]\!]$, and 2) the inner variable x is vacuous. However, in our settings, the free variable x in $[\![M]\!]$ gets bound by the inner lambda operator. This is due to the difference of operators binding the variable v; in the failed transformation, v is bound by a *normal* lambda operator, while in our transformation, the variable v is bound by a *meta-lambda* operator, thus v is actually a *meta-variable*, which keeps track of the free variables that it contains. This enables the inner lambda operator to bind the free variable contained in the meta-variable v. For a detailed definition of the objects in meta-lambda calculus, see Appendix A. Here we show only the typing of the transformation of lambda abstraction terms below.

Example 13

$$\small\cfrac{(MLAM)\cfrac{(MAPP)\cfrac{(MVAR)\cfrac{\kappa : (\Gamma \vdash \alpha \to \beta \Rightarrow \Gamma \vdash O), v : (\Gamma, x : \alpha \vdash \beta)}{\Vdash \kappa : (\Gamma \vdash \alpha \to \beta \Rightarrow \Gamma \vdash O)} \quad (LAM)\cfrac{(MVAR)\cfrac{\kappa : (\Gamma \vdash \alpha \to \beta \Rightarrow \Gamma \vdash O), v : (\Gamma, x : \alpha \vdash \beta)}{\Vdash v : (\Gamma, x : \alpha \vdash \beta)}}{\kappa : (\Gamma \vdash \alpha \to \beta \Rightarrow \Gamma \vdash O), v : (\Gamma, x : \alpha \vdash \beta) \Vdash \lambda x.v : (\Gamma \vdash \alpha \to \beta)}}{\kappa : (\Gamma \vdash \alpha \to \beta \Rightarrow \Gamma \vdash O), v : (\Gamma, x : \alpha \vdash \beta) \Vdash \kappa \triangleleft (\lambda x.v) : (\Gamma \vdash O)}}{\kappa : (\Gamma \vdash \alpha \to \beta \Rightarrow \Gamma \vdash O) \Vdash \zeta v.\kappa \triangleleft (\lambda x.v) : (\Gamma, x : \alpha \vdash \beta) \Rightarrow (\Gamma \vdash O)}$$

$$\small\cfrac{(MLAM)\cfrac{(MAPP)\cfrac{\kappa : (\Gamma \vdash \alpha \to \beta \Rightarrow \Gamma \vdash O) \atop \Vdash [\![M]\!]_c : ((\Gamma, x : \alpha \vdash \beta) \Rightarrow (\Gamma \vdash O)) \Rightarrow (\Gamma \vdash O) \quad \cfrac{\kappa : (\Gamma \vdash \alpha \to \beta \Rightarrow \Gamma \vdash O)}{\Vdash \zeta v.\kappa \triangleleft (\lambda x.v) : (\Gamma, x : \alpha \vdash \beta) \Rightarrow (\Gamma \vdash O)}}{\kappa : (\Gamma \vdash \alpha \to \beta \Rightarrow \Gamma \vdash O) \Vdash [\![M]\!]_c \triangleleft (\zeta v.\kappa \triangleleft (\lambda x.v)) : (\Gamma \vdash O)}}{\Vdash \zeta \kappa. [\![M]\!]_c \triangleleft (\zeta v.\kappa \triangleleft (\lambda x.v)) : (\Gamma \vdash \alpha \to \beta \Rightarrow \Gamma \vdash O) \Rightarrow (\Gamma \vdash O)}$$

4.3 Transforming Determiners

Determiners such as "every" and "only", which have proved to be problematic for Continuized Semantics, are successfully transformed in the following way.

$$
\begin{aligned}
[\![every]\!] &= [\![\lambda P.\lambda Q.\forall x(Px \to Qx)]\!] \\
&= \zeta\kappa.[\![\lambda Q.\forall x(Px \to Qx)]\!](\zeta v.\kappa(\lambda P.v)) \\
&= \zeta\kappa.(\zeta\kappa'.[\![\forall x(Px \to Qx)]\!](\zeta v'.\kappa'(\lambda Q.v')))(\zeta v.\kappa(\lambda P.v)) \\
&= \zeta\kappa.(\zeta\kappa'.(\zeta\kappa''.[\![Px \to Qx]\!](\zeta v''.\kappa''(\forall x.v'')))(\zeta v'.\kappa'(\lambda Q.v')))(\zeta v.\kappa(\lambda P.v)) \\
&= \zeta\kappa.(\zeta\kappa'.(\zeta\kappa''.\zeta\kappa'''.\kappa'''(Px \to Qx)(\zeta v''.\kappa''(\forall x.v'')))(\zeta v'.\kappa'(\lambda Q.v')))(\zeta v.\kappa(\lambda P.v)) \\
&= \zeta\kappa.(\zeta\kappa''.\zeta\kappa'''.\kappa'''(Px \to Qx)(\zeta v''.\kappa''(\forall x.v'')))(\zeta v'.(\zeta v.\kappa(\lambda P.v))(\lambda Q.v')) \\
&= \zeta\kappa.(\zeta v''.(\zeta v'.(\zeta v.\kappa(\lambda P.v))(\lambda Q.v'))(\forall x.v''))(Px \to Qx) \\
&= \zeta\kappa.(\zeta v'.(\zeta v.\kappa(\lambda P.v))(\lambda Q.v'))(\forall x.Px \to Qx) \\
&= \zeta\kappa.(\zeta v.\kappa(\lambda P.v))(\lambda Q.\forall x.Px \to Qx) \\
&= \zeta\kappa.\kappa(\lambda P.\lambda Q.\forall x.Px \to Qx)
\end{aligned}
$$

$$
\begin{aligned}
[\![everyone]\!] &= [\![\lambda Q.\forall x(Qx)]\!] \\
&= \zeta\kappa.[\![\forall x(Qx)]\!](\zeta v.\kappa(\lambda Q.v)) \\
&= \zeta\kappa.(\zeta\kappa'.[\![Qx]\!](\zeta v'.\kappa'(\forall x(v'))))(\zeta v.\kappa(\lambda Q.v)) \\
&= \zeta\kappa.(\zeta\kappa'.(\zeta\kappa''.\kappa''(Qx))(\zeta v'.\kappa'(\forall x(v'))))(\zeta v.\kappa(\lambda Q.v)) \\
&= \zeta\kappa.(\zeta\kappa''.\kappa''(Qx))(\zeta v'.(\zeta v.\kappa(\lambda Q.v))(\forall x(v'))) \\
&= \zeta\kappa.(\zeta v'.(\zeta v.\kappa(\lambda Q.v))(\forall x(v')))(Qx) \\
&= \zeta\kappa.(\zeta v.\kappa(\lambda Q.v))(\forall x(Qx)) \\
&= \zeta\kappa.\kappa(\lambda Q.\forall x(Qx))
\end{aligned}
$$

$$\llbracket only \rrbracket = \llbracket \lambda z.\lambda P.(Pz \wedge \forall x(Px \rightarrow x = z)) \rrbracket$$
$$= \zeta\kappa.\llbracket \lambda P.(Pz \wedge \forall x(Px \rightarrow x = z)) \rrbracket (\zeta v.\kappa(\lambda z.v))$$
$$= \zeta\kappa.(\zeta\kappa'.\llbracket Pz \wedge \forall x(Px \rightarrow x = z) \rrbracket (\zeta v'.\kappa'(\lambda P.v')))(\zeta v.\kappa(\lambda z.v))$$
$$= \zeta\kappa.(\zeta\kappa'.(\zeta\kappa''.\kappa''(Pz \wedge \forall x(Px \rightarrow x = z))(\zeta v'.\kappa'(\lambda P.v')))(\zeta v.\kappa(\lambda z.v))$$
$$= \zeta\kappa.(\zeta\kappa''.\kappa''(Pz \wedge \forall x(Px \rightarrow x = z))(\zeta v'.(\zeta v.\kappa(\lambda z.v))(\lambda P.v'))$$
$$= \zeta\kappa.(\zeta v'.(\zeta v.\kappa(\lambda z.v))(\lambda P.v')))(Pz \wedge \forall x(Px \rightarrow x = z))$$
$$= \zeta\kappa.(\zeta v.\kappa(\lambda z.v))(\lambda P.Pz \wedge \forall x(Px \rightarrow x = z)))$$
$$= \zeta\kappa.\kappa(\lambda z.\lambda P.Pz \wedge \forall x(Px \rightarrow x = z))$$

These transformations imply that our set of transformation rules does not result in the problem of types that we pointed out in Section 2.3 for a wide range of determiners.

4.4 Inverse Scope as `shift`/`reset`

In our analysis, inverse scope can be derived without covert movements. Instead, we define the *inverse scope operator* in terms of the `shift` operator.

Definition 14 (Inverse scope operator). *For any meta-lambda term ϕ :* $(e \rightarrow t) \rightarrow t$,

$$[\phi]_{INV} \overset{def}{\equiv} shift\, f.\phi(\lambda y.f \triangleleft (\lambda P.Py)) : (e \rightarrow e \rightarrow t) \rightarrow e \rightarrow t$$

For instance, the sentence (8) has a minor reading (called the "wife-reading") in which "every English man" over-scopes "a woman", that is distinct from a major reading (the "Queen-reading").

(8)   A woman loves [every English man]$_{INV}$.

The semantic representation of (8), where the inner generalized quantifer "every" is enclosed by the inverse scope operator, is interpreted in the following way.

$$\llbracket (\lambda P.some(woman)(P)([\lambda P.every(man)(P)]_{INV}\, love)) \rrbracket_c$$
$$= \zeta\kappa.\llbracket \lambda P.some(woman)(P) \rrbracket_c(\zeta s.\llbracket [\lambda P.every(man)(P)]_{INV} \rrbracket_c(\zeta e.\llbracket love \rrbracket_c(\zeta l.\kappa(s(e(l))))))$$
$$= \zeta\kappa.\llbracket [\lambda P.every(man)(P)]_{INV} \rrbracket_c(\zeta e.\kappa((\lambda P.some(woman)(P))(e\, love)))$$
$$= \zeta\kappa.\llbracket shift\, f.(\lambda P.every(man)(P))(\lambda y.f \triangleleft (\lambda P.Py)) \rrbracket_c(\zeta e.\kappa((\lambda P.some(woman)(P))(e\, love)))$$
$$= \zeta\kappa.(\zeta f.\llbracket (\lambda P.every(man)(P))(\lambda y.f \triangleleft (\lambda P.Py)) \rrbracket_c(\zeta x.x))(\zeta e.\kappa((\lambda P.some(woman)(P))(e\, love)))$$
$$= \zeta\kappa.(\zeta f.every(man)(\lambda y.f \triangleleft (\lambda P.Py)))(\zeta e.\kappa((\lambda P.some(woman)(P))(e\, love)))$$
$$= \zeta\kappa.every(man)(\lambda y.\kappa(some(woman)(love(y))))$$

This analysis predicts that any quantifier enclosed by an inverse scope operator cannot over-scope the lexical item that *resets*. For example, the inverse scope readings, *every > some*, in the sentences (9a) and (10a) are predicted to be available, while they are not in the sentences (9b) and (10b).

(9)   a.   Some woman introduced Bill to [every man]$_{INV}$.

   b.   Some woman only introduced [Bill]$_F$ to [every man]$_{INV}$.

(10)   a.   Some woman introduced [every man]$_{INV}$ to Sue.

   b.   Some woman only introduced [every man]$_{INV}$ to [Sue]$_F$.

5   Conclusion

In this paper, we have adopted, as a basic framework, meta-lambda calculus, as proposed in Bekki (2009), in which computational monads are represented as a triple of meta-lambda terms, called internal monads. Then we defined an internal monad for delimited continuations, which determines the translation rule of lambda terms, and defined two control operators, called *shift* and *reset*, under this setting.

We also showed that this setting properly serves as a continuation-based theory of formal semantics, and that it is free from the four crucial empirical/ theoretical problems of Continuized Semantics, and we have demonstrated how the use of control operators enables us to compute the meaning of sentences involving phenomena such as "focus movement" and inverse scope, without recourse to "covert movements."

References

Barker, C.: Introducing Continuations. In: Hastings, R., Jackson, B., Zvolenszky, Z. (eds.) SALT 11, CLC Publications, Ithaca (2001)

Barker, C.: Continuations and the Nature of Quantification. Natural Language Semantics 10(3), 211–241 (2002)

Barker, C.: Continuations in Natural Language. In: H. Thielecke (ed.): the Fourth ACM SIGPLAN Continuations Workshop (CW 2004). Technical Report CSR-04-1, School of Computer Science, University of Birmingham, Birmingham B15 2TT. United Kingdom, pp. 1–11 (2004)

Barker, C., Shan, C.-c.: Types as Graphs: Continuations in Type Logical Grammar. Journal of Logic, Language and Information 15(4), 331–370 (2006)

Bekki, D.: Monads and Meta-Lambda Calculus. In: Hattori, H. (ed.) JSAI 2008 Conference and Workshops. LNCS (LNAI), vol. 5447, pp. 193–208. Springer, Heidelberg (2009)

Danvy, O., Filinski, A.: Abstracting Control. In: LFP 1990, the 1990 ACM Conference on Lisp and Functional Programming, pp. 151–160 (1990)

Danvy, O., Filinski, A.: Representing control. Mathematical Structures in Computer Science 2(4) (1992)

de Groote, P.: Type raising, continuations, and classical logic. In: van Rooij, R., Stokhof, M. (eds.) The 13th Amsterdam Colloquium. Institute for Logic, Language and Computation, pp. 97–101. Universiteit van Amsterdam (2001)

Dybvig, R.K., Peyton-Jones, S., Sabry, A.: A Monadic Framework for Delimited Continuations. Journal of Functional Programming 17(6), 687–730 (2007)

Felleisen, M.: The Theory and Practice of First-Class Prompts. In: 15th ACM Symposium on Principles of Programming Languages, pp. 180–190 (1988)

Hayashishita, J.R.: 'Syntactic Scope and Non-Syntactic Scope'. In: Doctoral dissertation, University of Southern California (2003)

Kratzer, A.: Scope or pseudoscope? Are there wide scope indefinites? In: Rothstein, S. (ed.) Events and Grammar, pp. 163–196. Kluwer, Dordrecht (1998)

Otake, R.: Delimited continuation in the grammar of Japanese. Talk presented at Continuation Fest, Tokyo (2008)

Partee, B., Rooth, M.: Generalized conjunction and type ambiguity. In: Bauerle, R., Schwarze, C., Von Stechow, A. (eds.) Meaning, Use and Interpretation of Language, pp. 361–393. Walter De Gruyter Inc., Berlin (1983)

Plotkin, G.D.: Call-by-Name, Call-by Value and the Lambda Calculus. Theoretical Computer Science 1, 125–159 (1975)

Rooth, M.: A Theory of Focus Interpretation. Natural Language Semantics 1, 75–116 (1992)

Shan, C.-c.: A continuation semantics of interrogatives that accounts for Baker's ambiguity. In: Jackson, B. (ed.) Semantics and Linguistic Theory XII, Ithaca, pp. 246–265. Cornell University Press (2002)

Shan, C.-c.: Linguistic side effects. In: Barker, C., Jacobson, P. (eds.) Direct compositionality, pp. 132–163. Oxford University Press, Oxford (2007)

Shan, C.-c., Barker, C.: Explaining Crossover and Superiority as Left-to-right Evaluation. Linguistics and Philosophy 29(1), 91–134 (2006)

Sitaram, D.: Handling Control. In: The ACM Conference on Programming Language Design and Implementation (PLDI 1993). ACM SIGPLAN Notices, vol. 28, pp. 147–155. ACM Press, New York (1993)

Sitaram, D., Felleisen, M.: Control delimiters and their hierarchies. LISP and Symbolic Computation 3(1), 67–99 (1990)

Stratchey, C., Wadsworth, C.: 'Continuations: a mathematical semantics for handling full jumps'. Technical report, Oxford University, Computing Laboratory (1974)

Wagner, M.: NPI-Licensing and Focus Movement. In: Georgala, E., Howell, J. (eds.) SALT XV, CLC Publications, Ithaca (2006)

Winter, Y.: Flexibility Principle in Boolean Semantics: coordination, plurality and scope in natural language. MIT Press, Cambridge (2001)

Appendix

A Language of Meta-Lambda Calculus

A.1 Types and Meta-Types

The syntax of MLC is specified by the following definitions.

Definition 15 (Alphabet for MLC). *An alphabet for MLC is a sextuple $\langle \mathcal{GT}, \mathcal{Con}, \mathcal{Mcon}, \mathcal{Var}, \mathcal{Mvar}, \mathfrak{S} : \mathcal{Mvar} \to \mathcal{Pow}(\mathcal{Var}) \rangle$, which respectively represents a finite collection of ground types, constant symbols, meta-constant symbols, variables, meta-variables and an assignment function of free variables for each meta-variable.*

Definition 16 (Types). *The collections of types (notation $\mathcal{Typ}$) for an alphabet $\langle \mathcal{GT}, \mathcal{Con}, \mathcal{Mcon}, \mathcal{Var}, \mathcal{Mvar}, \mathfrak{S} \rangle$ is defined by the following BNF grammar (where $\gamma \in \mathcal{GT}$).*

$$\mathcal{Typ} := \gamma \mid \mathcal{Typ} \to \mathcal{Typ}$$

Definition 17 (Context). *A context is a finite list of pairs that are members of $\mathcal{Var} \times \mathcal{Typ}$ (notation $\Gamma = x_1 : \alpha_1, \ldots, x_n : \alpha_n$), where $x_1, \ldots, x_n$ are distinct variables.*

Definition 18 (Meta-types). *The collection of meta-types (notation $\mathcal{M}type$) for an alphabet $\langle \mathcal{GT}, \mathcal{C}on, \mathcal{M}con, \mathcal{V}ar, \mathcal{M}var, \mathfrak{S} \rangle$ is defined by the following BNF grammar (where Γ is a context and $\tau \in \mathcal{T}yp$).*

$$\mathcal{M}typ \;:=\; \Gamma \vdash \tau \mid \mathcal{M}typ \Rightarrow \mathcal{M}typ$$

Definition 19 (Meta-context). *A meta-context is a finite list of pairs that are members of $\mathcal{M}var \times \mathcal{M}Typ$ (notation $\Delta = X_1 : \sigma_1, \ldots, X_n : \sigma_n$), where $X_1, \ldots, X_n$ are distinct meta-variables.*

A.2   Raw-Terms

Definition 20 (Raw-terms). *The collection of raw-terms in Meta-Typed Lambda Calculus (notation Λ) is recursively defined by the following BNF notation, where $x \in \mathcal{V}ar$, $c \in \mathcal{C}on$, $X \in \mathcal{M}var$, and $C \in \mathcal{M}con$.*

$$\Lambda ::= x \mid c \mid \lambda x.\Lambda \mid \Lambda\Lambda$$
$$\mid X \mid C \mid \zeta X.\Lambda \mid \Lambda \triangleleft \Lambda$$

A.3   Judgment

Definition 21 (Judgment). *A judgment is a form $\Delta \Vdash M : \sigma$ where Δ is a meta-context, M is a raw-term, and σ is a meta-type, which is derived by the following rules.*

Variable
$$(VAR)\frac{}{\Delta \Vdash x : (\Gamma, x : \alpha, \Gamma' \vdash \alpha)}$$

Constant Symbol
$$(CON)\frac{}{\Delta \Vdash c : (\Gamma \vdash \alpha)} \quad where\ c \in \mathcal{C}on.$$

Lambda Abstraction
$$(LAM)\frac{\Delta \Vdash M : (\Gamma, x : \alpha \vdash \beta)}{\Delta \Vdash \lambda x.M : (\Gamma \vdash \alpha \to \beta)}$$

Functional Application
$$(APP)\frac{\Delta \Vdash M : (\Gamma \vdash \alpha \to \beta) \quad \Delta \Vdash N : (\Gamma \vdash \alpha)}{\Delta \Vdash MN : (\Gamma \vdash \beta)}$$

Substitution
$$(SUB)\frac{\Delta \Vdash M : (\Gamma, x : \alpha \vdash \beta) \quad \Delta \Vdash N : (\Gamma \vdash \alpha)}{\Delta \Vdash M[N/x] : (\Gamma \vdash \beta)}$$

Definition 22 (Judgment for meta-terms). *A judgment is a form $\Delta \Vdash M : \sigma$ where Δ is a meta-context, M is a meta-term, and σ is a meta-type, which is derived by the following rules.*

Meta-Variable
$$(MVAR)\frac{}{\Delta, X : \sigma, \Delta' \Vdash X : \sigma}$$

Meta-Constant Symbol
$$(MCON)\frac{}{\Delta \Vdash C : \sigma} \quad where\ C \in \mathcal{M}con.$$

Meta-Lambda Abstraction $\qquad (MLAM) \dfrac{\Delta, X : \sigma \Vdash M : \tau}{\Delta \Vdash \zeta X.M : \sigma \Rightarrow \tau}$

Meta-Functional Application $\qquad (MAPP) \dfrac{\Delta \Vdash M : \sigma \Rightarrow \tau \quad \Delta \Vdash N : \sigma}{\Delta \Vdash M \triangleleft N : \tau}$

Meta-Substitution $\qquad (MSUB) \dfrac{\Delta, X : \sigma \Vdash M : \tau \quad \Delta \Vdash N : \sigma}{\Delta \Vdash M[N/X] : \tau}$

A.4   Free Variables and Free Meta-Variables

Definition 23 (Free Variables and Meta-variables). *The set of free variables and free meta-variables are defined respectively by the following sets of rules.*

$$fv(x) = \{x\} \qquad\qquad fmv(x) = \{\}$$
$$fv(c) = \{\} \qquad\qquad fmv(c) = \{\}$$
$$fv(X) = \mathfrak{S}(X) \qquad\qquad fmv(\lambda x.M) = fmv(M)$$
$$fv(\lambda x.M) = fv(M) - \{x\} \qquad fmv(MN) = fmv(M) \cup fmv(N)$$
$$fv(MN) = fv(M) \cup fv(N) \qquad fmv(X) = \{X\}$$
$$fv(\zeta X.M) = fv(M) \qquad fmv(\zeta X.M) = fmv(M) - \{X\}$$
$$fv(M \triangleleft N) = fv(M) \qquad fmv(M \triangleleft N) = fmv(M) \cup fmv(N)$$

A.5   Substitution

Definition 24 (Substitution Rules for variables)

$$x[L/x] \overset{def}{\equiv} L$$
$$y[L/x] \overset{def}{\equiv} y \quad where \; y \not\equiv x.$$
$$c[L/x] \overset{def}{\equiv} c \quad where \; c \in \mathcal{C}on.$$
$$(\lambda x.M)[L/x] \overset{def}{\equiv} \lambda x.M$$
$$(\lambda y.M)[L/x] \overset{def}{\equiv} \lambda y.(M[L/x]) \quad where \; x \notin fv(M) \vee y \notin fv(L).$$
$$(\lambda y.M)[L/x] \overset{def}{\equiv} \lambda w.(M[w/y])[L/x] \quad where \; x \in fv(M) \wedge y \in fv(L).$$
$$(MN)[L/x] \overset{def}{\equiv} (M[L/x])(N[L/x])$$
$$C[L/x] \overset{def}{\equiv} C \quad where \; C \in \mathcal{M}con.$$
$$(\zeta X.M)[L/x] \overset{def}{\equiv} \zeta X.(M[L/x])$$
$$(M \triangleleft N)[L/x] \overset{def}{\equiv} (M[L/x]) \triangleleft (N[L/x])$$

Remark 25. $X[L/x]$ where $x \in fv(X)$ should be treated independently.

Definition 26 (Substitution Rules for meta-variables)

$$x[L/X] \overset{def}{\equiv} x$$

$$c[L/X] \overset{def}{\equiv} c \quad where \; c \in \mathcal{C}on.$$

$$(\lambda x.M)[L/X] \overset{def}{\equiv} \lambda x.(M[L/X])$$

$$(MN)[L/X] \overset{def}{\equiv} (M[L/X])(N[L/X])$$

$$X[L/X] \overset{def}{\equiv} L$$

$$Y[L/X] \overset{def}{\equiv} Y \quad where \; Y \not\equiv X.$$

$$C[L/X] \overset{def}{\equiv} C \quad where \; C \in \mathcal{M}con.$$

$$(\zeta X.M)[L/X] \overset{def}{\equiv} \zeta X.M$$

$$(\zeta Y.M)[L/X] \overset{def}{\equiv} \zeta Y.(M[L/X]) \quad where \; X \notin fmv(M) \vee Y \notin fmv(L).$$

$$(\zeta Y.M)[L/X] \overset{def}{\equiv} \zeta W.(M[W/Y])[L/X] \quad where \; X \in fmv(M) \wedge Y \in fmv(L).$$

$$(M \triangleleft N)[L/X] \overset{def}{\equiv} (M[L/X])(N[L/X])$$

B ζ-Theory

Axiom 27 (Permutation and Meta-Permutation)

$$\frac{\Delta \Vdash M = N : (\Gamma, x : \alpha, y : \beta, \Gamma' \vdash \delta)}{\Delta \Vdash M = N : (\Gamma, y : \beta, x : \alpha, \Gamma' \vdash \delta)} \qquad \frac{\Delta, X : \nu, Y : \upsilon, \Delta' \Vdash M = N : \sigma}{\Delta, Y : \upsilon, X : \nu, \Delta' \Vdash M = N : \sigma}$$

Axiom 28 (Weakening and Meta-Weakening)

$$\frac{\Delta \Vdash M = N : (\Gamma \vdash \beta)}{\Delta \Vdash M = N : (\Gamma, x : \alpha \vdash \beta)} \qquad \frac{\Delta \Vdash M = N : \sigma}{\Delta, X : \nu \Vdash M = N : \sigma}$$

Axiom 29 (Equivalence)

$$(=R)\frac{\Delta \Vdash M : \sigma}{\Delta \Vdash M = M : \sigma} \qquad (=S)\frac{\Delta \Vdash M = N : \sigma}{\Delta \Vdash N = M : \sigma}$$

$$(=T)\frac{\Delta \Vdash L = M : \sigma \quad \Delta \Vdash M = N : \sigma}{\Delta \Vdash L = N : \sigma}$$

Axiom 30 (Replacement)

$$(=\lambda)\frac{\Delta \Vdash M = N : (\Gamma, x : \alpha \vdash \beta)}{\Delta \Vdash \lambda x : \alpha.M = \lambda x : \alpha.N : (\Gamma \vdash \alpha \to \beta)}$$

$$(=F)\frac{\Delta \Vdash M = N : (\Gamma \vdash \alpha)}{\Delta \Vdash F(M) = F(N) : (\Gamma \vdash \beta)} \qquad (=A)\frac{\Delta \Vdash F = G : (\Gamma \vdash \alpha \to \beta)}{\Delta \Vdash F(M) = G(M) : (\Gamma \vdash \beta)}$$

Axiom 31 (Meta-Replacement)

$$(=M\lambda)\frac{\Delta, X : \sigma \Vdash M = N : \tau}{\Delta \Vdash \zeta X : \sigma.M = \zeta X : \sigma.N : \sigma \Rightarrow \tau}$$

$$(=MF)\frac{\Delta \Vdash M = N : \sigma}{\Delta \Vdash F \triangleleft M = F \triangleleft N : \tau} \qquad (=MA)\frac{\Delta \Vdash F = G : \sigma \Rightarrow \tau}{\Delta \Vdash F \triangleleft M = G \triangleleft M : \tau}$$

Axiom 32 (Function Equations)

$$(\alpha)\frac{\Delta \Vdash M : (\Gamma, x : \alpha \vdash \beta)}{\Delta \Vdash \lambda x.M = \lambda y.M[y/x] : (\Gamma \vdash \alpha \rightarrow \beta)} \qquad where \; y \notin fv(M).$$

$$(\beta)\frac{\Delta \Vdash F : (\Gamma, x : \alpha \vdash \beta) \qquad \Delta \Vdash M : (\Gamma \vdash \alpha)}{\Delta \Vdash (\lambda x : \alpha.F)M = F[M/x] : (\Gamma \vdash \beta)}$$

$$(\eta)\frac{\Delta \Vdash M : (\Gamma \vdash \alpha \rightarrow \beta)}{\Delta \Vdash \lambda x : \alpha.(Mx) = M : (\Gamma \vdash \alpha \rightarrow \beta)} \qquad where \; x \notin fv(M).$$

Axiom 33 (Meta-Function Equations)

$$(A)\frac{\Delta, X : \sigma \Vdash M : \tau}{\Delta \Vdash \zeta X.M = \zeta Y.M[Y/X] : \sigma \Rightarrow \tau} \qquad where \; y \notin fmv(M).$$

$$(B)\frac{\Delta, X : \sigma \Vdash F : \tau \qquad \Delta \Vdash M : \sigma}{\Delta \Vdash (\zeta X : \sigma.F) \triangleleft M = F[M/X] : \tau}$$

$$(H)\frac{\Delta \Vdash M : \sigma \Rightarrow \tau}{\Delta \Vdash \zeta X : \sigma.(MX) = M : \sigma \Rightarrow \tau} \qquad where \; X \notin fmv(M).$$

Problems with Intervention and Binding into Relations*

Alastair Butler[1] and Kei Yoshimoto[2]

[1] Japan Society for the Promotion of Science
6 Ichibancho, Chiyoda-ku, Tokyo 102-8471, Japan
ajb129@hotmail.com
[2] Center for the Advancement of Higher Education, Tohoku University
Kawauchi 41, Aoba-ku, Sendai 980-8576, Japan
kyoshimoto@mail.tains.tohoku.ac.jp

Abstract. In this paper we describe a formal system in which constraints on the interaction of operations involved in creating and supporting operator-variable dependencies during interpretive evaluation match effects of intervention and constraints prohibiting binding into relations that are observed in natural languages. The derived constraints are found to form two sides of the same coin, the one occurring when steps are taken to avoid the other. This formal result is of interest because both types of constraints afflict all forms of operator-variable dependencies in natural languages, suggesting that the wide range of cross-linguistic variation languages exhibit, especially when encoding longer distance dependencies, stems from there being no optimal way to encode operator-variable dependencies.

1   Introduction

Cross-linguistic data reveals that languages differ systematically when encoding operator-variable dependencies, especially those that need to span long distances. An archetypal case is that of WH arguments in constituent questions, where a long distance dependency can be created with the WH argument taking a wide semantic scope with the question while having to be appropriately integrated into the clause. In English one and only one WH constituent must occur sentence initially.

(1)   Who did John give what?

In Slavic languages all WH constituents receive a frontmost placement, as (2) from Bulgarian illustrates.

(2)   Koj kakvo e kupil?
      who what is bought
      'Who bought what?'

---

* We would like to thank the participants of LENLS 2009 and the reviewers for the helpful and challenging remarks we received.

K. Nakakoji, Y. Murakami, and E. McCready (Eds.): JSAI-isAI, LNAI 6284, pp. 181–196, 2010.

Motivation for WH constituents favouring frontmost positions follows from their requiring outermost scope with the question. Yet WH constituents can have embedded placements while somehow maintaining outermost scope with the question. For example, languages such as Japanese show an opposite pattern to Bulgarian by allowing canonical 'in-situ' placements for all WH constituents.

(3)   John-wa   dare-ni   nani-o   ageta ka?
      John-TOP who-DAT what-ACC gave Q
      'Who did John give what?'

Motivation for WH constituents favouring canonical argument placements can likely be traced to the need for the WH constituents to be appropriately integrated into the clause. However, on the face of it, such narrow syntactic placements of WH constituents are puzzling, since the grammar will require a mechanism whereby such WH constituents are able to take wider scopes than their (overt) syntactic positioning.

There are also "mixed" languages such as French, which, for questions with a single WH argument, has the option of having the WH argument placed frontmost, (4a), or placed in a canonical embedded argument position, (4b). Notably, from the perspective of interpretation (4a) and (4b) are identical: both are equally acceptable as nonecho single constituent questions.

(4)   a.   Qui' est-ce que un homme voit?
           who is-this that a   man   sees
           'Who does a man see?'
      b.   Un homme voit qui?
           a   man   sees who

In this paper we aim to provide a rational for why there is such diversity in how natural languages encode operator-variable dependencies. We will do this by utilising fine grained operations for manipulating operator-variable dependencies, the formal details of which are presented in an appendix. The paper is structured as follows. Section 2 presents the account as a series of results regarding the possible distribution of operations used to create and thereafter manage the allocation of operator-variable dependencies. Section 3 provides an overall summary.

2   The Account

The account we provide for scope interactions will be phrased in terms of Scope Control Theory (SCT) (Butler 2007). This provides a small logical language, defined in the appendix, with operations of scope manipulation that combine the static reformulations of Dynamic Semantics by Dekker (2002) and Cresswell (2002) with the Sequence Semantics of Vermeulen (1993). Five operations are of relevance here:

- Close($oper$,x,e) creates a number n ($n \geq 0$) of $oper$ quantifications that bind x, where n is determined by the x usage count over e;

- $\text{Use}(x,e)$ supports binding by incrementing the x usage count;
- $\text{Hide}(x,e)$ terminates the x usage count;
- $\text{Lam}(x,y,e)$ shifts a binding from x to y;
- $\text{Rel}(\boldsymbol{x},\boldsymbol{y},r,e)$ builds a relation from relation name r with sequence of expressions $\boldsymbol{e}$ as arguments, potentially changing the assignment with respect to which each argument is evaluated based on binding name sequences $\boldsymbol{x}$ and $\boldsymbol{y}$.

Using the definitions of (5), the combination of operations Close and Lam allows for the adoption of a scope picture like (6) as the basis for an account of how a WH constituent can receive a wide scope interpretation while being syntactically embedded.

(5) $\text{Cq} = \lambda \text{f}.\text{Close}(?,"q",\text{f})$
   $\text{Lqx} = \lambda \text{f}.\text{Lam}("q","x",\text{f})$

(6)

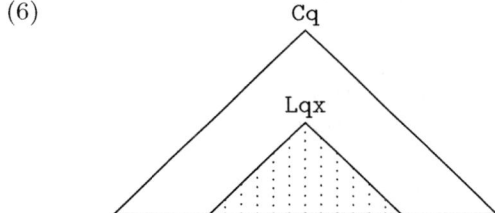

Notably (6) suggests at least two distinct manifestations of the same scope: (i) as a *semantic scope* signalled by the closure operation, Close, that marks the point from where the scope is actually created and so takes semantic effect—specifically in (6) as a "q" binding; and (ii) as a *syntactic scope* signalled by Lam, marking the point from where an already open scope gets integrated into the clause as a binding argument—specifically in (6) as an "x" binding. The one constraint (6) shows is that, in order to pick up on the binding created by Cq, the action of Lqx needs to occur embedded under Cq.

We can apply (6) to both (4a) and (4b) by supposing that the WH constituent *qui* 'who' has the argument binding role of Lqx, and that the question act of utterance contributes Cq, to ensure that Cq receives a widest scope placement. It follows that *un homme* 'a man' in (4a) is in the dotted region of (6), which is under both Cq and Lqx, as is illustrated by (7). In contrast, *un homme* in (4b) is in the region of (6) without dots that is under Cq, but outside the scope of Lqx, as is illustrated by (8).

(7)

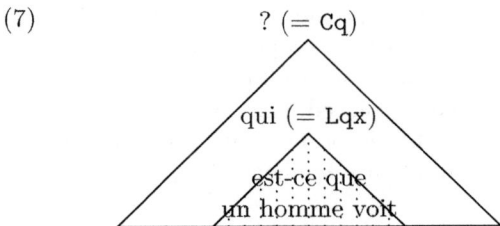

(8)

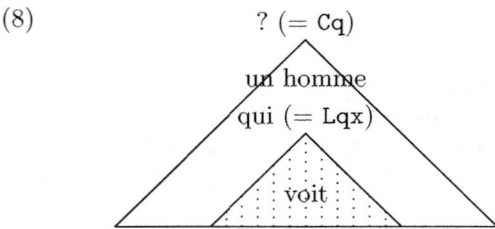

The operations of Cq and Lqx are mutually licensing: Cq relies on their being as many occurrences of Lam("q",#,#) as it creates bindings, else its presence will lead to vacuous "q" bindings left without integration into the clause; and Lqx relies on their being an available binding to shift to an argument binding, else its action will fail.

However, the picture of (6) cannot be the complete story, since Cq by itself has no idea how many "q" bindings should be created. With the SCT system, Cq gets this information by counting occurrences of Uq, as defined in (9).

(9) $Uq = \lambda f.Use("q",f)$

An instance of Uq needs to be under the scope of the Cq to which it provides its supporting information, but can be anywhere else, e.g., positioned above Lqx, (10), or below, (11).

(10)

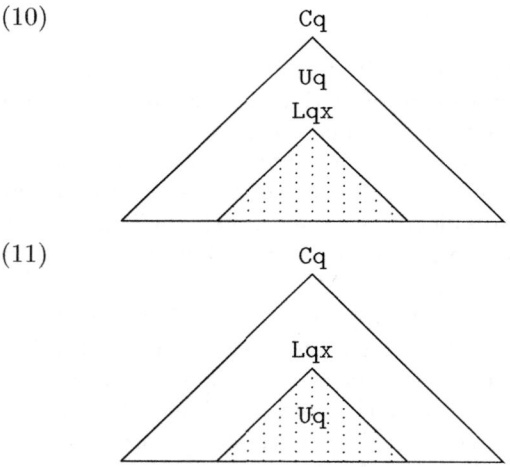

(11)

There is however a fly in the ointment for the approach of counting instances of Uq, which comes as the possibility of Uq falling under the scope of Hq, as defined in (12). Presence of Hq has the effect of ending the count for instances of Uq, and so any instance of Uq that is under the scope of Hq will no longer be providing its countable presence outside the scope of Hq. As (13) pictures, this makes an expression invalid wherever Lqx is placed, since a binding will not be created by Cq to support Lqx. It follows that in order to avoid the possibility of falling under the scope of any Hq, one wants a placement of Uq to be as high in the structure as possible.

(12)   Hq = λf.Hide("q",f)

(13)   *

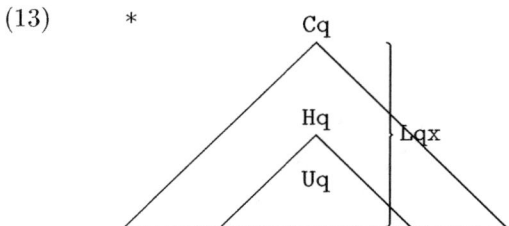

Interestingly this constraint can be matched to what happens with natural languages. For example, in French (see e.g., Mathieu 1999), while an operation like negation leaves the licensing of the fronted WH constituent in (14a) unhindered, negation will block the licensing of a WH constituent embedded within its scope, as in (14b).

(14)   a.   **Qui'** est-ce que tu   ne vois pas?
            who   is-this that you NE see   not
            'Who didn't you see?'

       b. *Tu ne vois pas **qui**?
            you NE see   not who

Supposing we expect the WH constituent *qui* 'who' to now be contributing the combination of Lqx and Uq and that any definition of negation involves the introduction of a closure operation that triggers the hiding of all quantificational usage information and so includes the presence of Hq, then we find that (14a) leads to the structure of (15) in which the instance of Uq is placed outside the scope of Hq and so conforms to the valid (11). In contrast, (14b) leads to the structure of (16) in which the instance of Uq is placed inside the scope of Hq, and so manifests the invalid structure of (13).

(15)   ? (= Cq)

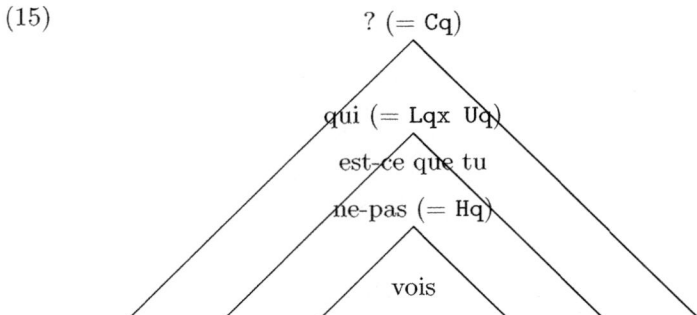

(16)   *

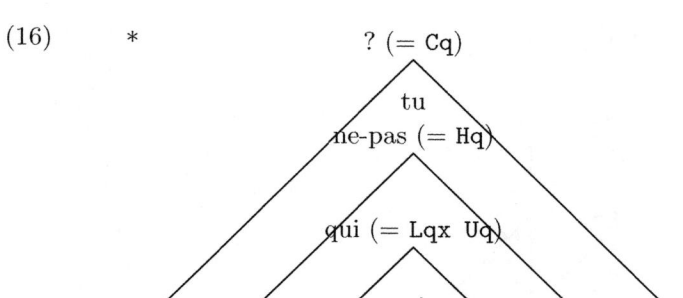

Note that SCT cannot do away with the `Hide` operation, since the presence of `Hide(`x`,#)` is essential to allowing for the embedding of instances of `Close(#,`x`,#)` for some value of x. For example, the placement of operations in (17) is invalid, with the problem being that both instances of `Cq` introduce a binding for the single `Lqx` operation.

(17)   *

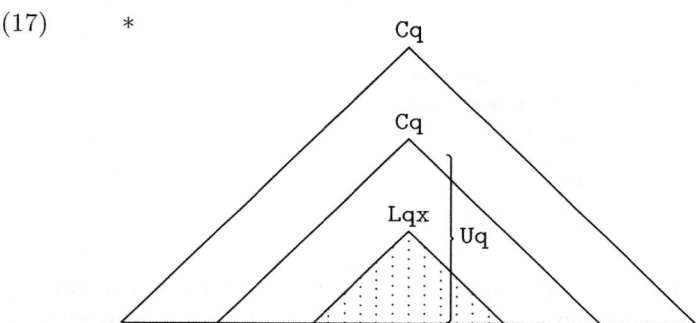

Placement of `Hq` in the portion of structure between the two instances of `Cq` in (17), as in (18), ensures that only the embedded instance of `Cq` is triggered to create a "q" binding.

(18)

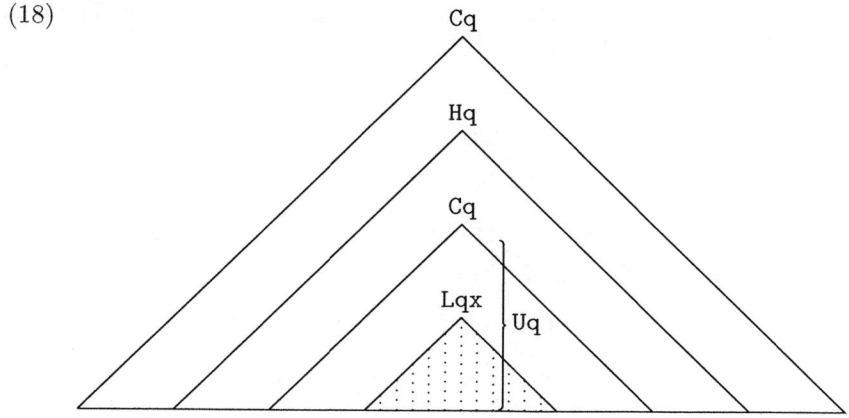

The intervention effect seen with (13) of Hq constraining the placement of Uq is not the only kind of restriction on where instances of the Use operation can be placed. Using (19), in addition to the definitions of (5), (9) and (12), we can illustrate with (20) a scenario where invalidity results from there needing to be a "q" binding (signalled by the presence of Lqx) that falls within an argument of an Rq relation without the support of Uq within the argument.

(19) Rq = λ1.Rel([..."q"...],#,#,1)

(20)

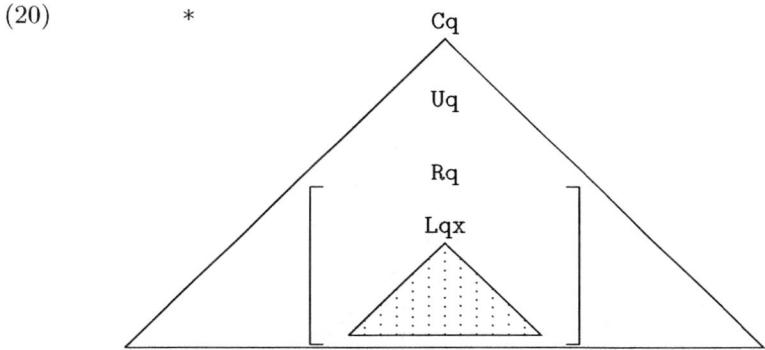

In contrast to (20) the scenario (21) is valid when there is placement of Uq anywhere within the argument of Rq.

(21)

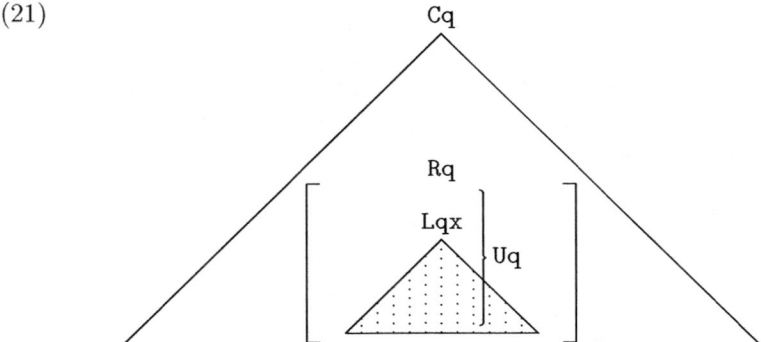

When the placement of Lqx scopes over Rq it becomes essential that Uq is outside the scope of Rq as shown by the contrast between the valid (22) and invalid (23). The reason for this contrast is that Lqx will shift the "q" binding to an "x" binding, so there will be no "q" binding present within the scope of Rq, thus requiring the absence of any Uq.

(22)

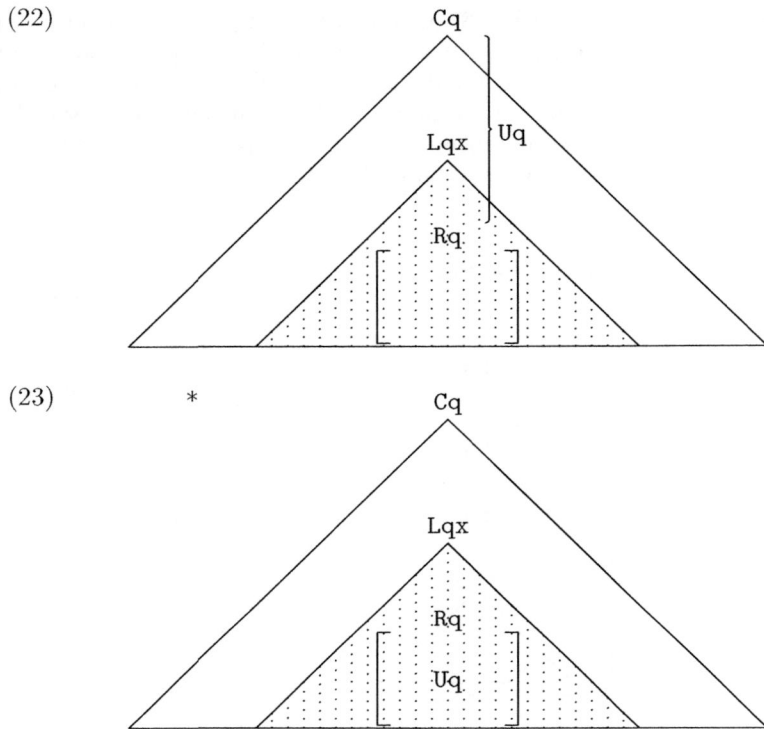

(23)   *

Combining the results of (20) and (21) on the one hand and (22) and (23) on the other we get the overall consequence that the placement of Uq should have as close a proximity as possible to the Lqx that it supports.

After an occurrence of Lqx what had been a "q" binding shifts to an "x" binding. Using (24) we can observe the contrast between the invalid (25) and valid (26). That is, just as it was necessary for a "q" binding that binds into an argument of an Rq relation to find Uq support within the scope of the argument, it is necessary for an "x" binding that binds into an argument of an Rx relation to find Ux support within the scope of the argument.

(24)   Ux = λf.Use("x",f)
       Rx = λl.Rel([..."x"...],#,#,l)

(25)   *

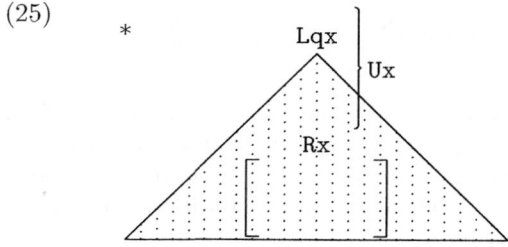

(26)

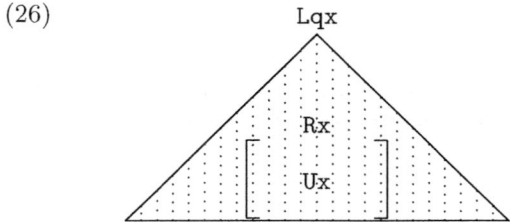

Combining the result of (22) together with (25) and (26), we can, for example, observe the contrast between the valid (27), in which both of the arguments of Rq involve Rx relations that contain arguments that support the "x" binding opened by Lqx, and the invalid (28), which is due to one of the arguments of Rq involving an Rx relation that fails to contain an argument that is able to support the "x" binding opened by Lqx.

(27)

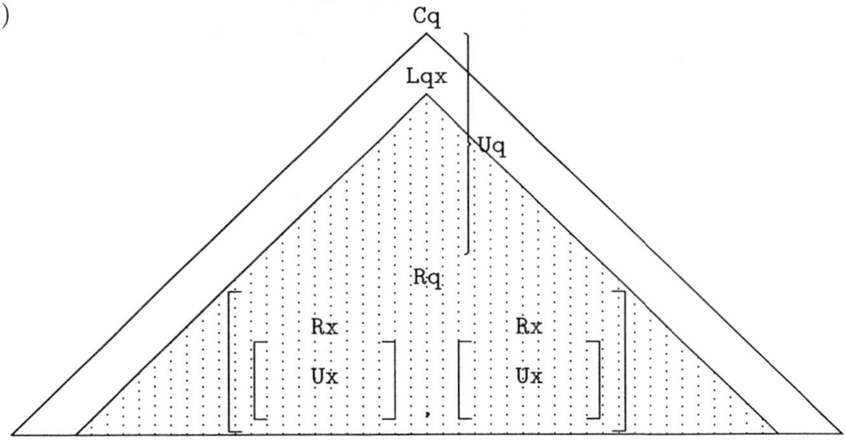

(28)   *

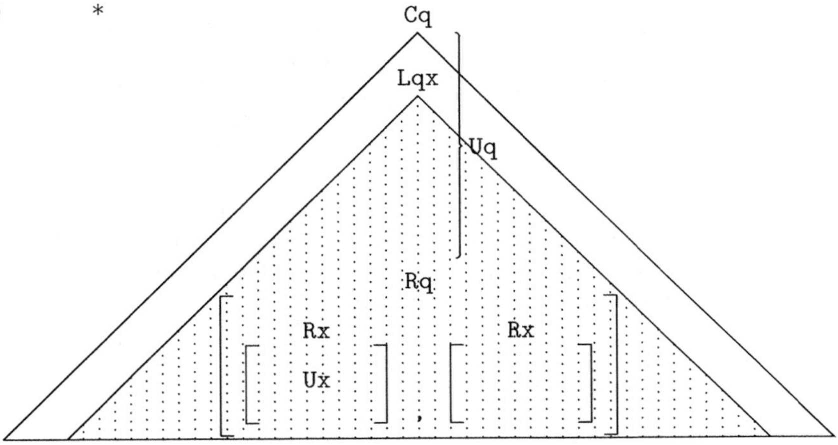

That (28) is invalid can be used to derive the observation that (29) is ungrammatical.

(29) *Who does John see and Bill like Mary?

Example (29) illustrates what has come to be known as the *coordinate structure constraint* of Ross (1967), which was originally phrased in terms of restrictions on movement:

(30)   *The Coordinate Structure Constraint*
        In a coordinate structure, no conjunct may be moved, nor may any element contained in a conjunct be moved out of that conjunct.

(Ross's (4.84))

From the insight of (28), the relevant empirical consequences of (30) follow from the inability for an x binding to bind into an argument e_i of a relation $\mathtt{Rel}([\ldots x \ldots],\#,\#,e)$ because e_i contains no $\mathtt{Use}(x,\#)$ able to support the binding. To see this applied, we can capture the invalidity of (29) with the placement of operations in (31).

(31)   *

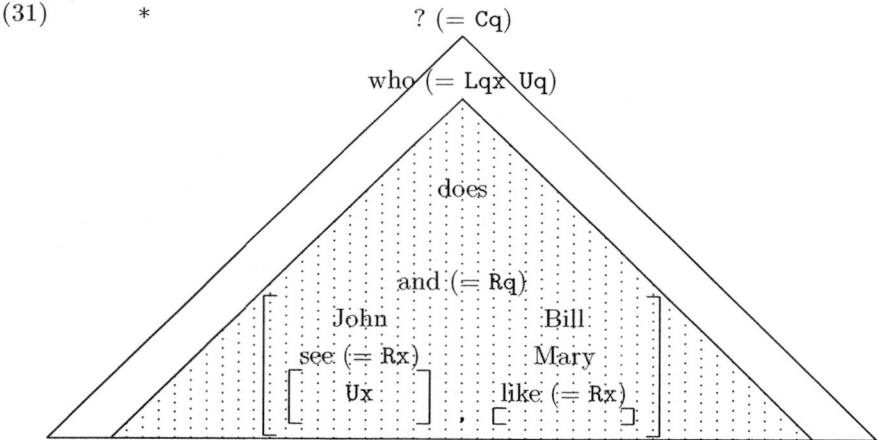

That the structure of (27) is valid demonstrates what was taken to be an 'across-the-board' exception to the coordinate structure constraint. We can illustrate an application of this structure with (32), with both conjuncts containing predicates that support the binding from *who*.

(32)   Who does John see and Bill like?

We can expect that (32) leads to the allocation of operations seen in (33), thereby matching (27).

(33)

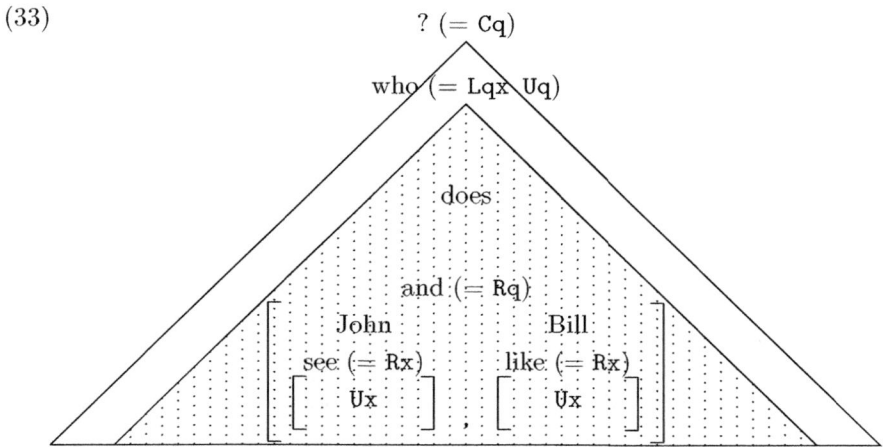

We have seen with the invalid structure of (28) and example (29/31) that the high placement of a binding operation runs the risk of falling foul of the coordinate structure constraint, which amounts to accidentally binding into a relation sensitised to check for the binding that fails to contain an argument that is able to support the binding. The way to avoid this constraint is to maintain a low placement for the binding operation. This amounts to requiring a low `Lam(#, x,#)` placement (and so also a low `Use(x,#)` placement, following the results of (20) and (21)), as in the valid (34).

(34)

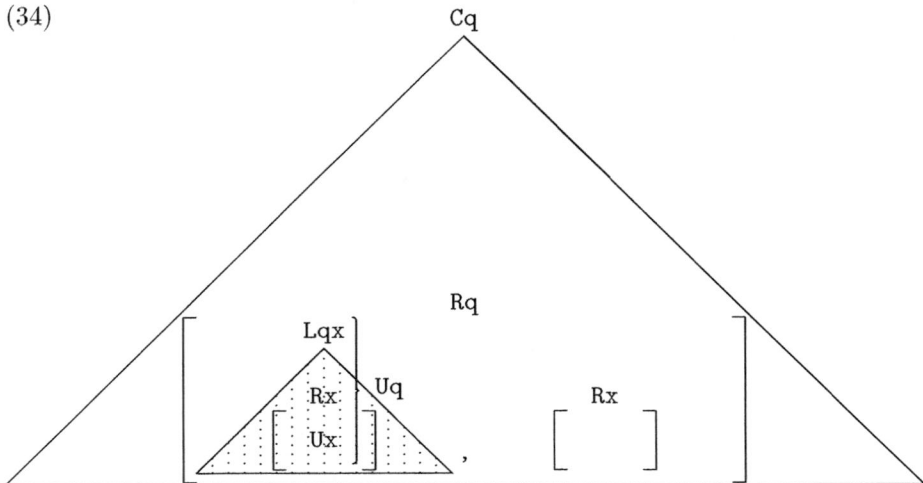

With (34) valid as a result of Lqx being kept low so as not to scope over a relation sensitised to and failing to support an "x" binding, we are also in a position to note the invalid (35). This shows the reverse consequence that when there is a relation that requires a binding, the binding should not fail to scope over the relation.

(35)   *

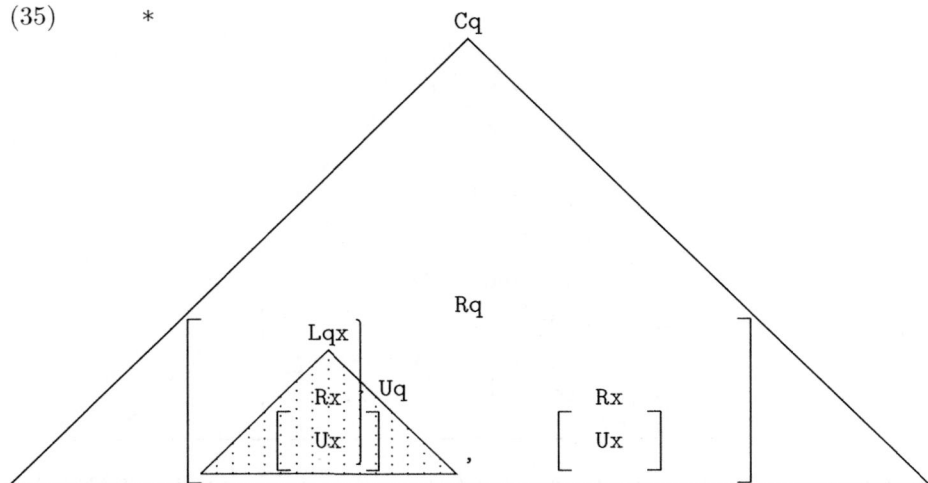

We can see a match for the contrast between the valid (34) and the invalid (35) in the data of (36).

(36) a. Who does John see and does Bill like her?
      b. *Who does John see and does Bill like?

We can expect that (36a) leads to the allocation of operations seen in (37), thereby matching (34).

(37)   ? (= Cq)

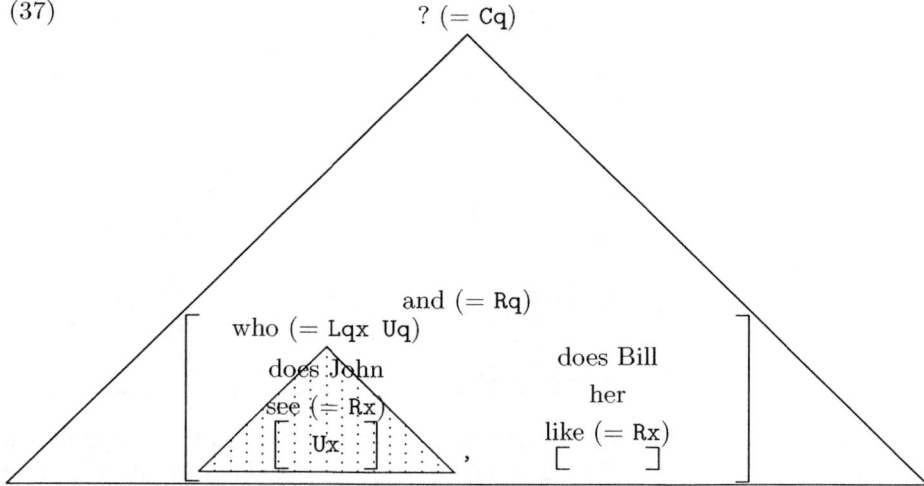

In contrast, (36b) leads to the allocation of operations seen in (38), thereby matching (35).

(38)   *

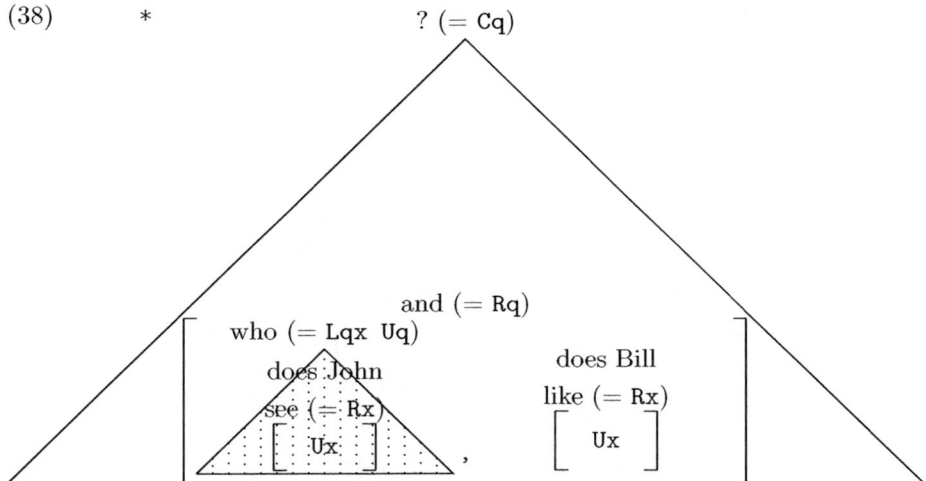

The results of binding into relations that we have seen are not limited to examples involving coordination, but rather the effects will apply whenever there is the combining of expressions. For example, the invalidity of the structure of (28) will apply to an example involving an attached adjunct clause, as (39/40) demonstrates.

(39) *Who did Mary cry after John hit?

(40)

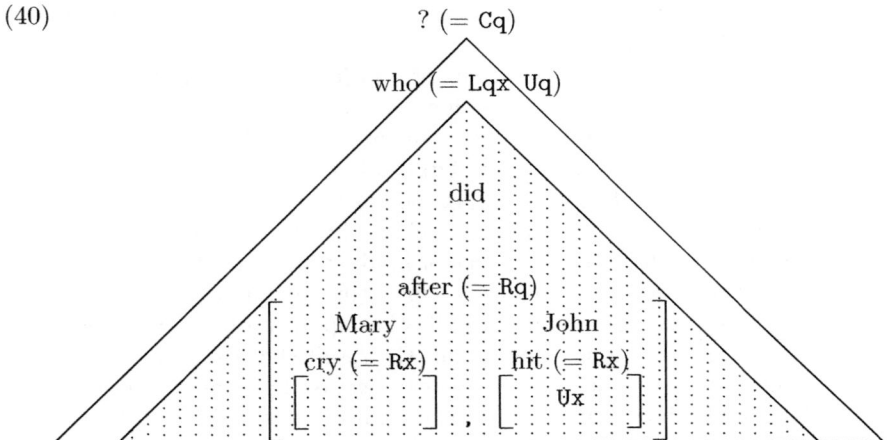

In contrast, the low placement of a WH constituent inside an adjunct clause leads to a valid structure, as demonstrated by (41/42).

(41)   Who cried after John hit who?

(42)

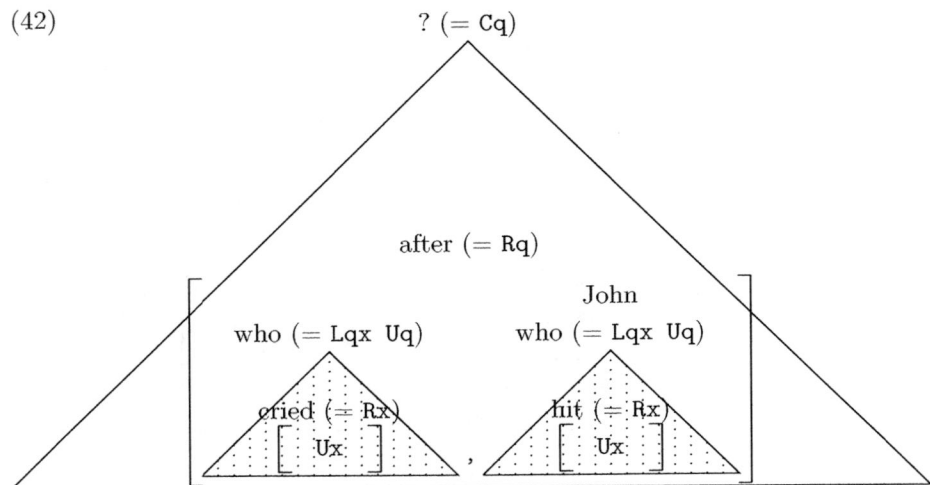

3   Summary

To sum up, the approach for capturing long distance operator-variable dependencies examined in this paper suggests that a binding operation can be factored into the components of: Close, to provide the closure that creates new scopes for bindings; Lam, to provide the instruction for when a scope should be integrated into the clause as a clause binding; and Use, to say how many scopes should be created by Close, as well as sustain binding into a sensitised Rel. These operations work together from potentially disparate locations. However the range of possible interactions is constrained.

Confining attention to the operations of Cq, Uq, Hq, Lqx, Rq, Rx and Ux: Cq must have widest placement; Uq must have as high a placement as possible to avoid intervention, that is, falling under an instance of Hq; Uq must be placed as close as possible to the instance of Lqx that it supports to avoid problems with binding into Rq relations (e.g., relations of coordination and adjunction); and finally Lqx which opens a binding for clause integration must have as low a placement as possible to avoid problems of binding into Rx relations (e.g., main predicates) that may or may not contain supporting instances of Ux. The totality of these constraints which generalise to all other applications of the Close, Use, Hide, Lam and Rel operations brings about the result that there can be no satisfactory systematic placement of operations to ensure that effects of intervention and problems with binding into relations are always avoided. With there being no optimal 'solution' to the placement of scope operations it is reasonable to expect the wide range of cross-linguistic variation languages exhibit, especially when coding longer distance dependencies, with any one systematic approach to operator placement having advantages as well as short comings.

References

Butler, A.: Scope control and grammatical dependencies. Journal of Logic, Language and Information 16, 241–264 (2007)

Cresswell, M.J.: Static semantics for dynamic discourse. Linguistics and Philosophy 25, 545–571 (2002)

Dekker, P.: Meaning and use of indefinite expressions. Journal of Logic, Language and Information 11, 141–194 (2002)

Mathieu, E.: French WH in situ and the Intervention Effect. In: Iten, C., Neeleman, A. (eds.) UCL Working Papers in Linguistics, pp. 441–472. University College, London (1999)

Ross, J.R.: Constraints on Variables in Syntax. Ph.D. thesis, MIT, Cambridge, Mass (1967)

Vermeulen, C.F.M.: Sequence semantics for dynamic predicate logic. Journal of Logic, Language and Information 2, 217–254 (1993)

Appendix: SCT Evaluation

In this appendix we present the evaluation procedure of Scope Control Theory we have assumed in this paper by way of a translation routine from expressions of a language with operators T, Use, Hide, Close, Lam and Rel, into formulas of a predicate logic notation. The idea is that translation returns a 'snapshot' of the dependencies evaluation establishes.

Extensive use is made of sequences and operations on sequences. Notably:

- $[x_0,...,x_{n-1}]$: a sequence with n elements, x_0 being frontmost.
- $\boldsymbol{x}$: abbreviation for a sequence.
- $\boldsymbol{x}_i$: the i-th element of a sequence, e.g. $[x_0,....,x_{n-1}]_i = x_i$, where $0 \leq i < n$.
- $|\boldsymbol{x}|$: the sequence length, e.g. $|[x_0,...,x_{n-1}]| = n$.
- $\mathrm{cons}(y,[x_0,...,x_{n-1}]) = [y,x_0,...,x_{n-1}]$.
- $\mathrm{snoc}(y,[x_0,...,x_{n-1}]) = [x_0,...,x_{n-1},y]$.

Translation is with respect to an assignment of a (possibly empty) sequence of predicate logic variables (scopes) to each binding name: $g : Name \longrightarrow Var^\star$. We employ relations on sequence assignments taking us from g to h or vice versa. For $\mathrm{shift}(op)$ the operation op needs to be specified, with suitable candidates being cons and snoc, to give shift(cons) and shift(snoc).

- $(g,h) \in \mathrm{pop}_x$ iff h is just like g, except that, $g(x) = \mathrm{cons}((g(x))_0, h(x))$.
- $(g,h) \in \mathrm{shift}(op)_{x,y}$ iff $\exists k : (h,k) \in \mathrm{pop}_y$ and k is just like g, except that, $g(x) = op((h(y))_0, k(x))$.

Relations are iterated when augmented with a positive superscript, e.g., pop_x^3 iterates pop_x three times.

We now define a 'usage count' operation $x(e)$. This formally defines the contribution of Use and Hide, returning a count of the number of times Use$(x,\#)$ occurs in expression e outside the scope of any Hide$(x,\#)$.

- $x(\texttt{Use}(y,e)) = \begin{cases} x(e) + 1 & \text{if } x = y \\ x(e) & \text{otherwise} \end{cases}$
- $x(\texttt{There}(y,e)) = \begin{cases} x(e) - 1 & \text{if } x = y \\ x(e) & \text{otherwise} \end{cases}$
- $x(\texttt{Hide}(y,e)) = \begin{cases} 0 & \text{if } x = y \\ x(e) & \text{otherwise} \end{cases}$
- $x(\texttt{T}(y,i)) = 0$
- $x(\texttt{Lam}(y,z,e)) = x(e)$
- $x(\texttt{Close}(oper,y,e)) = x(e)$
- $x(\texttt{Rel}(\boldsymbol{y},\boldsymbol{z},r,e)) = \sum_{i=0}^{|e|-1} x(e_i)$

A formal definition for the other operators is given below in terms of a recursive evaluation procedure with respect to a sequence assignment, g, returning either a translation into a formula of predicate logic notation or '$*$' to indicate failure of the evaluation.

- $(g,\texttt{T}(x,i))^\circ$ return $(g(x))_i$, provided $0 \le i < |g(x)|$; otherwise return $*$.
- $(g,\texttt{Use}(x,e))^\circ$ return $(g,e)^\circ$.
- $(g,\texttt{Hide}(x,e))^\circ$ return $(g,e)^\circ$.
- $(g,\texttt{Close}(oper,x,e))^\circ$ if $x(e) = 0$ return $(g,e)^\circ$ else $\exists h : (h,g) \in \text{pop}_x^{x(e)}$ return $oper(h(x))_0...(h(x))_{x(e)-1}(h,e)^\circ$, provided $(h,e)^\circ \ne *$; otherwise return $*$.
- $(g,\texttt{Lam}(x,y,e))^\circ$ given $\exists h : (g,h) \in \text{shift}(\text{cons})_{x,y}$ return $(h,e)^\circ$; otherwise return $*$.
- $(g,\texttt{Rel}(\boldsymbol{x},\boldsymbol{y},r,e))^\circ$ return $r((0,g,\boldsymbol{x},\boldsymbol{y},e)^\bullet,...,(|e| - 1,g,\boldsymbol{x},\boldsymbol{y},e)^\bullet)$, provided for $0 \le i < |e|$, $(i,g,\boldsymbol{x},\boldsymbol{y},e)^\bullet \ne *$; otherwise return $*$.

where:

- $(n,g,\boldsymbol{x},\boldsymbol{y},e)^\bullet$ if $|\boldsymbol{x}| = 0$ return $(g,e_n)^\circ$ else $\exists h_0...h_{|\boldsymbol{x}|} : h_0 = g$ and for $0 \le i < !|\boldsymbol{x}|$, $(h_i,h_{i+1}) \in (\text{pop}_{\boldsymbol{x}_i}^{|h_i(\boldsymbol{x}_i)|-\sum_{k=0}^n \boldsymbol{x}_i(e_k)}; \text{shift}(\text{snoc})_{\boldsymbol{x}_i,\boldsymbol{y}_i}^{|h_i(\boldsymbol{x}_i)|-\sum_{k=n}^{|e|-1} \boldsymbol{x}_i(e_k)})$ return $(h_{|\boldsymbol{x}|},e_n)^\circ$; otherwise return $*$.

A Translation from Logic to English with Dynamic Semantics*

Elizabeth Coppock and David Baxter

Cycorp, Inc., Austin TX 78731, USA

Abstract. We present a procedure for translating predicate logic into English, which generates both referring and non-referring expressions using a dynamically updated context representation. The system treats referring and non-referring expressions within a unified framework, capturing their common properties – both bound and referential anaphora require an accessible antedecent – and the special properties of non-referring expressions: Non-referring expressions are introduced with quantificational determiners, and correspond to short-term discourse referents.

Keywords: natural language generation, dynamic semantics, predicate logic, quantification, anaphora.

1 Introduction

The goal of the present work is to define an algorithm for translating formulas of predicate logic into concise, natural-sounding English, with quantificational expressions, proper names, indefinites, definite descriptions, and pronouns, wherever appropriate (examples are given in §2). Because predicate logic contains both constants and variables, this algorithm should generate both referring and non-referring expressions.

Work in the field of generating referring expressions [1–16] is designed for the task of providing an appropriate means of referring to a given object in a domain. For example, the input might be a situation in which there is more than one book, and the book in question is on a unique table. An appropriate output for this situation would be *the book on the table*. Because of the nature of the input and the task, these systems only generate expressions that genuinely refer to objects or groups of objects. Even those systems within this tradition of research that generate "quantificational" expressions [7, 12] such as *those women who have fewer than two children* or *the people who work for exactly 2 employers* really only generate referential expressions, referring to groups of objects. The field 'generating referring expressions' is thus appropriately named, so far, because it deals only with the generation of genuinely referring expressions.

---

* Thanks to the LENLS organizing committee and audience, David Beaver, Cleo Condoravdi, Nicholas Asher, Lucas Champollion and Elias Ponvert for feedback. This work was partially supported under the DARPA Rapid Knowledge Formation program.

K. Nakakoji, Y. Murakami, and E. McCready (Eds.): JSAI-isAI, LNAI 6284, pp. 197–216, 2010.
© Springer-Verlag Berlin Heidelberg 2010

A separate line of research, known as 'tactical generation' or 'realization' concentrates on sentence generation based on formal grammatical theories such as Head-Driven Phrase Structure Grammar (HPSG) [17–20], Lexical Functional Grammar (LFG) [21–24] and Combinatory Categorial Grammar (CCG) [25–27]. Some systems in this category take as input a logical semantic representation that contains quantifiers and variables. For example, the HPSG generation system described in [17] takes as input the typed feature structure corresponding to the CONTENT value of the top-level HPSG sign using Pollard and Yoo's HPSG analysis of quantification [28], which makes the input a notational variant of quantificational logic. Systems in this category may potentially generate non-referring expressions, such as *no man* and *himself* in *no$_i$ man likes himself$_i$*.

Tactical realization systems based purely on existing grammatical formalisms like [17] lack a representation of the discourse and a theory of antecedent-accessibility that could be used to decide, for example, when a pronoun, definite description, or indefinite description, would be an appropriate way to realize a given discourse referent. They are incomplete without algorithms for generating referring expressions. But the problem cannot be solved simply by adjoining a tactical realization system to a system for generating referring expressions, because there are common constraints between referential and bound variable anaphora. For example, both types of anaphora require a cognitively accessible antecedent. The fields of tactical realization and generating referring expressions should not be kept separate; rather, referring and non-referring expressions should be treated within a unified framework.

The idea that referential and bound variable anaphora have certain commonalities is one of the insights underlying dynamic theories of semantics, such as Discourse Representation Theory (DRT; [29]) and File Change Semantics [30]. Under such theories, both referring and non-referring expressions correspond to 'discourse referents' [31], which are not actual referents but elements of the discourse. According to Karttunen's definition (p. 4), "the appearance of an indefinite noun phrase [or any noun phrase, for our purposes] establishes a 'discourse referent' just in case it justifies the occurrence of a coreferential pronoun or a definite noun phrase later in the text." Pronouns can express either bound variables or constants, so a discourse referent may or may not correspond a genuine referent.

A representation containing discourse referents such as DRT's Discourse Representation Structures (DRSs) might therefore seem to be a natural starting point for a system designed to capture the commonalities between referential and bound variable anaphora. Combinations of HPSG and DRT (see [32] and references cited therein) could potentially be used in a practical natural language generation system; indeed, Minimal Recursion Semantics (MRS; [19]) is a form of Underspecified Discourse Representation Theory [33], and MRS-based natural language generation systems exist [6]. However, this strategy does not capture phenomena that reflect changes to the discourse context as the discourse proceeds, because if the input is a fully-formed DRS, then the discourse representation will remain fixed throughout the generation procedure. The dynamic nature of the framework is not utilized under such an approach.

The framework described in the present paper is inspired by some of the ideas underlying DRT, including the notion of the discourse referent. However, rather than using a complete Discourse Representation Structure as an input, a (somewhat more minimal) representation of the discourse is built up as the sentence is generated. Using this, the generation system keeps track of the changes that take place as the sentence is in progress, including the introduction of new discourse referents. It is thus dynamic, in the sense that it updates the discourse as it goes along.

2   Problems with Non-referential 'Discourse Referents'

Strictly speaking, it is not necessary to be able to generate quantificational determiners and bound variable anaphora, if the goal is to produce an English sentence for any given formula of predicate calculus. For an input like this:

(1) $\forall x[\mathbf{isa}(x, \mathbf{Man}) \rightarrow \mathbf{loves}(x, x)]$

one could give a direct translation like this:

(2)   For every x, if x is a man, then x loves x.

This output counts as English as long as explicit variables like 'x' count as English. But it is more desirable to produce the following kind of output:

(3)   Every man loves himself.

If the goal is to produce *concise* English translations of first-order logic formulas, then it is necessary to produce these kinds of non-referring expressions. This leads to special challenges having to do with determiner selection and capturing lifespan limitations.

2.1   Determiner Selection

In simple cases, universally quantified variables as in (4) are introduced with *every,* as in (5), and existentially quantified variables as in (6) are introduced with *some,* as in (7):

(4) $\forall x[\mathbf{loves}(\mathbf{Mary}, x)]$

(5)   Mary loves everything.

(6) $\exists x[\mathbf{loves}(\mathbf{Mary}, x)]$

(7)   Mary loves something.

A first pass at a determiner-selection algorithm would be then: If the variable is universally quantified, use *every*; if it is existentially quantified, use *some*. But this simple algorithm would fail at cases where a universally quantified variable is introduced with an indefinite determiner, as in donkey sentences. The universally quantified variables in the formula in (8) are introduced as indefinites in (9):

(8) $\forall x \forall y [[\text{isa}(x, \textbf{Donkey}) \wedge \text{isa}(x, \textbf{Farmer}) \wedge \textbf{owns}(x, y)] \rightarrow \textbf{beats}(x, y)]$

(9)   If a farmer owns a donkey, then he beats it.

Using *every* instead would not express the same idea:

(10)   If every farmer owns every donkey, then he beats it.

Thus, universally quantified variables are not always introduced with *every*.

One might then refine the algorithm to say that when the variable occurs in the antecedent of a conditional, then it is introduced with an indefinite. But there are cases of this type in which the variable is introduced by *every*:

(11) $\forall x [\text{isa}(x, \textbf{Donkey}) \rightarrow \textbf{loves}(\textbf{Mary}, x)]$

(12)   Mary loves every donkey.

When the *if-then* structure of the logical input formula is lost in the English translation, the variable is introduced with *every*.

Universally quantified variables can also be introduced with determiners other than *every* in the presence of negation. A formula like (13) has two concise renditions, (14) and (15), and neither uses *every*.

(13) $\forall x [\text{isa}(x, \textbf{Donkey}) \rightarrow \neg \textbf{loves}(\textbf{Mary}, x)]$

(14)   Mary doesn't love any donkey(s).

(15)   Mary loves no donkey(s).

When negation is expressed on the verb, the universally quantified variable is expressed with *any*; in the other case, the determiner *no* expresses both negation and universal quantification.

The facts are slightly different when the variable is in subject position. Consider an example in which x is the first argument of **loves**:

(16) $\forall x [\text{isa}(x, \textbf{Donkey}) \rightarrow \neg \textbf{loves}(x, \textbf{Mary})]$

In such a case, only *no* is possible; verbal negation with *any* is not possible:

(17)   No donkeys love Mary.

(18)   *Any donkeys don't love Mary.

Thus, syntactic considerations appear to play a role in determiner selection as well.

Negation does not always cause a universally-quantified variable to be realized with *any* or *no*; when the negation outscopes the universal quantifier, then *every* is used, as in the following example:

(19) $\neg\forall x[\mathbf{isa}(x, \mathbf{Donkey}) \rightarrow \mathbf{loves}(\mathbf{Mary}, x)]$

(20)   Mary doesn't love every donkey.

Thus, whether a universally quantified variable should be realized as *every, some/a, any,* or *no* depends (at least) on whether it is in the protasis of a conditional, the presence and relative scope of negation, and syntactic position.

Existentially bound variables are also sensitive to negation. When the existential quantifier outscopes negation, *some* is used, as usual, whether in subject or object position:

(21) $\exists x[\neg\mathbf{loves}(x, \mathbf{Mary})]$

(22)   Someone doesn't love Mary.

(23) $\exists x[\neg\mathbf{loves}(\mathbf{Mary}, x)]$

(24)   Mary doesn't love someone.

However, when negation outscopes the existential quantifier, *no* and *any* become appropriate, following the pattern observed with universal quantifiers:

(25) $\neg\exists x[\mathbf{loves}(x, \mathbf{Mary})]$

(26)   Noone loves Mary. / *Anyone doesn't love Mary.

(27) $\neg\exists x[\mathbf{loves}(\mathbf{Mary}, x)]$

(28)   Mary doesn't love anyone. / Mary loves noone.

Thus, whereas existential quantifiers give rise to *any* and *no* only when they are in the scope of negation, universal quantifiers give rise to them only when they outscope negation.

Since negation of an existential is equivalent to universal quantification over a negation, one might argue that *any* never really corresponds to a universal quantifier; to get an output with *any* one would convert a formula with a universal quantifier outscoping a negation to an equivalent formula with an existential quantifier outscoped by it. So (29) would be converted to (30) before producing a sentence with *any*:

(29) $\forall x[\neg\mathbf{loves}(\mathbf{Mary}, x)]$

(30) $\neg\exists x[\mathbf{loves}(\mathbf{Mary}, x)]$

This would allow one maintain the generalization that *any* corresponds to an existential quantifier, as is sometimes assumed (e.g. [34]). We assume, however, that *any* can correspond to universally quantified variables as well as existentially quantified ones. One argument for this view is that when multiple universal quantifiers outscope negation as in (31), both may correspond to an *any* phrase as in (32).

(31) $\forall x[\forall y[\neg \mathbf{loves}(x, y)]]$

(32)   It is not the case that anyone loves anyone (else).

Only one of the quantifiers in (31) can be "swapped" with a negation in order to produce an existential quantifier, yet there are two *any* phrases. At least one of them must correspond to a universal quantifier. Since we must allow universally quantified variables to be realized with *any* in such cases, we might as well allow it in general.

The facts listed above show that the choice of determiner to realize a variable is a non-trivial function of the logical operators present in the input formula and their relative scope. The determiner selection algorithm given in §3.3 captures these patterns.

2.2   Lifespan Limitations

Like referring expressions, non-referring expressions can be anaphoric, when there is an accessible antecedent in the discourse. The second realization of a given discourse referent should, barring potential ambiguity, take the form of a pronoun, whether the discourse referent corresponds to a constant or a variable:

(33) $\mathbf{loves}(\mathbf{Mary}, \mathbf{Mary})$

(34)   Mary loves herself.

(35) $\forall x \neg \mathbf{loves}(x, x)$

(36)   No woman loves herself.

However, unlike discourse referents corresponding to constants, discourse referents corresponding to variables (realized with non-referring expressions) exist in the discourse only temporarily, and thus have a limited 'lifespan' [31]. This is exemplified in (37), from Heim [30], and (38).

(37)   Everyone found a cat and kept it. #Then it ran away.

(38)   No$_i$ self-respecting lady will give you her$_i$ phone number. #I know her$_i$.

On the reading of (37) on which the universal quantifier outscopes the existential quantifier, the *it* in the second sentence cannot corefer with the *it* in the first sentence. Similarly, the discourse referent introduced by *no self-respecting lady* in (38) in is a *short-term referent* [31], whose lifespan ends with the end of the first sentence. From a parsing perspective, the challenge is to assign the right interpretations to pronouns. From a generation perspective, the challenge is to avoid generating pronouns that are anaphoric to discourse referents that are no longer active.

This is a challenge that does not arise with referential expressions. Contrast (38) with the following example:

(39) $Jane_i$ will give you her_i phone number. I know her_i.

In (39), the discourse referent survives into the second sentence, because proper names are referential, and the lifespan of a discourse referent introduced with a referential expression is in principle unlimited. If referring and non-referring expressions are not distinguished, then this difference between them cannot be captured. Thus, although bound variable and referential anaphora should be treated in a unified fashion, the treatment should not be so unified as to blur the distinction between them.

The lifespan of a short-term discourse referent does not always correspond to the enclosing tensed sentence. Whereas a discourse referent introduced by *every* is limited to the protasis of a conditional, a discourse referent introduced with an indefinite in the protasis may extend to the apodosis [29, 30]:

(40)   *If $every_i$ farmer owns $every_j$ donkey, then he_i beats it_j.

(41)   If a_i farmer owns a_j donkey, then he_i beats it_j.

The lifespan of the indefinite is not indefinite, however:

(42)   If a_i farmer owns a_j donkey, then he_i beats it_j. #I know him_i.

The lifespan of an indefinite introduced in the protasis of a conditional ends with the apodosis.

3   The Cyc NLG System

We now present a procedure for translating predicate calculus into English, which treats referring and non-referring expressions in a unified framework, and captures all of the facts described in the previous section, regarding both determiner selection and lifespan limitations. The system we present is the natural language generation (NLG) system for Cyc [35], a large-scale commonsense knowledge base and reasoning engine. Cyc is based on CycL, a logic that subsumes first order logic [36],[1] and the system we describe translates from CycL to English. Here, we concentrate on the first order portion of CycL, making limited use of Cyc-specific ontological distinctions, in order to maximize the applicability of our model. The input is described in detail in §3.1.

---

[1] The majority of the assertions in the Cyc Knowledge Base are statement of first-order logic; the majority of the remaining assertions can be transformed into statements first-order logic [36].

Our procedure uses two forms of dynamically updated context: the *discourse context,* which lists the discourse referents that have been mentioned, and the *operator context,* which stores information that is stripped away from the input formula. The discourse context is discussed in §3.2; the operator context will be discussed in §3.3.

3.1   Input: First-Order Predicate Calculus Part of CycL

The input to the Cyc NLG system is a formula of CycL, which is a higher-order logic built on first-order predicate calculus [37]. CycL has a number of fancy features, such as quoting, meta-assertions, lambda expressions (forming terms through variable abstraction), and kappa expressions (forming predicates through variable abstraction), most of which will not concern us here. In this paper, we will focus on inputs from the first-order portion of CycL. The set of expressions within this first-order portion contains variables (e.g. x, y, z), and atomic constants denoting individuals (e.g. **Mary**), collections (e.g. **Donkey**) predicates, (e.g. **loves**), and functions (e.g. **MotherOf**). The predicates and functions can in principle be of any arity. In CycL, arguments of predicates and functions can in principle be any other CycL expression, so CycL is higher order, but we can restrict our attention to first-order predicates and functions. The set of non-atomic expressions contains:[2]

- Non-Atomic Terms: if γ is a function and $\xi_1...\xi_n$ a sequence of arguments matching γ's argument constraints, then $\gamma(\xi_1...\xi_n)$ is a term.
- Atomic Sentences: if π is a predicate and $\xi_1...\xi_n$ a sequence of arguments matching π's argument constraints, then $\pi(\xi_1...\xi_n)$ is a sentence.
- Negations: if ϕ is a sentence then $\neg\phi$ is a sentence.
- Conjunctions: if ϕ is a sentence and ψ is a sentence than $\phi \wedge \psi$ is a sentence.
- Disjunctions: if ϕ is a sentence and ψ is a sentence then $\phi \vee \psi$ is a sentence.
- Implications: if ϕ is a sentence and ψ is a sentence then $\phi \rightarrow \psi$ is a sentence.
- Universals: if ϕ is a sentence and ξ is a variable then $\forall\xi\phi$ is a sentence.
- Existentials: if ϕ is a sentence and ξ is a variable then $\exists\xi\phi$ is a sentence.

This logic can be given a standard model-theoretic semantics for predicate calculus.

Other than the distinction between individuals and collections, this logic is perfectly standard. In Cyc, collections are often used in place of one-place predicates, so rather than **farmer**(x), we will have **isa**$(x, \textbf{Farmer})$, where **Farmer** represents the collection of all farmers, and **isa** is a predicate relating an individual to a collection, which holds if the individual is an instance of the collection.

---

[2] Regarding notation: We use the standard way of using parentheses in logic, rather than using the Lisp-style notation that is normally used for CycL. Constants are indicated with bold face (whereas in CycL they are prefixed with '#$') and variables with italics (rather than being prefixed with '?' as in CycL). Following CycL conventions, however, we use initial lowercase letters for predicates and initial uppercase letters for individuals and collections. Variables of all types are lowercase.

(We use the term *instance* for collections, rather than *member,* which we reserve for sets; the idea is that collections represent concepts, while sets are merely extensionally defined.) To give a more standard logic, one could replace all collections with single-arity predicates, so the distinction between individuals and collections is not completely crucial. However, it does happen to play a role in the grammar, so the grammar would have to be adapted if that distinction were eliminated.

The Cyc NLG system has full access to the Cyc Knowledge Base (KB), which contains an English lexicon. The lexicon includes a set of *generation templates,* which describe an English sentence or phrase corresponding to a function or predicate, with open slots for the arguments. For example, the predicate **likes** is associated with a template describing a transitive sentence in which the subject is the realization of the first argument, the verb is a form of *like* that agrees with the subject, and the object is a realization of the second argument. These templates thus accomplish argument linking. (Rather than stipulating the syntactic realization of arguments on a case-by-case basis, one could derive these templates from more general principles, so the present system is not crucially tied to a stipulative linking theory; we just take linking as given.) Aside from what is specified in generation templates, the grammatical structure of a generated utterance is determined procedurally. Therefore, the system that we describe here is not only a natural language generation system, but also a grammar.

Because it was developed for the purpose of generation rather than parsing, the theoretical constructs that this system uses are different from the ones that have been developed under the parsing perspective. In particular, there are two forms of context: *discourse context* and *operator context.* These are described in the following two subsections.

3.2   Discourse Context

Definition. A discourse context D is a set of *discourse referents* [31]. Like Heim's 'file cards' [30, 38] and the elements of DRT universes [29], these discourse referents need not correspond to any particular entity in the situation described by the sentence. Each discourse referent r is associated with a logical expression α, which can be either a variable or a closed term, composed entirely of constants (either atomic, e.g. **Mary**, or non-atomic, e.g. **Mother(Mary)** 'Mary's mother').

Insofar as the logical expression associated with these discourse referents can be either a variable or a closed term, they are similar to Muskens's 'registers' [39], and unlike the elements of DRT universes, which correspond only to variables. As Muskens points out [39], allowing proper names to be translated with constants eliminates the need for DRT's 'external anchoring' device. From a generation perspective, this design choice is quite natural; it would be a waste to convert constants in the input formula to variables before listing them among the discourse referents.

Discourse referents are also associated with *index features*: person, number and gender [40, 41]. Index features are computed on the basis of morphosyntactic information if it is available, or encyclopedic knowledge in the Cyc Knowledge Base (KB) otherwise. For example, the individual **Mary** is asserted to be a human female in the KB, so corresponding discourse referents will have a feminine gender feature. These index features determine the form that pronouns take.

Side effects. We recursively define a generation function $G(\alpha)$, where α can be any expression of the logic, which depends not only on α, but also on a global discourse D and a global operator context O. We subdivide the definition of $G(\alpha)$ into cases based on the logical expression type of α. A fundamental case is when α is an *atomic formula*, as in (43).

(43)   **loves(Mary, Mary)**

The most appropriate method for atomic formulas uses the generation template for the predicate – a mapping from a logical predicate to a partial specification of a sentence in natural language. The generation template for **loves** specifies a template from which a syntax tree is built. A syntax tree is like an HPSG sign [40], with "phonological", semantic, and syntactic information, including daughters for phrases. (Since we are computing textual output, the value of the so-called PHON feature is an orthographic string.)

A simplified rendition of the phrase that is ultimately generated for (43) is in Fig. 1. The tree is traversed left to right, depth first, and may be expanded during the traversal. Each time a node is visited, the value of the PHON feature is computed. We say that a subexpression of the logical formula has been *realized* if the PHON value of the phrase it corresponds to has been set. The PHON value of the mother is the concatenation of the PHON values of the daughters. The value of G is the PHON value of the top-level phrase.

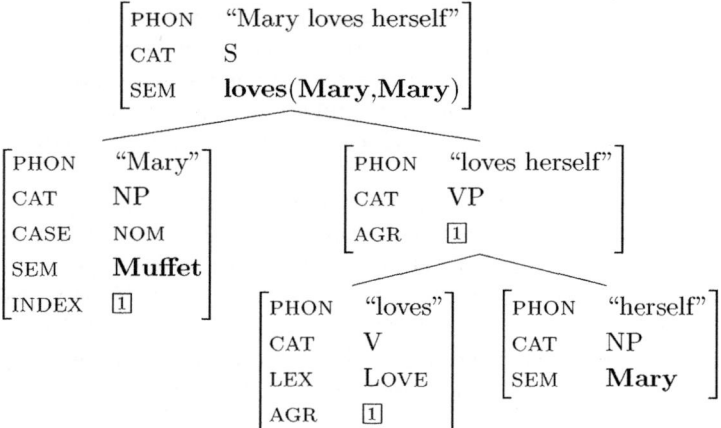

Fig. 1. Phrase generated for the input '**loves(Mary, Mary)**' (simplified)

What makes the semantics dynamic is that the computation of G can have *side effects* in the form of updates to D and O. The discourse context D is updated when a discourse referent is realized, as part of the algorithm for generating a constant or a variable. If α is an individual-denoting constant, and is not listed among the discourse referents, then $G(\alpha)$ returns a proper name, *and the discourse context is updated.* For example, the expression **Mary** in (43) is added to D after its first instance in the input formula is realized.

If α is listed in D, then an anaphoric expression is used. The algorithm for generating an anaphor is: If a pronoun would be ambiguous, then use a definite description or name; otherwise, use a pronoun. There is a great deal of sophistication and subtlety this treatment could acquire [42–48], but this is not our focus here. A (reflexive) pronoun is appropriate for the second instance of **Mary** in (43), so the formula is rendered, "Mary loves herself."

Like other discourse referents, variables are added to D after being mentioned, so as to be made eligible for subsequent rendering with anaphora. Thus, after *everything* is generated in (44), the variable x is added to the set of discourse referents, so it can be realized with a pronoun on its second mention:

(44)   Everything likes itself.
$\forall x[\textbf{likes}(x, x)]$

Lifespan limitations. Constants added to D during the execution of an instance of G may remain in D. However, it is necessary to remove variables from D when their "lifespan" is exhausted. Thus, if the argument to G is a quantificational sentence binding a variable ξ, then the following line must be executed before G exits:

$$D \leftarrow D - R(\xi)$$

where $R(\xi)$ stands for the discourse referent associated with ξ. Suppose the input formula were:

(45) $\forall x \neg[\textbf{loves}(\textbf{Doug}, x)] \wedge \forall x[\textbf{loves}(\textbf{Mary}, x)]$

This would be rendered correctly as: "Doug loves nothing and Mary loves everything" rather than "Doug loves nothing and Mary loves it."

Removal of variables from D is the key to accounting for the limited lifespan of the discourse referent introduced by *no self-respecting woman* in (38) ("No self-respecting woman will give you her phone number. #I know her.") At the end of the sentence, the variable bound by the universal quantifier is removed from the discourse context, making it impossible for there to be subsequent anaphoric references to it.[3] If a pronoun is generated in the following sentence, it will have to be associated with some other discourse referent.

---

[3] Some of Karttunen's observations in [31] suggest that discourse referents corresponding to existentials with maximally wide scope should not be removed from the discourse context. For example, consider the sentence "I have a proof of this theorem but it won't fit in this margin." One is tempted to analyze the first sentence using an existential quantifier; but then how can there be subsequent anaphoric reference to it? One possibility is to introduce a Skolem constant for wide-scope existentials.

3.3   Operator Context

$G(\alpha)$ depends not only on a discourse context D, but also on an operator context O. As we describe futher below, logical operators and negations are stripped away from the input formula, leaving a "clausal skeleton," and the information stripped away is stored in the operator context.

Definition. We define an operator context O as a tuple $\langle V, S, n \rangle$ where V is a set of variable type entries, S is a stack of logical symbols, and n is an integer representing the number of negations remaining to be expressed.

A variable type entry $v \in V$ is a tuple $\langle \alpha, \theta, \tau \rangle$, where α is a variable over individuals, θ is a quantifier symbol, and τ is a type. Types are Cyc *collections*, such as **Donkey**, the collection of all donkeys. As mentioned above, Cyc collections are like sets except that they are meant to represent concepts and have instances, rather than members [49].[4] Associating a type with a variable is conceptually similar to identifying the type as the 'restriction' of the quantifier, even though formally, quantifiers in first order logic do not specify restrictions, unlike generalized quantifiers.

The stack of symbols S contains the variables and logical operators with scope over the element of the formula that is currently being realized. We represent stacks as tuples $\langle \alpha_1 \dots \alpha_n \rangle$, where α_n is at the top of the stack. The wider the scope of the operator, the deeper on the stack it is. For the formula $\forall x \neg [\mathbf{loves}(\mathbf{Mary}, x)]$, at the point when **Mary** is being generated, $S = \langle x, \neg \rangle$. We use the variable x rather than the quantifier $\forall$ to indicate the scope of the associated universal quantifier, because the variable uniquely identifies the quantifier in question, while there may be many universal quantifiers in a formula. The quantifier associated with the variable is computable from the variable's type entry in V.

Finally, n is used to keep track of unexpressed negations. As we will see in the next section, negations can be removed from formulas in the course of constructing the clausal skeleton, and this counter helps to ensure that every negation is expressed exactly once. (Note that the value of n is not derivable from S, as any negation on S may be either expressed or unexpressed.)

Clausal skeletons. In the generation of some formulas, parts of the input formula are stripped away, leaving a *clausal skeleton*. For example, the clausal skeleton of (46) is (47).

(46) $\forall x \forall y [[\mathbf{isa}(x, \mathbf{Man}) \wedge \mathbf{isa}(y, \mathbf{Donkey}) \wedge \mathbf{owns}(x, y)] \rightarrow \mathbf{loves}(x, y)]$

(47) $\mathbf{owns}(x, y) \rightarrow \mathbf{loves}(x, y)$

The two **isa** statements in the antecedent of the conditional in (46) are *variable typing clauses*. The binary predicate **isa** relates an individual to a collection,

---

[4] Collections can also be complex; CycL contains collection-forming functions with which concepts like "the collection of farmers who own a donkey" can be expressed.

and signifies that the individual is an instance of the collection. Variable typing clauses are removed, along with the universal quantifiers (as mentioned in [50]).

For any input formula of the form $\forall \xi \alpha$, where ξ is a variable and α is a sentence, a simplified version of the algorithm for constructing clausal skeletons σ and new variable type entries v is as follows (where **Thing** is the most general collection, and $\sim$ stands for "is of the form" or "matches"):

(48)   − If $\alpha \sim [\psi \rightarrow \phi]$:
          • If $\psi \sim [\mathbf{isa}(\xi, \gamma)]$:
            $\sigma = \phi$
            $v = \langle \xi, \forall, \gamma \rangle$
          • Else if $\psi \sim [\delta_1 \wedge ... \wedge \delta_n]$ where $\delta_i \sim [\mathbf{isa}(\xi, \gamma)]$:
            $\sigma = [\delta_1 \wedge ... \wedge \delta_{i-1} \wedge \delta_{i+1} \wedge ... \wedge \delta_n] \rightarrow \phi$
            $v = \langle \xi, \forall, \gamma \rangle$
          • Else:
            $\sigma = \alpha$
            $v = \langle \xi, \forall, \mathbf{Thing} \rangle$
    − Else:
       $\sigma = \alpha$
       $v = \langle \xi, \forall, \mathbf{Thing} \rangle$

For a formula of the form $\exists \xi \alpha$, there are two cases:

(49)   − If $\alpha \sim [\delta_1 \wedge ... \wedge \delta_n]$ where $\delta_i \sim [\mathbf{isa}(\xi, \gamma)]$:
       $\sigma = [\delta_1 \wedge ... \wedge \delta_{i-1} \wedge \delta_{i+1} \wedge ... \wedge \delta_n]$
       $v = \langle \xi, \exists, \gamma \rangle$
    − Else:
       $\sigma = \alpha$
       $v = \langle \xi, \exists, \mathbf{Thing} \rangle$

The variable type entry v is added to the set V of variable type entries in the operator context O. After the operator context is updated, $G(\sigma)$ is computed; in other words, the clausal skeleton is realized. Then, importantly, the operator context is restored to its previous state. The information stored in the operator context surfaces when the variable is expressed, as described in §3.3.

The clausal skeleton is isomorphic, clause for clause, to the resulting English sentence. Thus the realization of (46) has two clauses, just as its clausal skeleton (47) has:

(50)   If a farmer owns a donkey, then he loves it.

When the antecedent of the clause consists entirely of a variable typing clause, all that remains in the clausal skeleton is the consequent. An input formula such as (51) will be realized as in (52), a single clause.

(51) $\mathbf{isa}(x, \mathbf{Man}) \rightarrow \mathbf{loves}(x, \mathbf{Mary})$

(52)   Every man loves Mary.

Thus, all of the content that is expressed below the clause level (in quantified noun phrases) comes from the operator context, and every clause in the English translation is part of the clausal skeleton.[5]

Negation stripping. When α is of the form $\neg\phi$, the clausal skeleton of α is ϕ. No variable type entries are produced in this case, of course, but the counter representing the number of unexpressed negations, n, is incremented. "Stripping" the negation in this way makes it possible for negations to be expressed sub-clausally, using a member of the *no*-series (*nobody, nothing,* etc.). We use clausal negation as a "back-up strategy" when expressing negation sub-clausally fails; if $n > 0$ after ϕ is realized, either the verb is negated or negation is expressed periphrastically with, for example, "It is false that...".

Updating the operator stack. The main purpose of the operator stack is to determine when NPIs are licensed. The generation procedure $G(\alpha)$ updates the operator stack whenever α is a formula whose operator is in the set $\{\forall, \exists, \neg, \rightarrow\}$. If α is a quantificational formula such as $\forall x\phi$, then the variable (x) is pushed onto the operator stack, and popped off of it at the completion of $G(\alpha)$. If α is a negative formula, then a $\neg$ symbol is pushed onto the operator stack and again, popped off of it after $G(\alpha)$ is computed. If α is an implication, then the symbol $\rightarrow$ is pushed onto the stack and the stack is popped at the completion of the generation of the *antecedent,* because implications only license NPIs in the antecedent.

Determiner selection algorithm. The definition of $G(\alpha)$ where α is a variable (*variable realization*) involves the operator context as well as the discourse context. As mentioned in §3.2, variables are realized as pronouns when they are listed in D and a pronoun would not be ambiguous. If a pronoun is not appropriate for realizing a variable, a lexical noun phrase containing a determiner and a noun is used. The noun is the realization of the variable's type, the τ such that $\langle \alpha, \theta, \tau \rangle \in V$. τ represents a collection, e.g. **Man** (the collection of all men), and can be realized as, for example, *man*. There are several types of determiner that may accompany the noun: DEFINITE (*the*), NO (*no*), UNIVERSAL (*every*), INDEFINITE (*a, some*), NPI (*any, a* or *some*).[6]

The first step in the algorithm for choosing a determiner type is to compute whether or not a variable could be expressed as an NPI, i.e., with *any.* Both universally and existentially bound variables can be realized with *any,* but the

---

[5]   If the variable typing clauses were not removed, the variables would be registered with type **Thing** and the output would contain more clauses: "If something is a farmer and some other thing is a donkey then the thing loves the other thing." If the variables were not registered in the operator context at all, then the universal quantifier would not be stripped from the formula, and the output would be as follows, with explicit variables: "For every x, for every y, if x is a farmer and y is a donkey and x owns y, then x loves y."

[6]   Another determiner type is WH (*what, which*). Wh- determiners are used for unbound variables in formulas to be generated with interrogative force. We ignore questions here for the sake of simplicity.

two quantifier types differ with respect to the scope that they must have relative to an NPI licenser. In order to qualify for being realized with *any*, an instance of a variable must be bound by either a universally quantified variable outscoping an NPI licenser, or an existentially bound quantifier outscoped by one. For example, (29), repeated below is rendered as in (54).

(53) $\forall x[\neg \mathbf{loves}(\mathbf{Mary}, x)]$

(54)   Mary doesn't love anything.

In contrast, when the universal quantifier is inside the scope of negation, the output should be *Mary doesn't love everything*.

The algorithm for determining whether or not NPI *any* is licensed is as follows: First, look up the quantifier of the variable in question; then if the quantifier is $\forall$, the question is whether the variable is deeper on the stack than an NPI licenser; if the quantifier is $\exists$, the question is whether the NPI licenser is deeper than it. Call the NPI licenser π; if the variable has no NPI licenser, then π is null.

Given π, a variable type entry $\langle \alpha, \theta, \tau \rangle$ for variable ξ, and an unexpressed negation counter n, the determiner is chosen according to the following algorithm:

(55)   – If $R(\chi) \in D$ (i.e., a previous instance of the variable has been realized), return DEFINITE.
　　　– If π is non-null:
　　　　• If $\pi = \neg$ and $n > 0$, then return NO and decrement n by one.
　　　　• Otherwise, return NPI.[7]
　　　– If $\theta = \forall$, return UNIVERSAL.
　　　– Otherwise, return INDEFINITE.

Note at this point that the set of variable type entries V is not redundant with the discourse context D, despite the fact that they may simultaneously contain the same variable as an entry. The discourse context is used for referents that have already been realized, but variable type entries are used in the formulation of the first mention of the variable.

Now consider example (46) again, repeated here as (56).

(56) $\forall x \forall y [[\mathbf{isa}(x, \mathbf{Man}) \wedge \mathbf{isa}(y, \mathbf{Donkey}) \wedge \mathbf{owns}(x, y)] \rightarrow \mathbf{loves}(x, y)]$

Right before x is generated for the first time, $\langle x, \mathbf{Man}, \forall \rangle \in V$ (i.e. x is universally quantified, and has type $\mathbf{Man}$), $S = \langle x, \rightarrow \rangle$ (i.e., the current expression is inside the antecedent of a conditional), $n = 0$ (there are no unexpressed negations), and x is not in D (so x has not previously been realized). Therefore $G(x)$ will realized with an NPI-type determiner, as either *any man* or with an indefinite (*a* or *some*). After x is realized, it will be in D, so it will be realized as a

---

[7] In the generation of variables with NPI-type determiners, we make the stylistic choice to use *any* when there is only one mention of the variable, and an indefinite otherwise.

masculine pronoun or as *the man*. If instead, S were merely $\langle x \rangle$ (so the current expression is not in the scope of an NPI licenser), then π would be null, and the appropriate determiner type would be EVERY.

Another case where NPI-type determiners are chosen is in the scope of negation, when there is no negation to be expressed, i.e., when $\pi = \neg$ and $n \not> 0$. Such a situation arises when negation is expressed on the verb, as in *Mary doesn't love anything*. When $n > 0$ (there are unexpressed negations) and $\pi = \neg$ (the NPI licenser for the variable is negation), the variable can be used to express negation, as in *Mary loves nothing*.

We are now in a position return to the contrast between indefinite determiners and *every* with respect to the lifespan of the discourse referents they introduce. Recall the fact that whereas discourse referents introduced by indefinites in the protasis of a conditional extend to the apodosis, those introduced by *every* are limited to the protasis:

(57)   If a$_i$ farmer owns a$_j$/*every$_j$ donkey, then he$_i$ loves it$_j$.

We have just seen how the acceptable variant of (57), with an indefinite determiner in the antecedent, is generated for an input like (46). The only way for *every* to be generated in the antecedent of a conditional is for the universally quantified variable to be outscoped by $\rightarrow$, as in the following example:

(58) $\forall x[\mathbf{isa}(x, \mathbf{Farmer}) \wedge \forall y[\mathbf{isa}(x, \mathbf{Donkey}) \rightarrow \mathbf{owns}(x, y)]] \rightarrow \mathbf{loves}(x, \mathbf{Mary})$

The formula in (58) would be rendered as follows:

(59)   If a farmer owns every donkey, then he loves Mary.

When it is time to realize the variable y for the first time, $S = \langle x, \rightarrow, y \rangle$, so the computation of π for y will yield a null value, since y is universally quantified and universally quantified variables must be outside the scope of an NPI licenser. The determiner selection algorithm will therefore correctly choose *every*. But in this case, the scope of the quantifier will have ended by the time the consequent of the conditional is reached. Thus, discourse referents introduced by *every* in the protasis of a conditional will never extend to the apodosis.

4   Summary and Outlook

We have presented an algorithm for translating predicate logic to English that uses dynamically updated information states. It deals with referring and non-referring expressions in a unified framework, capturing the fact that both require an accessible antecedent. This is formalized using the discourse context, where discourse referents are placed after they are introduced. The system also captures special features of non-referring expressions, which correspond to logical variables. Discourse referents associated with variables have limited lifespans and are introduced with quantificational determiners, whose use is governed by a complex set of factors, modelled with the operator context.

Just as general frameworks for generating semantic representations from English sentences (e.g. [51], [29]) are semantic theories, the framework presented here is in a sense a theory of semantics (or what we might call 'inverse semantics'). It differs in its use of theoretical primitives from semantic theories that were developed from a parsing perspective, as the theoretical constructs that we found useful in generation are slightly different from those that were found to be useful in parsing. Given such differences, the generation perspective could potentially shed light on the theory of semantics more generally, and provide more elegant or even more empirically adequate accounts of certain phenomena.

One area where the generation perspective may shed light is in NPI licensing. NPI *any* is not always licensed in the semantic scope of negation:

(60)   *Anyone doesn't love me.

On our analysis, this configuration is blocked by *Noone loves me*. Thus, it is not necessary to associate syntactic constraints with NPI *any* to rule (60) out. In general, implicit in the notion of a 'determiner selection algorithm' is the idea of blocking, an idea that is natural from the generation perspective, and we believe it may be worthwhile to pursue this view of quantificational determiners further.

Secondly, the notion of *accessibility* between pronouns and their antecedents receives quite a different treatment here from the one in Discourse Represenation Theory. Whereas the accessibility relationship is characterized in DRT as a structural relationship within Discourse Representation Structures, accessibility is formalized here as existence in the discourse context, a potentially transient state that ends for variables when the logical expression corresponding to the quantifier that binds them has been realized. In DRT, proper names always "float to the top" of a DRS, so they are always available; this corresponds to the fact that constants are never removed from the discourse context. We feel that the present view on accessibility has a certain intuitive appeal, and it would be worthwhile to compare the empirical predictions of the two approaches to see if they differ.

Natural language generation also puts presupposition in a new light. Definite descriptions and pronouns, for example, are usually analyzed as containing uniqueness presuppositions. The association of these items with uniqueness is encoded in the present system via conditions on the choice of referring expression type. Presuppositions in general need not be represented declaratively, but can be implicitly encoded in a procedural generation algorithm. This view on presupposition would capture the fact that pronouns are quite easy to process, and would therefore seem to carry a very simple message, contrary to what one would expect if they came associated with complex presuppositional content.

References

1. Horacek, H.: An algorithm for generating referential descriptions with flexible interfaces. In: Proceedings of the 35th Annual Meeting of the Association for Computational Linguistics, pp. 206–213 (1988)
2. Dale, R.: Cooking up referring expressions. In: Proceedings of the 27th Annual Meeting of the Association for Computational Linguistics (1989)

214 E. Coppock and D. Baxter

3. Reiter, E.: The computational complexity of avoiding false implicatures. In: Proceedings of the 28th Annual Meeting of the Association for Computational Linguistics (1990)
4. Reiter, E., Dale, R.: A fast algorithm for the generation of referring expressions. In: Proceedings of the 14th International Conference on Computational Linguistics, Nantes, pp. 232–238 (1992)
5. Dale, R., Reiter, E.: Computational interpretations of the Gricean maxims in the generation of referring expressions. Cognitive Science 19, 233–263 (1994)
6. Copestake, A., Flickinger, D., Malouf, R., Riehemann, S., Sag, I.: Translation using minimal recursion semantics. In: Proceedings of the Sixth International Conference on Theoretical and Methodological Issues in Machine Translation, Leuven, Belgium (1995)
7. Shaw, J., McKeown, K.: Generating referring quantified expressions. In: Proceedings of the first international conference on natural language generation, Mitzpe Ramon, Israel, pp. 100–107 (2000)
8. Krahmer, E., Van Erk, S., Verleg, A.: Graph-based generation of referring expressions. Computational Linguistics (2003)
9. Van Deemter, K.: Generating referring expressions: Boolean extensions of the incremental algorithm. Computational Linguistics 28, 37–52 (2002)
10. Siddharthan, A., Copestake, A.: Generating anaphora for simplifying text. In: Proceedings of the 4th Discourse Anaphora and Anaphor Resolution Colloquium, pp. 199–204 (2002)
11. Siddharthan, A., Copestake, A.: Generating referring expressions in open domains. In: Proceedings of the 42nd Annual Meeting of the Association for Computational Linguistics, Barcelona, Spain, pp. 408–415 (2004)
12. Varges, S., Van Deemter, K.: Generating referring expressions containing quantifiers. In: Proceedings of the 6th International Workshop on Computational Semantics, pp. 1–13 (2005)
13. Kelleher, J.D., Kruijff, G.J.M.: Incremental generation of spatial referring expressions in situated dialog. In: Proceedings of COLING/ACL 2006 (2006)
14. Paraboni, I., Van Deemter, K., Masthoff, J.: Generating referring expressions: Making referents easy to identify. Computational Linguistics 33, 229–254 (2007)
15. Van Deemter, K., Krahmer, E.: Graphs and booleans: on the generation of referring expressions. In: Bunt, H., Muskins, R. (eds.) Computing Meaning, vol. 3, pp. 397–422. Springer, Dordrecht (2008)
16. Areces, C., Koller, A., Striegnitz, K.: Referring expressions as formulas of description logic. In: White, M., Nakatsu, C., McDonald, D. (eds.) Proceedings of the Fifth International Natural Language Generation Conference, Salt Fork, Ohio, pp. 42–49. Association for Computational Linguistics (2008)
17. Wilcock, G., Matsumoto, Y.: Head-driven generation with HPSG. In: Proceedings of COLING-ACL 1998: Workshop on Usage of WordNet in Natural Language Processing Systems, pp. 1393–1397 (1998)
18. Carroll, J., Flickinger, D., Copestake, A., Poznanski, V.: An efficient chart generator for (semi-)lexicalist grammars. In: Proceedings of the 7th European Workshop on Natural Language Generation, Toulouse, France (1990)
19. Copestake, A., Flickinger, D., Pollard, C., Sag, I.A.: Minimal recursion semantics: An introduction. Research on Language and Computation 3, 281–332 (2005)
20. Carroll, J., Oepen, S.: High efficiency realization for a wide-coverage unification grammar. In: Dale, R., Wong, K.F. (eds.) Proceedings of the Second International Joint Conference on Natural Language Processing (IJNLP 2005), Springer, Heidelberg (2005)

21. Wedekind, J., Kaplan, R.M.: Ambiguity-preserving generation with LFG- and PATR-style grammars. Computational Linguistics 22, 555–558 (1996)
22. Wedekind, J.: Semantic-driven generation with LFG- and PATR-style grammars. Computational Linguistics 25, 277–281 (1999)
23. Kaplan, R.M., Wedekind, J.: LFG generation produces context-free languages. In: Proceedings of the 18th Conference on Computational Linguistics, Saarbrücken, Germany, pp. 425–431 (2000)
24. Cahill, A., van Genabith, J.: Robust pcfg-based generation using automatically acquired lfg approximations. In: Proceedings of the 21st International Conference on Computational Linguistics and 44th Annual Meeting of the ACL, Sydney, Australia, pp. 1033–1040. Association for Computational Linguistics (2006)
25. Calder, J., Reape, M., Zeevat, H.: An algorithm for generation in unification categorial grammar. In: Proceedings of the 4th Conference of the European Chapter of the Association for Computational Linguistics, Manchester, UK, pp. 233–240 (1989)
26. Phillips, J.D.: Generation of text from logical formulae. Machine Translation 8, 209–235 (1993)
27. White, M.: Reining in CCG chart realization. In: Belz, A., Evans, R., Piwek, P. (eds.) INLG 2004. LNCS (LNAI), vol. 3123, pp. 182–191. Springer, Heidelberg (2004)
28. Pollard, C., Yoo, E.J.: A unified theory of scope for quantifiers and *wh-* phrases. Journal of Linguistics 34(2), 415–445 (1998)
29. Kamp, H., Reyle, U.: From Discourse to Logic. Kluwer Academic Publishers, Dordrecht (1993)
30. Heim, I.: The Semantics of Definite and Indefinite Noun Phrases. PhD thesis, MIT (1982)
31. Karttunen, L.: Discourse referents. In: McCawley, J.D. (ed.) Syntax and Semantics 7: Notes from the Linguistic Underground, pp. 363–385. Academic Press, New York (1976)
32. Sailer, M.: Npi licensing, intervention and discourse representation structures in hpsg. In: Müller, S. (ed.) Proceedings of the HPSG 2007 Conference. CSLI Publications, Stanford (2007)
33. Reyle, U.: Dealing with ambiguities by underspecification: Construction, representation, and deduction. Journal of Semantics 10(2), 123–179 (1993)
34. De Swart, H.: Licensing of negative polarity items under inverse scope. Lingua 105, 175–200 (1998)
35. Lenat, D.: Cyc: A large-scale investment in knowledge infrastructure. Communications of the ACM 38 (1995)
36. Ramachandran, D., Reagan, P., Goolsbey, K.: First-orderized ResearchCyc: Expressivity and efficiency in a common-sense ontology. In: Papers from the AAAI Workshop on Contexts and Ontologies: Theory, Practice and Applications, Pittsburg, PA (2005)
37. Matuszek, C., Cabral, J., Witbrock, M., DeOliveira, J.: An introduction to the syntax and content of Cyc. In: Proceedings of the 2006 AAAI Spring Symposium on Formalizing and Compiling Background Knowledge and Its Applications to Knowledge Representation and Question Answering, Stanford, CA (2006)
38. Heim, I.: File change semantics and the familiarity theory of definiteness. In: Baurle, R., Schwarze, C., Von Stechow, A. (eds.) Meaning, Use, and the Interpretation of Language, pp. 164–189. Walter de Gruyter, Berlin (1983)
39. Muskens, R.: Combining Montague semantics and discourse representation. Linguistics and Philosophy 19, 143–186 (1996)

40. Pollard, C., Sag, I.A.: Head-Driven Phrase Structure Grammar. University of Chicago Press, Chicago (1994)
41. Wechsler, S., Zlatić, L.: The Many Faces of Agreement. Center for the Study of Language and Information, Stanford (2003)
42. Chafe, W.L.: Givenness, contrastiveness, definiteness, subjects, topics and point of view. In: Li, C.N. (ed.) Subject and topic, pp. 25–55. Academic Press, New York (1976)
43. Ariel, M.: Accessing NP antecedents. Routledge, London (1990)
44. Gundel, J.K., Hedberg, N., Zacharski, R.: Cognitive status and the form of referring expressions in discourse. Language 69, 274–307 (1993)
45. Brennan, S.: Centering attention in discourse. Language and Cognitive Processes 10, 137–167 (1995)
46. Grosz, B.J., Joshi, A.K., Weinstein, S.: Providing a unified account of definite noun phrases in discourse. In: Proceedings of the 21st Annual Meeting of the Association for Computational Linguistics, Cambridge, MA, pp. 44–49 (1983)
47. Grosz, B.J., Joshi, A.K., Weinstein, S.: Centering: A framework for modeling the local coherence of discourse. Computational Linguistics 21, 203–226 (1995)
48. Beaver, D.I.: The optimization of discourse anaphora. Linguistics and Philosophy 27, 3–56 (2004)
49. Lenat, D.B., Guha, R.V.: Building Large Knowledge-Based Systems. Addison-Wesley, Reading (1990)
50. Baxter, D., Shepard, B., Siegel, N., Gottesman, B., Schneider, D.: Interactive natural language explanations of cyc inferences. In: Proceedings of AAAI 2005: International Symposium on Explanation-aware Computing, Washington, D.C. (2005)
51. Heim, I., Kratzer, A.: Semantics in Generative Grammar. Blackwell, Oxford (1998)

Semantics of Possibility Suffix "(Rar)e"

Takashi Iida*

Department of Philosophy, Keio University
iida@flet.keio.ac.jp

Abstract. A semantical analysis of a Japanese verbal suffix "(rar)e" is given in the framework of event semantics. It is claimed that the primary sense of "(rar)e" is that of situational and volitional possibility. This sense is analyzed in terms of the concept of possible stages of history during a given time interval and the thematic role of agent. The other two senses of "(rar)e" are shown to be derivable from the primary sense. It is observed that the ability sense of "(rar)e" appears when it is used in an attribute sentence, and an analysis of a certain class of attribute sentences is given, which shows they are essentially generic sentences and hence, the predicate with "(rar)e" expressing ability is a generic predicate. Finally, the reason why "(rar)e" expresses a realized possibility in certain contexts are explained by the interaction of the primary sense of "(rar)e" and perfective aspect. Throughout, it is emphasized that the basic classification of Japanese sentences into states of affairs sentences and attribute sentences is essential for giving the semantical account of Japanese predicates.

1   Introduction

I have chosen this topic because it gives me a good opportunity to present a framework for Japanese semantics, which has been well known for some time among a number of Japanese grammarians. My work may be regarded as a formal reworking of the theoretical insights of those grammarians[1].

This framework mainly consists in making use of a broad classification of Japanese sentences. Such a classification of sentences has been a part of the tradition of Japanese grammatical studies; some people say it can be traced back even to Edo-era. According to it, Japanese sentences are classified into those that report a concrete state of affairs and those that ascribe an enduring property to a subject. These two types have been called differently by different scholars. Here I adopt the terminologies used by Masuoka Takashi. He classifies Japanese sentences into "jishou jyojyutsu bun (sentences reporting states of affairs)" and "zokusei jyojyutsu bun (sentences reporting attributes)" ([Ma1], p.22). I am going to abbreviate the former as "state of affairs sentences" and the latter as "attribute sentences".

---

* I thank Lajos Brons and Koji Mineshima for helpful comments and discussions.

[1] Among the books which present such a framework I would like to cite the following two. [Ko1] and [Ma1].

K. Nakakoji, Y. Murakami, and E. McCready (Eds.): JSAI-isAI, LNAI 6284, pp. 217–234, 2010.

Many authors make a subdivision of the former. According to Masuoka, state of affairs sentences are divided in turn into those express "dynamic states of affairs" and those express "static states of affairs". As he explains a "dynamic state of affairs" as "an event which happens at a particular time and place", the distinction between "dynamic" and "static" among the states of affairs corresponds to the one between events and states which is well-known in philosophy and plays an important role in the linguistic study of aspect.

In sum, there is a broad classification of Japanese sentences that can be set in the following table.

(A) state of affairs sentences
  (A–1) sentences expressing dynamic states of affairs (event sentences)
  (A–2) sentences expressing static states of affairs (state sentences)
(B) attribute sentences

If you wish to give a semantic account of a Japanese sentence, you should always be aware of the type of sentence which it belongs to, namely, whether it is an event sentence, a state sentence, or a sentence attributing a certain property to a subject. Different types of sentences should be given different semantic accounts, and sometimes the same sentence has different readings corresponding to the different types to which it is thought to belong. We have a good example of this in the case of sentences with the so-called possibility suffix "(rar)e". The same sentence with this suffix can have a state sentence reading, an attribute sentence reading , and sometimes even an event sentence reading.

Whether a sentence is an event sentence or a state sentence is determined by its main predicate. Let us call a predicate which makes an event sentence as an "event predicate", and a predicate which makes a state sentence as a "state predicate". Japanese verb phrases may be divided into two classes according to whether they make event predicates or state predicates. It is important to note that this is a classification of verb phrases and not a classification of verbs themselves. For, there are verb suffixes which make state verb phrases out of event verbs or event verb phrases. A typical example is a suffix "tei" which makes a state verb phrase out of an event verb phrase. Depending on the verb it is appended to, it may mean either the state that the event of the specified type is in progress, or the state resulting from the occurrence of the event. "Akeru" (open, transitive) and "aku" (open, intransitive) give a nice contrasting pair of sentences.

(1) Taro ga   mado   o   ake-tei-ru.
    Taro NOM window(s) ACC open PROG NON-PAST
    Taro is opening the window(s).

(2) Mado   ga   ai-tei-ru.
    window(s) NOM open RESU NON-PAST
    The window(s) is(are) open.

In order to see whether a predicate is an event predicate or a state predicate, it is enough to look at a non-past sentence which has it as the main predicate. If the sentence cannot refer to the present state of affairs, then it is an event sentence. On the other hand, if the sentence may refer to the present state of affairs, it is a state sentence.

In their admirable textbook of Japanese grammar [MT1], Masuoka Takashi and Takubo Yukinori say "the suffix '(rar)e' is appended to a dynamic verb to make a stative verb." The following two sentences are given as the examples ([MT1], p.106.).

(3) Taro wa   Hanako ni   a-u.
Taro TOP Hanako DAT meet NON-PAST
Taro is going to meet Hanako.

(4) Taro wa   Hanako ni   a-e-ru.
Taro TOP Hanako DAT meet can NON-PAST
Taro can meet Hanako.

Using the above test, it is easy to see that the sentence (3) is an event sentence and (4) is a state sentence. Thus, the verb phrase "a-e-ru" is a state verb phrase, whereas the verb it comes from is an event verb. As Masuoka and Takubo state, the suffix "(rar)e" makes a state verb phrase out of an event verb, or more generally, an event verb phrase.

However, it is not true that any event verb phrase can take the suffix "(rar)e". Although the passives are event verb phrases, we cannot put "(rar)e" to passives. In contrast, you can put "(rar)e" to causatives, as the following example shows.

(5) tabe-sase-rare-na-i.
eat CAUSE can not NON-PAST
I cannot make her (or him or them) to eat it.

Thus, it is not accurate to say simply that "the suffix '(rar)e' is appended to a dynamic verb to make a stative verb".

One promising lead is the idea which is also found in [MT1], namely that of classifying verbs into "volitional verbs" and "non-volitional verbs" (p.13). Shall we claim then that the suffix "(rar)e" is appended to a verb phrase whose core is a volitional verb to make a stative verb phrase? Unfortunately, most of the verbs which are classified as "non-volitonal" also can take "(rar)e". For example, the follwing sentence is perfectly all right, in spite of the fact that it has a non-volitional verb "nemuru" (sleep) as the core of the predicate.

(6) Nemur-e-na-i.
sleep can not NON-PAST
I cannot sleep.

Again, a non-volitional verb "korobu" (fall down) can occur with "(rar)e" in the sentences like the following.

(7) Taro wa   umaku korob-e-ru.
Taro TOP well   fall down can NON-PAST
Taro can fall down in the right way.

However, we can save the distinction, if we do not regard it as the one among the verbs themselves, but as the one among the uses of them. We do not have volitonal verbs or non-volitional verbs; what we have are the volitional and non-volitional uses of verbs.

Thus, we now have a rough characterization of the function of the suffix "(rar)e": it is appended to an event verb phrase whose core consists of a volitonal use of a verb to make a stative verb phrase. We may call our suffix that of "volitional possibility". Our semantics of "(rar)e" should explain what is involved in "volitional possibility". This should be our first task.

Another problem we should consider about the semantics of "(rar)e" is concerned with the fact that a sentence containing this suffix may have different interpretations. For example,

(8) Taro wa   oyog-e-ru.
    Taro TOP swim can NON-PAST
    Taro can swim.

means either (a) Taro is able to swim now, or (b) Taro has the ability to swim. Moreover, the sentence with the past tense particle "ta"

(9) Taro wa   oyog-e-ta.
    Taro TOP swim can PAST

has three different interpretations. It may mean that (a) Taro was able to swim at a particular time in the past, (b) Taro had an ability to swim in the past, or (c) Taro managed to swim. These different interpretations become explicit if we put appropriate adverbs.

(9a) Taro wa   sono toki   oyog-e-ta.
     Taro TOP at that time swim can PAST
     Taro could swim at that time.
(9b) Taro wa   mukashi oyog-e-ta.
     Taro TOP years ago swim can PAST
     Taro was able to swim years ago.
(9c) Taro wa   yatto   oyog-e-ta.
     Taro TOP somehow swim can PAST
     Taro somehow managed to swim.

Thus, our second task is to explain why a sentence with "(rar)e" may have such different interpretations and how they are related to each other.

2   "(Rar)e" of Volitional Possibility

2.1   Semantics of Event Verbs and State Verbs

My analysis of the possibility suffix "(rar)e" is in the framework of event semantics. Event semantics is particularly well suited to analyzing a Japanese verbal predicate, because a Japanese verbal predicate may have a complex structure consisting of various suffixes and they can be regarded as expressing certain operations in the domain consisting of events and states[2]. One of these suffixes is our possibility suffix, and as we saw before, it turns an event predicate into a state predicate.

---

[2] This is one of the reasons why I have not adopted a framework like Stit whose basic elements are propositions, not events or states.

Let me give an example of a Japanese verbal predicate with various suffixes. In the sentence

(10) Taro wa   eki   made ik-ase-rare-tei-ru.
　　 Taro TOP station to   go CAUSE PASS PROG NON-PAST
　　 Taro is now in the process of being made to go to the station.

the verbal predicate "ikaserareteiru" consists of a verb stem "ik", a tense particle "ru", and three suffixes in between, namely,

"(s)ase" causative,
"rare" passive,
"tei" progressive.

As was mentioned before, there are two main kinds of predicates, dynamic predicates which make event sentences, and static predicates which make state sentences. Each verb and verb phrase introduces a type of event or state. A dynamic verb phrase is assigned a set of events as its extension, and a static verb phrase is assigned as its extension a set of time intervals in which the state it expresses holds. In the present account, for the sake of simplicity, we assume that time is linear and discrete: it consists of unit intervals. We also assume that each time unit I has its immediate future I^+ and immediate past I^-. A period is a set of consecutive unit time intervals.

Although I am certain that we should countenance events as full-fledged individuals just like persons and stars, I don't think we should do the same with states[3]. So, a state verb may be regarded as expressing a certain relation between time intervals and the individuals involved in the state in question.

Moreover, each verb is thought to have a fixed number of arguments. For example, "oyogu" (swim) is an event verb (dynamic verb) with one argument, and "taberu" (eat) is an event verb with two arguments. Likewise, "iru" (is at, or, stay at) is a state verb with two arguments. Each argument of an event verb plays a definite role in the events the verb denotes. Thus, the one and only argument of "oyogu" is the agent of a swimming event, and the two arguments of "taberu" are the agent and the theme (patient) of an eating event. In contrast to this, a state verb "iru" expresses a relation between individuals like persons or animals, locations and time intervals.

The extensions of these three verbs are given in the following way. Here "e" is a variable that ranges over events, and "I" is the one over unit time intervals. I suppose it is obvious what "SWIM(e)", "AGENT(x, e)", etc. mean.

$$\|\mathbf{oyog}(x)\| = \{e|\text{SWIM}(e) \wedge \text{AGENT}(x, e)\}$$
$$\|\mathbf{taber}(x, y)\| = \{e|\text{EAT}(e) \wedge \text{AGENT}(x, e) \wedge \text{THEME}(y, e)\}$$
$$\|\mathbf{i}(x, y)\| = \{I|x \text{ is at } y \text{ during } I\}$$

A verb makes the core of the main predicate of an event sentence or state sentence. A verb cannot appear by itself in a sentence, but it should be accompanied

---

[3] In an expanded version of event semantics I developed in [Ii1], its ontology comprises not only individuals like persons or token events but also states and event types. Still, states are not individuals, but universals just as event types are.

with a tense particle. State of affairs sentences are sentences that report particular events or states, and these can be located in time only with the aid of a tense expression. Let us take a very simple sentence.

(11) Taro ga   oyog-u.
     Taro NOM swim NON-PAST
     Taro is going to swim.

A semantic theory should give us the truth condition of this sentence relative to the context C of its utterance. As (11) ends with the non-past tense particle "ru" and the verb "oyog" preceding it is an event verb, (11) is true if and only if there are some events which are of the type SWIM with Taro as their agents and happens at some time after its utterance. If we denote by "I_0" some time interval later than the utterance time I^C and E^{I_0} is the set of all the events that occur during I_0, then this condition can be written in this way.

$$\|\mathbf{oyog}(x)\|_{x=\text{Taro}} \cap E^{I_0} \neq \emptyset$$

or

$$\{e : \text{SWIM}(e) \wedge \text{AGENT}(\text{Taro}, e)\} \cap E^{I_0} \neq \emptyset$$

In contrast to this, a state sentence with non-past tense is true relative to C when the utterance time I^C is in the extension of its state predicate. For example, a state sentence

(12) Taro wa   Toukyou ni i-ru.
     Taro TOP Tokyo   at stay NON-PAST
     Taro is in Tokyo.

is true in C if and only if

$$I^C \in \|\mathbf{i}(x)\|_{x=\text{Taro}}$$

namely,

Taro is at Tokyo during I^C.

2.2   Analysis of Volitional Possibility

We already know how to give a simple sentence "Taro ga oyog-u (Taro is going to swim)" its truth condition relative to the context C. Let us consider a sentence with the possibility suffix "(rar)e"

(8) Taro wa   oyog-e-ru.
    Taro TOP swim can NON-PAST

As I mentioned before, this sentence has at least two intepretations, namely, it may mean that (a) Taro can swim now, or that (b) Taro has the ability to swim. My claim is that the reading (a) is more basic than (b) and the latter reading is derivable in some sense from (a). So, for the time being, let us interpret (8) as saying that Taro can swim now.

Please note that this sentence is about the present state of Taro, and not a future event involving Taro as in (11). (8) is a state sentence, and its predicate "oyog-e-" denotes a state of Taro. Our task is to show how the extension of "oyog-e-" is determined by the extension of the core event verb "oyog-".

In the following analysis the concept of action is taken for granted. We are not going to consider how this concept should be analyzed[4] except that we stipulate that

an event e is an action $\Leftrightarrow \exists x$ AGENT(x, e).

I make an assumption that for any time interval I there are a number of possibilities that might come to be actualized during its immediate future I^+. Let us call these possibilities "possible stages of history immediately after I" and designate it by

$$H^I$$

Let h be an element of such a set. h should satisfy the following conditions.

(i) h is a continuation of the actual history up to I.
(ii) Any general law that holds in the actual world also holds in h.

Given such apparatus, the first idea that might occur is to give an application condition of a predicate "V + (rar)e" as follows.

"V + (rar)e" is true of an agent a at $I \Leftrightarrow$ there are possible stages of history immediately after I in which there is an action by a which is of the type V.

This characterization judges correctly that the cases where a might suffer the V type event just after I are not the cases where "V + (rar)e" is true of a at I. For example, the fact there are possible stages of history immediately after I in which a might fall down unintentionally, does not make it true that "a wa korob-e-ru (a is able to fall down)". For the latter to be true, what is needed is not simply an event of a's falling down, but an action on a's part of falling down intentionally.

However, the present characterization is too simple for many cases[5]. Consider the various utterances of the following sentence.

---

[4] This is another difference between my analysis and that given by Stit theory like [Hol]. But, I don't think that the concept of action should be a primitive one. For an interesting attempt at analysis, see [Pil].

[5] A referee pointed out that the present analysis in terms of "possible stages of history" could not account for the sentences like the following.

(i) Taro wa   kaijuu   ni   datte kat-e-ru
    Taro TOP monster(s) DAT even win can NON-PAST

However, the issues raised by such a sentence are part of the general problem of how to give a semantic account of a discourse containing reference to a fictional object like *kaijuu*, and hence, tangential to our concern in this paper. Moreover, I strongly doubt the possible worlds and the like give us an appropriate framework to do semantics of such discourses. If my doubt is well founded, then the sentences like (i) do not give us a good reason to extend our "possible stages of history" to something further apart from the actual world.

(13) (Sore wa)   Tabe-rare-na-i.
     (that TOP) eat can not NON-PAST
     I can't eat (that).

1. "Tabe-rare-nai" uttered by a person who judges the offered food is poisoned.
2. "Tabe-rare-nai" uttered by a person whose religion prohibits eating a certain kind of food.
3. "Tabe-rare-nai" uttered by a person who is on diet.
4. "Tabe-rare-nai" uttered by a person who has a very strong moral feeling when he or she is offered food that is obtained by morally dubious procedure.

These cases remind us that there is a normative dimension in an action. An action is possible for an agent only if it is not against the norms the agent adopts[6]. There are many kinds of norms an agent may adopt; they might be prudential, religious, aesthetic or moral. In the above examples a speaker judges "Tabe-rare-nai" because taking the offered food involves the violation of the norm the speaker adopts.

Let us incorporate such considerations into our semantics of "(rar)e". What should be done is to make clear what "a possible stage of history immediately after I" is. It is no grand thing like a possible world. It is not even like a temporal stage of it. "A possible stage of history immediately after I" is more like a possible alternative open to an agent at I. Let us adopt this terminology and make it relative to an agent a and a norm ν[7] as well as a time interval I. Thus, the set of possible alternatives open to an agent a under a norm ν at I is designated by

$$H_{a,\nu}^{I}$$

If h is an element of this set, it should satisfy the following condition as well as (i) and (ii) above.

(iii) In h there is no violation of the norm ν.

For each possible alternative h in $H_{a,\nu}^{I}$, there is the set of all the events that occur during I^{+}. It is denoted by "$E_{h}^{I^{+}}$". We also suppose that the context C supplies the relevant norm $\nu(C)$ when it is necessary.

At last, we can define the extension relative to the context C of the predicate with the possibility suffix

$$(\mathbf{oyog}(x))\mathbf{e}$$

applied to a as follows.

$$\|(\mathbf{oyog}(x))\mathbf{e}\|_{x=a}^{C} = \{I | \exists h \exists e[h \in H_{a,\nu(C)}^{I} \wedge e \in E_{h}^{I^{+}} \wedge e \in \|\mathbf{oyog}(x)\|_{x=a}]\}$$

---

[6] However, this is an oversimplification. The relevant norm may not be the one the agent herself adopts. We may judge that some actions are not possible for an agent, not because they are against a norm the agent adopts, but because they are against the norm we ourselves adopt.

[7] The norm ν may not be the one the agent a adopts. See the previous note.

Using this definition, the truth condition of (8) relative to the context C can be easily deduced. As (8) is a state sentence, it is true in C if and only if the utterance time I^C is in the extension of

$$\|(\mathbf{oyog}(x))\mathbf{e}\|_{x=\text{Taro}}$$

and, it is true in turn if and only if

$$\exists h \exists e [h \in H^{I^C}_{\text{Taro},\nu(C)} \wedge e \in E_h^{(I^C)^+} \wedge \text{SWIM}(e) \wedge \text{AGENT}(\text{Taro}, e)],$$

that is, there are some possible alternative open to Taro at I^C in which occur some swimming events with Taro as an agent without any violation of the norm relevant to the context of the utterance.

3   "(Rar)e" of Ability

However, there are two senses (8) may express.

(a) Taro is in the state of being able to swim now.
(b) Taro has an ability to swim.

Even though our present analysis might be true of (a), it is definitely not true of (b). (8) read as (b) is concerned with Taro's ability, whereas (8) read as (a) is concerned with Taro's present situation. Why are there two different readings of (8)? How can we know which is the right reading on each occasion?

I claim that (8) can be interpreted either as a state of affairs sentence or an attribute sentence, and that the former way of taking (8) results in the volitional possibility reading and the latter way of taking it results in the ability reading. Our analysis so far has considered (8) only as a state of affairs sentence. In order to give an analysis of (8) as expressing an ability, we should know how the semantics of an attribute sentence is given.

3.1   Attribute Sentences as Generic Sentences

I do not claim that the following analysis applies to all sorts of attribute sentences in Japanese. It is intended to apply to those sentences which attribute an enduring property to a certain subject. These sentences are divided into those which attribute such a property to a particular individual like Taro and Hanako and those which attribute it to a certain kind. Here are the typical examples of both types of attribute sentences.

(14) Taro wa   kashikoi.
     Taro TOP wise
     Taro is wise.
(15) Zou   wa   hana   ga   nagai.
     elephants TOP trunks NOM long
     Elephants have long trunks.

In the following, I will be mainly concerned with the former type of an attribute sentence, namely, a sentence that attributes an enduring property to a particular individual.

(8) read as (b) is one of them. According to the present analysis, such a reading results when it contains an operator which will be written as "**Gen**". The choice of this symbol reflects the thought that a large class of Japanese attribute sentences are in fact generic sentences. In general, an attribute sentence consists of two distinct parts; the first part, which we call "subject-part", presents the subject of the sentence, and the second part, which we call "attribute-part", attributes a property to the subject. Both the subject-part and the predicate-part contain the generic operator "**Gen**".These two parts are typically connected by a particle "wa". Finally, a tense particle is an outermost part of the sentence. Hence, (8) read as an attribute sentence has the following structure.

$$\{ \mathbf{Gen}_{SP}(\text{Taro}) \text{ wa } \mathbf{Gen}_{AP}(\text{oyog-e})\} \text{ ru.}$$

Let us recall that we compared (8) read as (a) to an event sentence (11).

(11) Taro ga   oyog-u.
     Taro NOM swim NON-PAST
     Taro is going to swim.

If we change the particle "ga" to "wa" in (11), then the resulting sentence

(16) Taro wa   oyog-u.
     Taro TOP swim NON-PAST
     Taro swims.

can be construed as an attribute sentence, which says that Taro is a person who swims, in contrast to an event sentence which talks about a future event of Taro's swimming[8]. Such a reading of (16) has an analysis similar to (8) read as (b).

$$\{ \mathbf{Gen}_{SP}(\text{Taro}) \text{ wa } \mathbf{Gen}_{AP}(\text{oyog})\} \text{ u}$$

Attribute sentences can be classified again according to whether their attribute parts are event predicates or state predicates. (16) is an attribute sentence having an event predicate as its attribute part, and a sentence like the following is one having a state predicate as its attribute part[9].

(17) Hanako wa   gakusei da.
     Hanako TOP student COPULA
     Hanako is a student.

Our analysis is intended to apply to both kinds of attribute sentences. It is based on two ideas.

---

[8] (16) can be read in this way, too. But, for such an interpretation to be possible, "wa" should be pronounced with a stress.

[9] A state predicate can be a predicate derived from a noun like "gakusei" of (17) as well as a verbal predicate.

1. A property attributed to the subject of an attribute sentence is an enduring one, partly because the subject has it for the period that is much wider than the period that is the time of its utterance.
2. An untensed part of an attribute sentence has the form

$$\mathbf{Gen}(S) \text{ wa } \mathbf{Gen}(P).$$

The predicate "$\mathbf{Gen}(P)$" is an attributive predicate whose extension consists of individuals. It is the third kind of predicate which is different from an event predicate whose extension consists of events and a state predicate whose extension consists of time intervals[10].

Let me explain this, taking (17) as an example. In the case of attribute sentences with individual subjects, it is true in the context C if and only if the individual which is denoted by "$\mathbf{Gen}(S)$" is in the extension of the predicate "$\mathbf{Gen}(P)$" during a certain extended period of time determined by C and its tense. And, if "S" is a proper name as in (17), "$\mathbf{Gen}(S)$" denotes just the individual denoted by "S". Moreover, "$\mathbf{Gen}(\text{gakusei-da})$" is an attribute derived from a state predicate "gakusei-da" and it is simply a predicate whose extension consists of individuals which are students. Hence, (17) is true now if and only if Hanako is a student during a certain extended period of time which includes the present.

3.2 Attribute Predicates Derived from Event Predicates

"$\mathbf{Gen}(\text{oyog})$" in (16) is also a predicate whose extension consists of certain individuals, but it is an attribute derived from an event predicate "oyog" , and it is not so simple to say what kind of individuals are in it. This depends on what it means to say that a property is "enduring", because a property expressed by "$\mathbf{Gen}(\text{oyog})$" should be an enduring property of Taro.

When is a property "enduring"? I think there are two respects in endurity and both of them are necessary to it. First, for a property F to be an enduring property of an individual A, F should be true of A during some extensive period of time. We should be able to say that A continues to be F during that period. This does not necessarily mean that F should always be true of A in that period. Even in that period F may not be true of A occasionally. But, A should be F in the uniform manner during that period, and this is the second respect necessary to endurity. During the period in question, F may be true of A only occasionally, but F should be true of A occasionally all through the period. In short, a property attributed to a subject in an attribute sentence should be a property which is uniformly true of it during an extensive period of time. Let us say that an enduring property should be both extensive and uniform.

---

[10] Hence, we have three kinds of predicates corresponding to three types of sentence.

[sentence type]	[predicate type]	[extension of predicate]
event sentence	event predicate	consists of events
state sentence	state predicate	consists of time intervals
attribute sentence	attribute predicate	consists of individuals in general

Let us consider the sentence (16) again. One reading of this sentence is that Taro is a kind of person who swims, or a swimmer. When do we say that somebody is a swimmer? Suppose Taro has swam only once in the past. Is that enough to make him a swimmer? I suppose not. Then how many times should Taro have swum in the past to be a swimmer? But, if he swam a number of times many years ago, but has not swum for several years now, then can we count him as a swimmer now?

I am not sure that there are unique correct answers to these questions. The answers I give here are only tentative and subject to change in the course of a more detailed study of the data. With this proviso, let us try to answer them.

I think it is reasonable to refuse the title "swimmer" to a person who has swum only once during the period relevant to the conversation. I also think that for a person to be a swimmer it is not only necessary that he or she swims a number of times during the period in question but also that he or she swims regularly all through the period. So, let us divide the period in question into several parts, and see whether it contains the events of the person's swimming. If such events are found in most of them, may we conclude that the person is a swimmer?

However, this by itself does not guarantee that the events of that person's swimming are distributed uniformly during the period. For, if the period is divided into unequal parts, we will get the wrong result. Suppose that we divide the part of the entire period where the events of Taro's swimming occur into many parts, but never divide the part of the period where such events do not occur, and suppose also that Taro once did many swimmings many years ago, but has not swum since. Then, Taro would be counted as a swimmer now. So let us demand that the period in question should be divided into equal parts[11]. Still this is not enough. On one hand, there are some actions which are so difficult to perform that having successfully performed such an action just once or twice is enough to qualify the person who did it as the owner of the relevant ability. On the other hand, there are some actions which are so routinely done many times a day by a normal person that a few successful performances a day or a week is not enough to qualify a person as the owner of the ability. This means that the division of the period may be coarse or fine. For a kind of action which is performed rarely and only with difficulty, the division should be rough. In contrast to this, for a kind of action which is performed quite often and with no difficulty, the division should be fine. Thus, we should divide the period differently for different kinds of action.

Let us try to be a little more formal. First, we define "an equal division of the period".

Definition. Let Π be a time period. By "an equal division of Π", we mean a set Δ of the periods π_i which satisfies the following conditions:

    1. For each i, $\pi_i \subseteq \Pi$.
    2. $\bigcup \Delta = \Pi$.

---

[11] The divided parts need not be strictly equal. It is enough that there is no extreme difference between the duration of each part.

3. For any i and j, if $i \neq j$, then $\pi_i \cap \pi_j = \emptyset$.
4. For any i and j, $\mathrm{dur}(\pi_i) = \mathrm{dur}(\pi_j)$.
Here, "$\mathrm{dur}(\pi_i)$" indicates the duration of π_i.

As the above considerations suggest, only some of these divisions are appropriate for our purpose, and it is different with each predicate which divisions are appropriate. So, let us express the relation "an equal division Δ of a period Π is appropriate for a predicate 'P'"[12] by

$$\mathrm{AP}(\Delta, \Pi, \text{"}P\text{"}).$$

With this relation, we can express the condition for a person to be a swimmer in the period Π: a person is a swimmer during Π if and only if there is an equal division Δ of Π appropriate for the predicate "swim" and most of the elements of Δ contain the events of the person's swimming.

In general, when "$P(x)$" is an event predicate, the extension of $\|\mathbf{Gen}(P(x))\|$ relative to the period Π is given by the following.

$$\|\mathbf{Gen}(P(x))\|_\Pi = \{x \mid \exists \Delta[\mathrm{AP}(\Delta, \Pi, \text{"}P(x)\text{"}) \wedge$$
$$\mathbf{most}\, \pi(\pi \in \Delta \to \exists I \exists e(I \subseteq \pi \wedge e \in E^I \wedge e \in \|P(x)\|)]\}$$

3.3   Attribute Predicates Derived from State Predicates

Let us turn to the case of an attribute predicate derived from a state predicate. For such a predicate, it seems a simple matter to give its extension. Suppose "$\mathbf{Gen}(P(x))$" is such a predicate derived from a state predicate "$P(x)$". A state predicate has the property of uniformity in the sense that if it holds in a certain period, then it holds in any part of that period. Hence, if something has an attribute derived from a state predicate during Π, then doesn't it have the same attribute in any time interval I within Π? That is, can't its extension relative to Π be simply this?

$$\|\mathbf{Gen}(P(x))\|_\Pi = \{x \mid \forall I(I \in \Pi \to I \in \|P(x)\|)\}$$

But, even though a state predicate is admitted to possess such a property of uniformity, it is not necessarily true that an attribute predicate derived from it has also that property. Let us consider the following pair of sentences both of which have a state predicate "urusai" (noisy).

(18) Kuruma ga   urusa-i.
     Car(s)   NOM noisy NON-PAST
     The noise is coming out of the car.

(19) Kono kuruma wa   urusa-i.
     this   car   TOP noisy NON-PAST
     This car is noisy.

---

[12] In an extended framework which contains event types and states as universals, "P" may be replaced by an event type expressed by "P". See note 3 above.

(18) can be interpreted as is here translated, and in that case, it is a sentence reporting a particular state. Although (19) can be also interpreted as a sentence reporting a particular state, it has another interpretation according to which the property of noisiness is attributed to a particular car. In the latter interpretation, (19) is an attribute sentence.

Let us consider (19). For the car to have the attribute of noisiness during a certain period, is it necessary that the car never ceases to be noisy during the period? Not at all. If it should, then a noisy car should be always in the move or continue to emit some sound even when it is not moving. Hence, the above characterization of the extension of the attribute predicate derived from a state predicate is wrong.

Thus, it is obvious that we need the concept of an equal division of the period appropriate for a predicate also in the case of an attribute predicate derived from a state predicate. When "$P(x)$" is a state predicate, the extension relative to a period Π of an attribute predicate "$\mathbf{Gen}\ (P(x))$" is given as follows.

$$\|\mathbf{Gen}(P(x))\|_{\Pi} = \{x \mid \exists\Delta[\mathrm{AP}(\Delta, \Pi, \text{“}P(x)\text{”}) \wedge$$
$$\mathbf{most}\ \pi(\pi \in \Delta \to \exists I(I \subseteq \pi \wedge I \in \|P(x)\|))]\}$$

Before moving on, I would like to say something about the semantics of an attribute sentence with a subject which denotes a kind, not an individual. Such a sentence asserts of the kind that the individuals belonging to it typically have a certain property. For example, the sentence

(15) Zou wa hana ga nagai. (Elephants have long trunks.)

ascribes a typical property of having a long trunk to the kind *elephant*. How should we explain this "typicality"? I think that this "typicality" of the property is very much similar to the endurity of the property. A typical property also should be extensive and uniform. But, this time what should be distributed extensively and uniformly are not events, but individuals of the kind. That is, a typical property of the kind is one found extensively and uniformly among its instances. So, the rough idea is like the following.

> Let "$\mathbf{Gen}(S)$ wa $\mathbf{Gen}(P)$" be an attribute sentence in which "S" denotes a certain kind K. Ignoring tense, this sentence is true if and only if there is a set Σ of individuals which belong to K such that (i) it forms an extenseive part of the entire individuals of K, and (ii) there is an appropriate division of Σ into subsets most of which contain an individual which has the property expressed by "$\mathbf{Gen}(P)$". As we know what the property expressed by "$\mathbf{Gen}(P)$" already, we will have the desired truth condition.

It goes without saying that this is only a rough idea and there are still many problems to be solved before this idea can give a feasible account of those Japanese attribute sentences which can be regarded as generic sentences.

3.4   Analysis of Ability

After all this, we now have everything we need to state the truth condition of (8) as read as a claim about Taro's ability to swim. As we said before, (8) read in this way is an attribute sentence and has the following form.

(20) { **Gen(Taro)** wa **Gen((oyog(x))e)**} **ru.**

This sentence is true in the context C if and only if the following holds for a certain extended period Ω^C determined by C.

$$\text{Taro} \in \|\mathbf{Gen((oyog}(x))\mathbf{e})\|_{\Omega^C}$$

As "**Gen((oyog(x))e)**" is an attribute predicate derived from a state predicate, this is equivalent to

$$\text{Taro} \in \{x | \exists \Delta [AP(\Delta, \Omega^C, \text{``}(\mathbf{oyog}(x))\mathbf{e}\text{''}) \wedge$$
$$\mathbf{most}\, \pi(\pi \in \Delta \rightarrow \exists (I \in \pi \wedge I \in \|(\mathbf{oyog}(x))\mathbf{e})\|))]\}$$

Only thing left for us to do now is to spell out the last part of this, namely, to figure out what the following means.

$$I \in \|(\mathbf{oyog}(x))\mathbf{e})\|)$$

But it is nothing but the volitional possibility we have dealt with before. What we get at the end is the following.

> (20) is true in the context C if and only if there is an equal division Δ of the period Ω^C appropriate to the predicate "oyog-e", and in most of the parts thus divided, there are time intervals at which some possible alternatives to Taro are open where the swimming events with Taro as an agent might occur without any violation of the relevant norm.

In short, to say that Taro has the ability to swim for a certain period of time is to say that for most of the parts of that period Taro is in the state of being able to swim at that period.

The sentence (8) "Taro wa oyog-e-ru" has two readings not because the suffix "(rar)e" has two different meanings such as volitional possibility and ability. It is just an instance of the widely observed phenomena in Japanese. (8) can be read either as a sentence reporting a particular state involving Taro or as a sentence attributing a property to Taro. The following two have also such two readings.

(21) Taro wa   okubyou dat-ta.
     Taro TOP cowardly COPULA NON-PAST
     Taro behaved cowardly, vs. Taro was a coward.

(22) Hanako wa   ie   kara de-na-i.
     Hanako TOP home from go_out not NON-PAST
     Hanako does not go out today, vs. Hanako never goes out.

Obviously, it is not a good policy to suppose that the phrases like "okyobyou dat-ta" and "ie kara de-nai" have two meanings. The ambiguity we see in such cases are not lexical, but structural. Our analysis about "(rar)e" given here is concerned with such structural features rather than the specific meaning of it. Thus it can be easily adopted to account the ambiguity found in (21) or (22).

4   "(Rar)e" of Realized Possibility

Let us consider the sentence we encountered before.

> (9) Taro wa   oyog-e-ta.
> Taro TOP swim can PAST

Suppose (9) is uttered now. If we are watching Taro in the pool and have just seen Taro swim, then the most natural interpretaion of it is that Taro has just swum successfully. Let us call such an interpretation the realized possibility interpretation. According to [MT1] p.107, when the suffix "(rar)e" is followed by "ta" which expresses perfective aspect as well as past tense, it may mean the realization of a possibility. So, the realized possibility interpretation of (9) implies the non-modal sentence.

> (23) Taro wa   oyoi-da.
> Taro TOP swim PAST
> Taro swam.

If we consider the sentences with adverbs like "yatto" (somehow) and "mou" (already), such an implication is still clearer.

> (24) [=(9c)] Taro wa   yatto   oyog-e-ta.
> Taro TOP somehow swim can PAST(PERF)
> Taro somehow managed to swim.

On the other hand, you may wonder what is the point of the modal expression "(rar)e" if (9) only reports the same fact (23) does.

In order to explain the existence of implications like the one from (9) to (23) and the role of the modal suffix in the realized possibility interpretation, I claim that if a state predicate with the modal suffix "(rar)e" is put just before the perfective particle "ta", it should be turned into an event predicate and that is done by an operator I refer to by "**R**". Thus, in the realized possibility interpretation of (9), the predicate has the following form.

R $(\mathbf{oyog}(x)\text{-}\mathbf{e})$

This is an event predicate, and hence, its extension consists of events. What events should be in the extension? The idea is that the extension consists of those events which actually occur during the period the state predicate "$\mathbf{oyog}(x)\text{-}\mathbf{e}$" holds. In general, the extension relative to the context C of

R $((P(x))\ \mathbf{e})$

is given by the following set.

$$\{e|\exists I[I \in \|(P(x))\,\mathbf{e}\| \wedge e \in \|P(x)\| \wedge e \in E^{I^+} \wedge \text{non-violation}(e, \nu(C))]\}$$

where "non-violation$(e, \nu(C))$" means that e involves no violation of the norm $\nu(C)$. As the first conjunct inside "$\exists I$" follows from the remaining three conjuncts, this reduces to

$$\{e|e \in \|P(x)\| \wedge \text{non-violation}(e, \nu(C))]\}$$

It is no wonder that (23) follows from (9). (9) is different from (23) in that it claims that there has not been any violation of the relevant norm in the event of Taro's swimming. For example, if the relevant norm is a prudential one, then the speaker may mean by the utterance of (9) that Taro swam without any untoward consequences to himself. But I think there is yet another difference between (9) and (23). It is the difference in the implicatures they have.

Let us go back to the characterization of the volitional possibility "(rar)e" expresses. According to it, if you wish to know whether a given interval belongs to the extension of the predicate with "(rar)e", you have to consider the possible stages of history just after that interval. If there exist those which contain an event satisfying the relevant condition, then the predicate is true, and if not it is false. Suppose that an event of certain type happens no matter what possible stages of history follow. For example, no matter how you and the others act and no matter what happens in general, you have to swim, isn't it misleading or inappropriate to say that "oyog-e-ru" (I can swim)? To say this in such circumstances is like saying something is possible when it is known to the speaker that it is necessary, and it violates one of the conversational maxims. Hence, it seems reasonable to suppose that the following condition is presupposed in the appropriate use of the predicate "$\mathbf{oyog}(x)$-$\mathbf{e}$" applied to an agent a in the context C.

$$\exists h \neg \exists e[h \in H_{a,\nu(C)}^{I^+} \wedge e \in E_h^{I^+} \wedge e \in \|\mathbf{oyog}(x)\|_{x=a}]$$

in other words,

$$\exists h \forall e[(h \in H_{a,\nu(C)}^{I^+} \wedge e \in E_h^{I^+}) \rightarrow e \notin \|\mathbf{oyog}(x)\|_{x=a}]$$

that is, there are some possible alternatives to the agent a where a does not swim.

In the realized possibility interpretation of (9), its truth condition is non-modal just as (23) is. However, the presupposition accompanying the use of the predicate "$\mathbf{oyog}(x)$-$\mathbf{e}$" remains, namely the presupposition that there exist some possibilities that Taro did not swim, and it results from the use of the modal suffix "(rar)e".

5   Conclusion

In conclusion, let me remind the fact that the sentence we have just considered, namely,

(9) Taro wa   oyog-e-ta.
Taro TOP swim can PAST

has three different interpretations, all of which we have accounted for in the foregoing.

First, it can be interpreted as expressing a volitional possibility, namely, it may mean that Taro was able to swim if he wished in a certain particular occasion in the past. According to this interpretation, (9) is a state sentence about Taro's particular state in the past.

Second, it can be interpreted as saying that Taro once had the ability to swim. Then, (9) is interpreted as an attribute sentence ascribing the property of being able to swim to Taro during some extended period in the past.

Third, it can be interpreted as expressing a realized possibility. According to this interpretation, (9) is an event sentence which says about Taro's accomplishment. It is non-modal in its truth condition, but has an implicature which is modal in character.

References

[Ho1] Horty, J.F.: Agency and Deontic Logic. Oxford University Press, Oxford (2001)
[Ii1] Takashi, I.: Nihon-go Keishiki Imi-ron no Kokoromi (2): Doushi-ku no Imi-ron (An Essay in Japanese Formal Semantics (2): Semantics of Verb Phrases.), ms. Keio University (2001)
[Ko1] Seiji, K.: Nihon-go wa Donna Gengo ka (What Kind of Language is Japanese?). Chikuma Shobou, Tokyo (1994)
[Ma1] Takashi, M.: Meidai no Bunpou (The Grammar of Propositions). Kuroshio Publishing, Tokyo (1987)
[MT1] Takashi, M., Yukinori, T.: Kiso Nihongo Bunpou. Kaiteiban (Basic Japanese Grammar, Revised Edition). Kuroshio Publishing, Tokyo (1992)
[Pi1] Pietroski, P.M.: Causing Actions. Oxford University Press, Oxford (2000)

An Adaptive Logic for the Formal Explication of Scalar Implicatures

Hans Lycke*

Centre for Logic and Philosophy of Science, Ghent University,
Blandijnberg 2, 9000 Gent, Belgium
Hans.Lycke@UGent.be
http://logica.ugent.be/hans

Abstract. Hearers get at the intended meaning of *uncooperative utterances* (i.e. utterances that conflict with the prescriptions laid down by the *Gricean maxims*) by pragmatically deriving sentences that reconcile these utterances with the maxims. Such pragmatic derivations are made according to pragmatic rules called *implicatures*. As they are pragmatic in nature, the conclusions drawn by applying implicatures remain uncertain. In other words, they may have to be withdrawn in view of further information. Because of this last feature, Levinson argued that implicatures should be formally modeled as *non–monotonic* or *default* rules of inference. In this paper, I will do exactly this: by relying on the *Adaptive Logics Programme*, I will provide a formal explication of implicatures as default inference rules. More specifically, I will do so for a particular kind of implicatures, viz *scalar implicatures*.

Keywords: Gricean pragmatics, scalar implicatures, linguistic scales, defeasible inference rules, adaptive logics.

1 Scalar Implicatures

In contemporary pragmatics, the *Gricean maxims* (see [6, pp. 26–27]) are interpreted not as actual maxims, but as *heuristic markers* for both speakers and hearers (see e.g. [1],[10]).

> Instead of thinking about them as rules (or rules of thumb) or behavioral norms, it is useful to think of them as primarily *inferential heuristics* which then motivate the behavioral norms. (sic, [10, p. 35])

The maxims provide speakers the guidelines to model their utterances in a way that best serves their communicative purposes (whatever these may be: information transfer, transfer of emotions,...). Moreover, they provide hearers the guidelines to decipher the intended meaning of utterances that are in conflict with the maxims (henceforth, these will be called *uncooperative utterances*). The latter is done by deriving sentences that reconcile uncooperative utterances with the maxims (obviously, hearers will only do so in case they are convinced the

---

* The author is a Postdoctoral Fellow of the Special Research Fund of Ghent University.

K. Nakakoji, Y. Murakami, and E. McCready (Eds.): JSAI-isAI, LNAI 6284, pp. 235–251, 2010.

speaker assumed they are capable to get at the actual meaning of the utterance in spite of its deviance from the prescriptions stated by the maxims). These derivations are obviously not deductive derivations, but pragmatic ones. Hence, the intrinsic features of such derivations are distinct from those of deductive ones. Most importantly, the consequences of pragmatic derivations are only accepted in a defeasible way, meaning that they might be withdrawn at some point, for example in case the speaker explicitly rejects them, or because they conflict with the background knowledge shared by speaker and hearer (see e.g. [7],[9],[10]).

The pragmatic rules that enable hearers to get at the intended meaning of uncooperative utterances, are called *implicatures*. As these rules yield defeasible consequences, Levinson [10, ch. 1] has argued convincingly that they should be captured formally as *non–monotonic* or *default* rules of inference. That is exactly what I will do in this paper: by relying on the *Adaptive Logics Programme* (see e.g. [2],[3]), I will provide a formal explication of implicatures as default inference rules. More specifically, I will do so for a particular kind of implicatures, viz *scalar implicatures*. The latter are based on *linguistic scales*,[1] which are partially ordered sets of sets of linguistic expressions $\langle \Delta_1, ..., \Delta_n \rangle$ (the partial ordering relation has to be defined over the sets of linguistic expressions "in a contextually salient way," see [10, p. 105]). The linguistic expressions in Δ_i are considered more high–ranked than those in Δ_j in case $i < j$.

Example 1. The following are all linguistic scales:[2] $\langle$ *and, or* $\rangle$, $\langle$ *all, most, many, some* $\rangle$, $\langle$ *succeed, try* $\rangle$, $\langle$ *book,* $\{chapter\ 1, chapter\ 2, ...\}$ $\rangle$,...

Scalar implicatures arise from linguistic scales in the following way: the assertion by a speaker of a sentence containing a low–ranked linguistic expression will force the hearer to implicate the negation of the corresponding sentences with more high-ranked linguistic expressions. For, so the reasoning goes, if the speaker would have been in a position to use a more high–ranked linguistic expression, he would have done so (in order to comply with the maxim of quantity that states that we should be as informative as our communicative purposes require us to be).

Example 2. "John ate *some* of the cookies" IMPLICATES THAT "John didn't eat *all* of the cookies"

Finally, remark that I am not concerned with the specific characteristics of linguistic scales (what Levinson called the diagnostics of linguistic scales, see [10, p. 81]), nor with how people recognize a linguistic scale in a particular conversational context (what I would like to call the psychology of linguistic scales). I am merely concerned with how scalar implicatures are used by hearers in order to get at the intended meaning of assertions made by speakers. Hence, I will simply

---

[1] Levinson [10, p. 105] called these scales *Hirschberg scales*. I have opted for the more neutral *linguistic scales* coined by Verhoeven [12, p. 9].

[2] For reasons of convenience, singletons occurring in linguistic scales are represented by their sole elements. Hence, where $l_1,...,l_n$ are linguistic expressions, $\langle l_1, ..., l_n \rangle$ is an abbreviation of $\langle \{l_1\}, ..., \{l_n\} \rangle$.

presuppose that some linguistic scales are available to hearers in a particular conversational context. Formally, this means that the information available to the hearer in a conversational context is taken to be a couple $\langle\ \varGamma^{\mathbf{u}}\cup\varGamma^{\mathbf{bk}}\ ,\ \varGamma^{\mathbf{ls}}\ \rangle$, where $\varGamma^{\mathbf{u}}$ represents the utterances made by the speaker (as they are heard by the hearer), $\varGamma^{\mathbf{bk}}$ represents the background knowledge shared by both speaker and hearer (as supposed by the hearer), and $\varGamma^{\mathbf{ls}}$ contains all linguistic scales that are available to the hearer in the particular context.

2   The Role of Standard Logic

The consequences obtained by means of pragmatic inference steps (in casu, scalar implicatures) are defeasible, which means that speakers might withdraw them at a certain point. The reasons for withdrawal can be twofold. First of all, new information might be acquired that is in conflict with the pragmatically derived conclusions (e.g. the speaker has made some new utterances). In formal terms, this comes down to non–monotonicity. Secondly, pragmatic consequences might also be withdrawn because the deductive consequences of some of the utterances made by the speaker contradict them. In practice, this comes down to the fact that people sometimes draw (wrong) pragmatic conclusions from utterances before they have full insight in what the speaker has actually said. If they obtain more insight (which, let's face it, might not happen at all), they will withdraw these conclusions. Formally, this corresponds to the fact that people are not logically omniscient (which, in the approach presented below, is a strictly proof theoretic feature).

The second reason for withdrawing pragmatic consequences clearly shows that scalar implicatures are always applied against a deductive background (i.e. they are ampliative inference rules). Traditionally, this deductive background is captured by means of a *standard logic* (**SL**),[3] which means that the logical symbols (the logical connectives, modal operators,...) are interpreted standardly. However, when trying to explicate implicatures formally, interpreting the logical symbols in the standard way leads to the so–called *implementation–problem* (for a discussion related to the *or*–implicature, see [8],[11],[13]). In short, this comes down to the fact that the implicatures either generate too many or too few pragmatic consequences (dependent on the way you determine when to withdraw pragmatic consequences). The problem is related to the fact that **SL** doesn't distinguish between sentences the hearer heard the speaker utter and sentences the hearer merely derived from those she heard the speaker utter. Obviously, the implicatures should only be applied to the former, not to the latter.

In this paper, the deductive background is captured by means of a non–standard logic, viz the logic $\mathbf{SL^u}$, a particular extension of $\mathbf{SL}$.[4] This logic is

---

[3] This *standard logic* is usually an extension of *classical logic* (**CL**). Consequently, the approach I will present below is not only applicable to the classical connectives and quantifiers, but also to a whole range of non–classical quantifiers (most, many,...) and modal operators (necessary, possible,...).

[4] A specific $\mathbf{SL^u}$ will be characterized in section 4.1. Moreover, the non–standard extension of propositional **CL** (called $\mathbf{CL^u}$) has been characterized in [11].

defined over the language $\mathcal{L}^u$ that not only contains the standard logical symbols, but also contains *utterance–symbols*. The latter are non–standard logical symbols that are used to formally represent the utterances made by the speaker. More specifically, utterances are represented by sentences that only contain utterance–symbols (these are called *utterance–sentences*). The other information available to the hearer in a conversational context (i.e. the shared background knowledge) is represented by sentences only containing standard symbols (these sentences are called *standard sentences*). In view of section 1, this means that the set Γ^u only contains utterance–sentences and that the set Γ^{bk} only contains standard sentences! In this way, the logic $\mathbf{SL}^u$ is able to formally make the distinction between sentences the hearer heard the speaker utter and sentences the hearer derived from those sentences. As a consequence, in the adaptive logics approach presented below, scalar implicatures will be captured as default inference rules that may only be applied to utterance–sentences. For, this avoids the implementation–problem in a way that resembles the actual reasoning process at hand.

A closing remark is necessary though. From an utterance–sentence A, it is always possible to derive the corresponding standard sentence B by means of the logic $\mathbf{SL}^u$.[5] As a consequence, despite the non–standard interpretation of the utterance–symbols, the hearer is still able to derive all standard deductive consequences from the utterances made by the speaker, as is shown by theorem 1.[6]

Theorem 1. *For Γ the set of standard sentences corresponding to the utterance–sentences in Γ^u and for A a standard sentence:*

$$\Gamma^u \cup \Gamma^{bk} \vdash_{\mathbf{SL}^u} A \ \textit{iff} \ \Gamma \cup \Gamma^{bk} \vdash_{\mathbf{SL}} A.$$

3   The Adaptive Logics Approach

The adaptive logic $\mathbf{SI}^s$ now captures the reasoning process of the hearer while trying to uncover the full intended meaning of the utterances made by the speaker in a conversational context. In line with the argumentation of Levinson [10, ch. 1], the adaptive logic $\mathbf{SI}^s$ characterizes scalar implicatures proof theoretically as non–monotonic inference rules. Below, only a general (and quite intuitive) characterization of $\mathbf{SI}^s$ will be given.

General Characterization of $\mathbf{SI}^s$. All standard adaptive logics are characterized completely by the following three elements: a *lower limit logic* (**LLL**), a *set of abnormalities* Ω (a set of formulas characterized by a logical form $\mathbf{F}$), and an *adaptive strategy*.[7] In case of the logic $\mathbf{SI}^s$, the **LLL** is the logic $\mathbf{SL}^u$

---

[5] For a good understanding, the standard sentence B corresponding to the utterance–sentence A is obtained by replacing all utterance–symbols in A by the corresponding standard symbols.

[6] For all $\mathbf{SL}$ and $\mathbf{SL}^u$, the proof of theorem 1 is completely analogous to the proof of theorem 3 in [11]. Hence, no proof will be given in this paper.

[7] For an elaborated characterization of the *standard format* of adaptive logics, see e.g. [2],[3].

(see section 2).[8] Given a conversational context $\langle \Gamma^{\mathbf{u}} \cup \Gamma^{\mathbf{bk}} , \Gamma^{\mathbf{ls}} \rangle$, the consequences derivable from the premise set $\Gamma^{\mathbf{u}} \cup \Gamma^{\mathbf{bk}}$ by means of the logic $\mathbf{SL^u}$ are called the *deductive consequences* of that premise set, which means that they are non–defeasible (i.e. they cannot be withdrawn!). In other words, the logic $\mathbf{SL^u}$ is the stable, deductive background against which some defeasible inference steps can be made.

Where $A[e]$ expresses that the linguistic expression e occurs in the formula A, the set of abnormalities Ω of $\mathbf{SI^s}$ is defined as follows:

Definition 1. $\Omega = \{A[e] \wedge B[e'] \mid \langle ..., \Delta \cup \{e'\}, ..., \Theta \cup \{e\}, ... \rangle \in \Gamma^{\mathbf{ls}};$ $A[e]$ *is an utterance–sentence; $B[e']$ is obtained from $A[e]$ by (1) replacing all utterance–symbols by the corresponding standard symbols and (2) replacing the linguistic expression e by $e'\}$.*

The *defeasible* consequences of the logic $\mathbf{SI^s}$ (in casu, those representing the consequences obtained by applying scalar implicatures) are yielded by treating the abnormalities (the elements of Ω) in a particular way. More specifically, the logic $\mathbf{SI^s}$ falsifies as many abnormalities as possible. In general, this comes down to the following: if a formula $A \vee Dab(\Delta)$ is an $\mathbf{SL^u}$–consequence of a premise set Γ (with $Dab(\Delta)$ a finite disjunction of abnormalities), the formula A is considered an $\mathbf{SI^s}$–consequence of Γ *on the condition* that none of the abnormalities in Δ can be interpreted as true.

The above implies that a formula A is a *possible* $\mathbf{SI^s}$–consequence of a premise set Γ in case A is either a deductive or a defeasible consequence of Γ. Formally, this is expressed as follows:

Definition 2. *The formula A is a possible $\mathbf{SI^s}$–consequence of the premise set Γ iff there is a finite $\Delta \subset \Omega$ such that $\Gamma \vdash_{\mathbf{SL^u}} A \vee Dab(\Delta)$.*

If $\Delta = \emptyset$, the formula A is a deductive consequence of Γ, while in case $\Delta \neq \emptyset$, A is a defeasible consequence of Γ. As deductive consequences of a premise set are derivable unconditionally, they necessarily enter the $\mathbf{SI^s}$–consequence set of a premise set. On the other hand, defeasible consequences are only derivable conditionally, so that some might have to be withdrawn from the $\mathbf{SI^s}$–consequence set of a premise set. Which of the defeasible consequences have to be withdrawn, is determined by the Dab–consequences of the premise set, together with the adaptive strategy. A Dab–consequence of a premise set Γ is a finite disjunction of abnormalities that is deductively derivable from Γ.

Definition 3. $Dab(\Delta)$ *is a Dab–consequence of Γ iff $\Gamma \vdash_{\mathbf{SL^u}} Dab(\Delta)$.*

As no abnormalities need to be falsified in order to derive a Dab–consequence from a premise set, a Dab–consequence of a premise set is true unconditionally. Hence, some of the disjuncts of a Dab–consequence have to be true. This

---

[8] Obviously, the logic $\mathbf{SI^s}$ will differ according to the particular logic $\mathbf{SL^u}$ that is chosen as its $\mathbf{LLL}$. Hence, one might say that there are multiple versions of the logic $\mathbf{SI^s}$. One of these will be characterized in section 4.2. In this section though, the logic $\mathbf{SI^s}$ is characterized in general.

implies that some (and possibly all) of the defeasible consequences obtained by presupposing the falsity of these disjuncts have to be withdrawn. In the end, the adaptive strategy is decisive, for the latter provides the criterion to determine which defeasible consequences have to be withdrawn in view of the Dab–consequences of a premise set. The adaptive strategy of the logic $\mathbf{SI^s}$ is the *normal selections* strategy.[9] In general, this strategy states that a defeasible consequence of a premise set Γ obtained by presupposing the falsity of the abnormalities $A_1, ..., A_n$, is withdrawn in case the formula $Dab(\{A_1, ..., A_n\})$ is a Dab–consequence of Γ. Hence, in view of definition 2, $\mathbf{SI^s}$–derivability is defined as follows:

Definition 4. $\Gamma \vdash_{\mathbf{SI^s}} A$ *iff there is a finite* $\Delta \subset \Omega$ *such that* $\Gamma \vdash_{\mathbf{SL^u}} A \vee Dab(\Delta)$ *and* $\Gamma \nvdash_{\mathbf{SL^u}} Dab(\Delta)$.

Example. To illustrate the logic $\mathbf{SI^s}$, consider the example below.

Example 3. Consider the conversational context $\langle \{A[e]\} \cup \emptyset, \{\langle e', e\rangle, ...\}\rangle$. From the premise set $\{A[e]\}$, the formula $\neg B[e'] \vee (A[e] \wedge B[e'])$ is deductively derivable. By interpreting the abnormality $A[e] \wedge B[e']$ as false, the formula $\neg B[e']$ may be derived defeasibly. At this point, the conversational context doesn't provide any reason to withdraw the formula $\neg B[e']$ from the adaptive consequence set of $\{A[e]\}$. Nevertheless, if at some later point, the speaker should utter $B[e']$, the conversational context is extended to $\langle \{A[e], B[e']\} \cup \emptyset, \{\langle e', e\rangle, ...\}\rangle$. Consequently, the abnormality $A[e] \wedge B[e']$ cannot be considered as false anymore (because it is now deductively derivable from the premise set $\{A[e], B[e']\}$). This implies that the formula $\neg B[e']$ has to be withdrawn from the adaptive consequence set of $\{A[e], B[e']\}$.

4   Applying the Adaptive Framework

Let's consider a particular application of the general approach set out in the previous sections. More specifically, consider the *cookie conversation* below which contains some applications of scalar implicatures based on the linguistic scale $\langle \texttt{All}, \texttt{Many}, \texttt{Some}\rangle$.

Example 4 (The Cookie Conversation). John's mother is talking to the nanny about John's eating behavior.

MOTHER   Did John eat something this afternoon?
NANNY   Yes, he ate some cookies.

    IMPLICATES THAT   John didn't eat many cookies.
    IMPLICATES THAT   John didn't eat all cookies.

---

[9] A lot of other strategies have been characterized in the adaptive logics literature (see e.g. [2],[3]), but these will not be considered here.

NANNY   In fact, he ate many.

FORCES WITHDRAWAL OF   John didn't eat many cookies.

MOTHER   He didn't eat them all, did he?

NANNY   No, he didn't.

In view of the linguistic scale present in the conversational context described above, viz the scale $\langle$All, Many, Some$\rangle$, the assertion of the nanny that John ate some cookies, yields two scalar implicatures. For, from the nanny's first assertion John's mother will pragmatically derive that John didn't eat all cookies, as well as that he didn't eat many of them. However, when the nanny afterwards asserts that John ate a lot of cookies, John's mother is forced to withdraw one of those pragmatic conclusions, viz the latter one.

Representation of Linguistic Expressions. To capture the implicatures involved in the cookie conversation, the language $\mathcal{L}$ of *classical logic* isn't satisfactory, for not all linguistic expressions in the linguistic scale $\langle$All, Many, Some$\rangle$ can be expressed by classical means. Hence, the standard logic **SL** capturing the deductive background against which the scalar implicatures are performed (see section 2) cannot be classical logic. Consequently, I will take **SL** to be a straightforward extension of classical logic, viz the logic $\mathbf{CL}_{\exists 10}$.

The logic $\mathbf{CL}_{\exists 10}$ is based on the language $\mathcal{L}_{\exists 10}$, obtained by adding the *generalized quantifier* $\exists^{10}$ to the language $\mathcal{L}$ of classical logic.[10] This newly added quantifier expresses that there are at least ten objects in the domain for which something is the case. Consequently, the quantifier $\exists^{10}$ is semantically characterized as follows:[11]

$$v_M((\exists_\alpha^{10})A_\alpha) = 1 \quad \text{iff} \quad \text{there are } \beta_1, ..., \beta_{10} \in \mathcal{C} \cup \mathcal{O} \text{ such that } v(\beta_1) \neq v(\beta_2),$$
$$v(\beta_1) \neq v(\beta_3), ..., v(\beta_9) \neq v(\beta_{10}), \text{ and } v_M(A_{\beta_1}) = ... =$$
$$v_M(A_{\beta_{10}}) = 1.$$

Proof theoretically, the characterization of $\exists^{10}$ is obtained by means of the following axiom:

A$\exists^{10}$ $(\exists_\alpha^{10})A_\alpha \equiv (\exists_{\alpha_1})...(\exists_{\alpha_{10}})(A_{\alpha_1} \wedge ... \wedge A_{\alpha_{10}} \wedge \neg(\alpha_1 = \alpha_2) \wedge \neg(\alpha_1 = \alpha_3) \wedge$
     $... \wedge \neg(\alpha_9 = \alpha_{10}))$

Soundness and completeness proofs for $\mathbf{CL}_{\exists 10}$ are obtained by standard means. As a consequence, these are left to the reader.

Besides the quantifier $\exists^{10}$, the language $\mathcal{L}_{\exists 10}$ also contains a number of defined quantifiers, viz the generalized quantifiers All, Many and Some. These are *relational* quantifiers, which means that they express a relation between two formulas A and B. For example, the quantifier All expresses that *all* objects that are A are B as well (the other quantifiers are explicated analogously). Formally, the quantifiers All, Many and Some are defined as follows:[12]

---

[10] For more on generalized quantifiers, see e.g. [5],[15].

[11] Any member of the domain is taken to be named by a member of $\mathcal{C} \cup \mathcal{O}$, with $\mathcal{C}$ the set of individual constants and $\mathcal{O}$ a set of pseudo–constants (see also section 4.1).

[12] By using brackets and commas, I follow the notational conventions of [5].

Definition 5. *For α an individual variable:*

$$(\text{All}_\alpha)(A_\alpha, B_\alpha) \quad =_{df} \quad (\forall_\alpha)(A_\alpha \supset B_\alpha)$$
$$(\text{Many}_\alpha)(A_\alpha, B_\alpha) \quad =_{df} \quad (\exists_\alpha^{10})(A_\alpha \wedge B_\alpha)$$
$$(\text{Some}_\alpha)(A_\alpha, B_\alpha) \quad =_{df} \quad (\exists_\alpha)(A_\alpha \wedge B_\alpha)$$

Some remarks concerning these defined quantifiers are necessary. First of all, the quantifier Many is generally considered to be context–dependent (see e.g. [15]). Hence, in view of the conversational context provided by the cookie conversation, I have arbitrarily taken Many to be at least ten. Secondly, the introduction of the defined quantifiers is necessary to capture the real meaning of the scalar implicatures occurring in the cookie example. More specifically, to capture the scalar implicature from *some* to *not all*. For, remember that a scalar implicature is obtained by negating a sentence in which a low–ranked linguistic expression (in casu, *some*) is replaced by a more high–ranked one (in casu, *all*). In spite of appearances, one cannot capture this formally by deriving the negation of a formula in which the logical expression $\exists$ is replaced by the logical expression $\forall$. For example, consider the cookie conversation: $(\exists_\alpha)(C_\alpha \wedge E_{j\alpha})$ expresses that *John ate some cookies* (literally, the formula states that there are objects that are cookies and are eaten by John). Moreover, suppose that one would (pragmatically) derive the formula $\neg(\forall_\alpha)(C_\alpha \wedge E_{j\alpha})$ from the formula $(\exists_\alpha)(C_\alpha \wedge E_{j\alpha})$. Obviously, that doesn't capture the intended meaning of the scalar implicature at all, for the derived formula doesn't state that John didn't eat all cookies, but states that not everything is a cookie and is eaten by John. The problem resides in the fact that this formula does not only refer to cookies, but might also refer to tables, chairs,... Hence, John may well have eaten all cookies, as long as there is something that is not a cookie, the sentence still applies (which is obviously not what was intended). The quantifiers All, Many and Some are introduced to avoid this kind of mix up between linguistic and logical expressions.

Overview. In the remaining of this paper, a particular version of the adaptive logic $\mathbf{SI^s}$ will be characterized, viz the one that is able to capture the scalar implicatures occurring in the cookie conversation. This particular version of the logic $\mathbf{SI^s}$ will be called $\mathbf{CL_{\exists 10}^s}$. Well now, given the adaptive logics approach outlined in section 3, the lower limit logic of the logic $\mathbf{CL_{\exists 10}^s}$ is a particular extension of the logic standardly taken to capture the deductive background against which the scalar implicatures are performed. For the logic $\mathbf{CL_{\exists 10}^s}$, this is the logic $\mathbf{CL_{\exists 10}^u}$, an extension of the logic $\mathbf{CL_{\exists 10}}$ discussed above. Below, the logic $\mathbf{CL_{\exists 10}^u}$ will be characterized first (in section 4.1). Next, a characterization of the adaptive logic $\mathbf{CL_{\exists 10}^s}$ will be provided (in section 4.2). At the end, the cookie conversation will be reconsidered (in section 4.3).

4.1   The Lower Limit Logic $\mathbf{CL_{\exists 10}^u}$

The logic $\mathbf{CL_{\exists 10}^u}$ is based on the language $\mathcal{L}_{\exists 10}^u$. The latter is obtained by adding to the language $\mathcal{L}_{\exists 10}$ of $\mathbf{CL_{\exists 10}}$ an utterance–symbol $\dot{s}$ for each standard logical symbol s. As a consequence, the utterance–symbols of the language $\mathcal{L}^{\mathbf{u}}_{\exists 10}$ are the following:

$$\dot{\neg}, \dot{\wedge}, \dot{\vee}, \dot{\supset}, \dot{\equiv}, \dot{\exists}, \dot{\exists}^{10}, \dot{\forall}, \dot{=}, \dot{\mathtt{All}}, \dot{\mathtt{Many}}, \dot{\mathtt{Some}}$$

As their standard counterparts, the utterance–symbols $\dot{\mathtt{All}}, \dot{\mathtt{Many}}$ and $\dot{\mathtt{Some}}$ are defined connectives.

Definition 6. *For α an individual variable:*

$$
\begin{aligned}
(\dot{\mathtt{All}}_\alpha)(A_\alpha, B_\alpha) &=_{df} (\dot{\forall}_\alpha)(A_\alpha \dot{\supset} B_\alpha) \\
(\dot{\mathtt{Many}}_\alpha)(A_\alpha, B_\alpha) &=_{df} (\dot{\exists}^{10}_\alpha)(A_\alpha \dot{\wedge} B_\alpha) \\
(\dot{\mathtt{Some}}_\alpha)(A_\alpha, B_\alpha) &=_{df} (\dot{\exists}_\alpha)(A_\alpha \dot{\wedge} B_\alpha)
\end{aligned}
$$

In the remaining of this paper, also the connectives $\supset, \equiv, \dot{\supset}$ and $\dot{\equiv}$ will be treated as defined connectives (defined in the standard way). Consequently, only the most essential logical symbols are taken to be primitive.

Finally, let $\mathcal{S}, \mathcal{P}^r, \mathcal{C}, \mathcal{V}$, and $\mathcal{W}^{\mathbf{u}}_{\exists 10}$ be respectively the set of sentential letters, the set of predicative letters of rank r, the set of individual constants, the set of individual variables, and the set of well–formed formulas of the language $\mathcal{L}^{\mathbf{u}}_{\exists 10}$. All are defined in the usual way.

Semantics. The semantics of the logic $\mathbf{CL}^{\mathbf{u}}_{\exists 10}$ isn't defined for the language $\mathcal{L}^{\mathbf{u}}_{\exists 10}$, but for the language $\mathcal{L}^{\mathbf{u+}}_{\exists 10}$. The latter is obtained by adding the set of pseudo–constants $\mathcal{O}$ to the language $\mathcal{L}^{\mathbf{u}}_{\exists 10}$. In the semantics of $\mathbf{CL}^{\mathbf{u}}_{\exists 10}$, the set $\mathcal{C} \cup \mathcal{O}$ plays the role usually played by $\mathcal{C}$, with this distinction that it is required that any element of the domain is named by at least one element of $\mathcal{C} \cup \mathcal{O}$. As a consequence, the introduction of $\mathcal{O}$ greatly simplifies the semantic characterization of the quantifiers.[13]

Let $\mathcal{F}^{\mathbf{u+}}_{\exists 10}$ be the set of all formulas of $\mathcal{L}^{\mathbf{u+}}_{\exists 10}$ (both open and closed ones), and let $\mathcal{W}^{\mathbf{u+}}_{\exists 10}$ be the set of all well–formed (closed) formulas of $\mathcal{L}^{\mathbf{u+}}_{\exists 10}$. Both are defined in the standard way. Moreover, let $\mathcal{W}^{\dot{s}+}_{\exists 10}$ be the set of well–formed formulas of $\mathcal{L}^{\mathbf{u+}}_{\exists 10}$ of which the main logical symbols are utterance–symbols (see definition 7), and let $\mathcal{W}^{\dot{\neg}+}_{\exists 10}$ be the set of well–formed formulas $\dot{\neg}A$ of $\mathcal{L}^{\mathbf{u+}}_{\exists 10}$ such that the main logical symbol of the formula A is a standard symbol (see definition 8).

Definition 7. $\mathcal{W}^{\dot{s}+}_{\exists 10} = \bigcup(\; \{\dot{\neg}A \mid A \in \mathcal{S}\},$
$\{\dot{\neg}\pi\beta_1...\beta_r \mid \pi \in \mathcal{P}^r \text{ and } \beta_1, ..., \beta_r \in \mathcal{C} \cup \mathcal{O}\},$
$\{\alpha \dot{=} \beta \mid \alpha, \beta \in \mathcal{C} \cup \mathcal{O}\},$
$\{\dot{\neg}(\alpha \dot{=} \beta) \mid \alpha, \beta \in \mathcal{C} \cup \mathcal{O}\},$
$\{\dot{\neg}\dot{\neg}A \mid A \in \mathcal{W}^{\mathbf{u+}}_{\exists 10}\},$
$\{A \dot{\wedge} B \mid A, B \in \mathcal{W}^{\mathbf{u+}}_{\exists 10}\},$
$\{\dot{\neg}(A \dot{\wedge} B) \mid A, B \in \mathcal{W}^{\mathbf{u+}}_{\exists 10}\},$
$\{A \dot{\vee} B \mid A, B \in \mathcal{W}^{\mathbf{u+}}_{\exists 10}\},$

---

[13] Obviously, the set $\mathcal{O}$ should have at least the cardinality of the largest model considered. If there is no such model, a suitable $\mathcal{O}$ has to be selected for each model.

$$\{ \dot{\neg}(A \dot{\vee} B) \mid A, B \in \mathcal{W}^{\mathbf{u}+}_{\exists 10} \},$$
$$\{ (\dot{\exists}_\alpha) A_\alpha \mid A_\alpha \in \mathcal{F}^{\mathbf{u}+}_{\exists 10} \},$$
$$\{ \dot{\neg}(\dot{\exists}_\alpha) A_\alpha \mid A_\alpha \in \mathcal{F}^{\mathbf{u}+}_{\exists 10} \},$$
$$\{ (\dot{\exists}^{10}_\alpha) A_\alpha \mid A_\alpha \in \mathcal{F}^{\mathbf{u}+}_{\exists 10} \},$$
$$\{ \dot{\neg}(\dot{\exists}^{10}_\alpha) A_\alpha \mid A_\alpha \in \mathcal{F}^{\mathbf{u}+}_{\exists 10} \},$$
$$\{ (\dot{\forall}_\alpha) A_\alpha \mid A_\alpha \in \mathcal{F}^{\mathbf{u}+}_{\exists 10} \},$$
$$\{ \dot{\neg}(\dot{\forall}_\alpha) A_\alpha \mid A_\alpha \in \mathcal{F}^{\mathbf{u}+}_{\exists 10} \} \ )$$

Definition 8. $\mathcal{W}^{\dot{\neg}+}_{\exists 10} = \bigcup (\ \{ \dot{\neg}(\alpha = \beta) \mid \alpha, \beta \in \mathcal{C} \cup \mathcal{O} \},$
$$\{ \dot{\neg}\neg A \mid A \in \mathcal{W}^{\mathbf{u}+}_{\exists 10} \},$$
$$\{ \dot{\neg}(A \wedge B) \mid A, B \in \mathcal{W}^{\mathbf{u}+}_{\exists 10} \},$$
$$\{ \dot{\neg}(A \vee B) \mid A, B \in \mathcal{W}^{\mathbf{u}+}_{\exists 10} \},$$
$$\{ \dot{\neg}(\exists_\alpha) A_\alpha \mid A_\alpha \in \mathcal{F}^{\mathbf{u}+}_{\exists 10} \},$$
$$\{ \dot{\neg}(\exists^{10}_\alpha) A_\alpha \mid A_\alpha \in \mathcal{F}^{\mathbf{u}+}_{\exists 10} \},$$
$$\{ \dot{\neg}(\forall_\alpha) A_\alpha \mid A_\alpha \in \mathcal{F}^{\mathbf{u}+}_{\exists 10} \} \ )$$

Characterizing Models. A $\mathbf{CL}^{\mathbf{u}}_{\exists 10}$–model M is a couple $\langle D, v \rangle$ with D a non-empty domain and v an assignment function. The latter is defined as follows:

C1.1 $v : \mathcal{S} \cup \mathcal{W}^{\dot{s}+}_{\exists 10} \cup \mathcal{W}^{\dot{\neg}+}_{\exists 10} \to \{0,1\}$

C1.2 $v : \mathcal{C} \cup \mathcal{O} \to D$ (where $D = \{ v(\alpha) \mid \alpha \in \mathcal{C} \cup \mathcal{O} \}$)

C1.3 $v : \mathcal{P}^r \to \rho(D)$ (the power set of the r–th Cartesian product of D)

The assignment function v of the model M is extended to a valuation function $v_M : \mathcal{W}^{\mathbf{u}+}_{\exists 10} \to \{0,1\}$ by the following semantic postulates:

C2.1   For $A \in \mathcal{S} \cup \mathcal{W}^{\dot{\neg}+}_{\exists 10}$, $v_M(A) = 1$ iff $v(A) = 1$.

C2.2   For $\alpha, \beta \in \mathcal{C} \cup \mathcal{O}$, $v_M(\alpha = \beta) = 1$ iff $v(\alpha) = v(\beta)$.

C2.3   For $\pi \in \mathcal{P}^r$ and $\beta_1, ..., \beta_r \in \mathcal{C} \cup \mathcal{O}$, $v_M(\pi \beta_1 ... \beta_r) = 1$ iff $\langle v(\beta_1), ..., v(\beta_r) \rangle \in v(\pi)$.

C2.4 $v_M(\neg A) = 1$ iff $v_M(A) = 0$.

C2.5 $v_M(A \wedge B) = 1$ iff $v_M(A) = 1$ and $v_M(B) = 1$.

C2.6 $v_M(A \vee B) = 1$ iff $v_M(A) = 1$ or $v_M(B) = 1$.

C2.7 $v_M((\exists_\alpha) A_\alpha) = 1$ iff $v_M(A_\beta) = 1$ for at least one $\beta \in \mathcal{C} \cup \mathcal{O}$.

C2.8 $v_M((\exists^{10}_\alpha) A_\alpha) = 1$ iff there are $\beta_1, ..., \beta_{10} \in \mathcal{C} \cup \mathcal{O}$ such that $v(\beta_1) \neq v(\beta_2)$, $v(\beta_1) \neq v(\beta_3), ..., v(\beta_9) \neq v(\beta_{10})$, and $v_M(A_{\beta_1}) = ... = v_M(A_{\beta_{10}}) = 1$.

C2.9 $v_M((\forall_\alpha) A_\alpha) = 1$ iff $v_M(A_\beta) = 1$ for all $\beta \in \mathcal{C} \cup \mathcal{O}$.

C2.10 For $A \in \mathcal{S}$, $v_M(\dot{\neg} A) = 1$ iff $v_M(A) = 0$, and $v(\dot{\neg} A) = 1$.

C2.11 For $\alpha, \beta \in \mathcal{C} \cup \mathcal{O}$, $v_M(\alpha \dot{=} \beta) = 1$ iff $v(\alpha) = v(\beta)$, and $v(\alpha \dot{=} \beta) = 1$.

C2.12 For $\alpha, \beta \in \mathcal{C} \cup \mathcal{O}$, $v_M(\dot{\neg}(\alpha \dot{=} \beta)) = 1$ iff $v(\alpha) \neq v(\beta)$, and $v(\dot{\neg}(\alpha \dot{=} \beta)) = 1$.

C2.13 For $\pi \in \mathcal{P}^r$ and $\beta_1, ..., \beta_r \in \mathcal{C} \cup \mathcal{O}$, $v_M(\dot{\neg}\pi \beta_1 ... \beta_r) = 1$ iff $v_M(\pi \beta_1 ... \beta_r) = 0$, and $v(\dot{\neg}\pi \beta_1 ... \beta_r) = 1$.

C2.14 $v_M(\dot{\neg}\neg A) = 1$ iff $v_M(A) = 1$, and $v(\dot{\neg}\neg A) = 1$.

C2.15 $v_M(A \dot{\wedge} B) = 1$ iff $v_M(A) = 1$ and $v_M(B) = 1$, and $v(A \dot{\wedge} B) = 1$.

C2.16 $v_M(\dot{\neg}(A \dot{\wedge} B)) = 1$ iff $v_M(\dot{\neg} A) = 1$ or $v_M(\dot{\neg} B) = 1$, and $v(\dot{\neg}(A \dot{\wedge} B)) = 1$.

C2.17 $v_M(A \dot{\vee} B) = 1$ iff $v_M(A) = 1$ or $v_M(B) = 1$, and $v(A \dot{\vee} B) = 1$.

C2.18 $v_M(\dot{\neg}(A\dot{\vee}B)) = 1$ iff $v_M(\dot{\neg}A) = 1$ and $v_M(\dot{\neg}B) = 1$, and $v(\dot{\neg}(A\dot{\vee}B)) = 1$.

C2.19 $v_M((\dot{\exists}_\alpha)A_\alpha) = 1$ iff $v_M(A_\beta) = 1$ for at least one $\beta \in \mathcal{C} \cup \mathcal{O}$, and $v((\dot{\exists}_\alpha)A_\alpha) = 1$.

C2.20 $v_M(\dot{\neg}(\dot{\exists}_\alpha)A_\alpha) = 1$ iff $v_M(\dot{\neg}A_\beta) = 1$ for all $\beta \in \mathcal{C} \cup \mathcal{O}$, and $v(\dot{\neg}(\dot{\exists}_\alpha)A_\alpha) = 1$.

C2.21 $v_M((\dot{\exists}_\alpha^{10})A_\alpha) = 1$ iff there are $\beta_1, ..., \beta_{10} \in \mathcal{C} \cup \mathcal{O}$ such that $v(\beta_1) \neq v(\beta_2)$, $v(\beta_1) \neq v(\beta_3),..., v(\beta_9) \neq v(\beta_{10})$, $v_M(A_{\beta_1}) = ... = v_M(A_{\beta_{10}}) = 1$, and $v((\dot{\exists}_\alpha^{10})A_\alpha) = 1$.

C2.22 $v_M(\dot{\neg}(\dot{\exists}_\alpha^{10})A_\alpha) = 1$ iff for all $\beta_1,...,\beta_{10} \in \mathcal{C} \cup \mathcal{O}$: if $v_M(A_{\beta_1}) = ... = v_M(A_{\beta_{10}}) = 1$ then $v(\beta_1) = v(\beta_2)$ or $v(\beta_1) = v(\beta_3)$ or ... or $v(\beta_9) = v(\beta_{10})$, and $v(\dot{\neg}(\dot{\exists}_\alpha^{10})A_\alpha) = 1$.

C2.23 $v_M((\dot{\forall}_\alpha)A_\alpha) = 1$ iff $v_M(A_\beta) = 1$ for all $\beta \in \mathcal{C} \cup \mathcal{O}$, and $v((\dot{\forall}_\alpha)A_\alpha) = 1$.

C2.24 $v_M(\dot{\neg}(\dot{\forall}_\alpha)A_\alpha) = 1$ iff $v_M(\dot{\neg}A_\beta) = 1$ for at least one $\beta \in \mathcal{C} \cup \mathcal{O}$, and $v(\dot{\neg}(\dot{\forall}_\alpha)A_\alpha) = 1$.

Semantic Consequence. Remember that pseudo–constants were introduced merely as a semantic aid, to simplify the characterization of the quantifiers. However, pseudo–constants are not allowed in the premises nor the conclusion of arguments. Hence, semantic consequence is defined over formulas that do not contain any pseudo–constants. In other words, semantic consequence is defined over well–formed formulas of the language $\mathcal{L}^{\mathbf{u}}_{\exists 10}$.

A well–formed formula A of the language $\mathcal{L}^{\mathbf{u}}_{\exists 10}$ is verified by a model M iff $v_M(A) = 1$. Moreover, a model M is a model of a premise set Γ iff M verifies all elements of Γ. Finally, semantic consequence is defined as follows:

Definition 9 (Semantic Consequence). $\Gamma \vDash_{\mathbf{CL}^{\mathbf{u}}_{\exists 10}} A$ iff A is verified by all $\mathbf{CL}^{\mathbf{u}}_{\exists 10}$–models of the premise set Γ.

Proof Theory. Proof theoretically, the logic $\mathbf{CL}^{\mathbf{u}}_{\exists 10}$ is characterized completely by adding the axioms in table 1 to the axiom system of $\mathbf{CL}_{\exists 10}$ (as described above). Proofs are defined in the standard way, as sequences of well–formed formulas each of which is either an axiom, a premise or a formula derived from earlier ones by application of a rule of inference. Consequently, derivability is defined as follows:

Definition 10 (Derivability). $\Gamma \vdash_{\mathbf{CL}^{\mathbf{u}}_{\exists 10}} A$ iff there is a proof of A from $B_1, ..., B_n \in \Gamma$.

Soundness and Completeness. Soundness and completeness for the logic $\mathbf{CL}^{\mathbf{u}}_{\exists 10}$ is easily obtained by extending the proofs of theorems 1 and 2 in [11]. As the extensions are completely straightforward, this is left to the reader.

Theorem 2. $\Gamma \vdash_{\mathbf{CL}^{\mathbf{u}}_{\exists 10}} A$ iff $\Gamma \vDash_{\mathbf{CL}^{\mathbf{u}}_{\exists 10}} A$.

Table 1. Additional Axioms of $\mathbf{CL}^{\mathrm{u}}_{\dot{\exists}10}$

A$\dot{\neg}$	$\dot{\neg}A \supset \neg A$	A$\dot{\neg}\dot{\neg}$	$\dot{\neg}\dot{\neg}A \supset A$
A$\dot{\wedge}$	$(A\dot{\wedge}B) \supset (A \wedge B)$	A$\dot{\neg}\dot{\wedge}$	$\dot{\neg}(A\dot{\wedge}B) \supset (\dot{\neg}A \vee \dot{\neg}B)$
A$\dot{\vee}$	$(A\dot{\vee}B) \supset (A \vee B)$	A$\dot{\neg}\dot{\vee}$	$\dot{\neg}(A\dot{\vee}B) \supset (\dot{\neg}A \wedge \dot{\neg}B)$
A$\dot{=}$	$(\alpha\dot{=}\beta) \supset (\alpha = \beta)$	A$\dot{\neg}\dot{=}$	$\dot{\neg}(\alpha\dot{=}\beta) \supset \neg(\alpha = \beta)$
A$\dot{\exists}$	$(\dot{\exists}_\alpha)A_\alpha \supset (\exists_\alpha)A_\alpha$	A$\dot{\neg}\dot{\exists}$	$\dot{\neg}(\dot{\exists}_\alpha)A_\alpha \supset (\forall_\alpha)\dot{\neg}A_\alpha$
A$\dot{\exists}^{10}$	$(\dot{\exists}^{10}_\alpha)A_\alpha \supset (\exists^{10}_\alpha)A_\alpha$	A$\dot{\neg}\dot{\exists}^{10}$	$\dot{\neg}(\dot{\exists}^{10}_\alpha)A_\alpha \supset \neg(\exists^{10}_\alpha)A_\alpha$
A$\dot{\forall}$	$(\dot{\forall}_\alpha)A_\alpha \supset (\forall_\alpha)A_\alpha$	A$\dot{\neg}\dot{\forall}$	$\dot{\neg}(\dot{\forall}_\alpha)A_\alpha \supset (\exists_\alpha)\dot{\neg}A_\alpha$

4.2   The Adaptive Logic $\mathbf{CL}^{\mathrm{s}}_{\dot{\exists}10}$

The lower limit logic (**LLL**) of the logic $\mathbf{CL}^{\mathrm{s}}_{\dot{\exists}10}$ is the logic $\mathbf{CL}^{\mathrm{u}}_{\dot{\exists}10}$ described in section 4.1, and the adaptive strategy of $\mathbf{CL}^{\mathrm{s}}_{\dot{\exists}10}$ is the normal selections strategy. Hence, before I can move on to the semantics and the proof theory, the set of abnormalities Ω still needs to be defined (in section 3 only a general characterization of Ω has been given — see definition 1).

Definition 11. $\Omega = \Omega_1 \cup \Omega_2 \cup \Omega_3$, with

$\Omega_1 = \{(\dot{\mathrm{Some}}_\alpha)(A_\alpha, B_\alpha) \wedge (\mathrm{Many}_\alpha)(A'_\alpha, B'_\alpha) \mid A, B$ only contain utterance–symbols; A', B' are obtained from respectively A and B by replacing all utterance–symbols by the corresponding standard symbols$\}$

$\Omega_2 = \{(\dot{\mathrm{Some}}_\alpha)(A_\alpha, B_\alpha) \wedge (\mathrm{All}_\alpha)(A'_\alpha, B'_\alpha) \mid A, B$ only contain utterance–symbols; A', B' are obtained from respectively A and B by replacing all utterance–symbols by the corresponding standard symbols$\}$

$\Omega_3 = \{(\dot{\mathrm{Many}}_\alpha)(A_\alpha, B_\alpha) \wedge (\mathrm{All}_\alpha)(A'_\alpha, B'_\alpha) \mid A, B$ only contain utterance–symbols; A', B' are obtained from respectively A and B by replacing all utterance–symbols by the corresponding standard symbols$\}$

By defining Ω in this way, the logic $\mathbf{CL}^{\mathrm{s}}_{\dot{\exists}10}$ is only able to capture scalar implicatures based on the linguistic scale $\langle \mathrm{All}, \mathrm{Many}, \mathrm{Some}\rangle$. Obviously, Ω can easily be extended in order to capture more scalar implicatures. To keep things as simple as possible, I will not do so here.

Semantics. The $\mathbf{CL}^{\mathrm{s}}_{\dot{\exists}10}$–semantics is based on the **LLL**–models of a premise set Γ. More specifically, to generate more consequences than the **LLL**, the $\mathbf{CL}^{\mathrm{s}}_{\dot{\exists}10}$–consequences are defined by reference to one or multiple *selected sets* of **LLL**–models of Γ, i.e. sets of *preferred* **LLL**–models of Γ. Hence, the $\mathbf{CL}^{\mathrm{s}}_{\dot{\exists}10}$–semantics is a so–called *preferential semantics*. As the **LLL** of $\mathbf{CL}^{\mathrm{s}}_{\dot{\exists}10}$ is the logic $\mathbf{CL}^{\mathrm{u}}_{\dot{\exists}10}$, semantic consequence for the logic $\mathbf{CL}^{\mathrm{s}}_{\dot{\exists}10}$ is defined as follows:

Definition 12 (Semantic Consequence). $\Gamma \vDash_{\mathbf{CL}^{\mathrm{s}}_{\dot{\exists}10}} A$ iff A is verified by all elements of some selected sets of $\mathbf{CL}^{\mathrm{u}}_{\dot{\exists}10}$–models of Γ.

Defining Selected Sets. Whether a particular $\mathbf{CL}^{\mathrm{u}}_{\dot{\exists}10}$–model M of Γ will make it to a selected set Σ, depends on the *abnormal part* of M and on the adaptive strategy of the logic $\mathbf{CL}^{\mathrm{s}}_{\dot{\exists}10}$. The abnormal part of a model M is the set of abnormalities M verifies.

Definition 13. *Where M is a $\mathbf{CL}^{\mathbf{u}}_{\exists 10}$–model, the abnormal part of M is the set* $Ab(M) = \{A \in \Omega \mid v_M(A) = 1\}$.

The adaptive strategy makes the actual selection among the $\mathbf{CL}^{\mathbf{u}}_{\exists 10}$–models. This is done by comparing their abnormal parts. As the adaptive strategy of the logic $\mathbf{CL}^{\mathbf{s}}_{\exists 10}$ is the normal selections strategy, a selected set Σ is defined by means of a two–step procedure. First, the *minimally abnormal models* of a premise set Γ are defined.

Definition 14. *A $\mathbf{CL}^{\mathbf{u}}_{\exists 10}$–model M of Γ is* minimally abnormal *iff there is no $\mathbf{CL}^{\mathbf{u}}_{\exists 10}$–model M' of Γ such that $Ab(M') \subset Ab(M)$.*

Secondly, all minimally abnormal models that verify the same abnormalities, are grouped together into distinct sets. These sets are the selected sets of $\mathbf{CL}^{\mathbf{u}}_{\exists 10}$–models of a premise set Γ.

Definition 15. $\Phi(\Gamma) = \{Ab(M) \mid M$ *is a minimally abnormal model of Γ*$\}$.

Definition 16. *A set Σ of $\mathbf{CL}^{\mathbf{u}}_{\exists 10}$–models of Γ is a* selected set *iff for some $\phi \in \Phi(\Gamma)$, $\Sigma = \{M \mid Ab(M) = \phi\}$.*

Proof Theory. As the logic $\mathbf{CL}^{\mathbf{s}}_{\exists 10}$ is a standard adaptive logic, its proof theory has some characteristic features shared by all adaptive logics (see also [2],[3]). First of all, a $\mathbf{CL}^{\mathbf{s}}_{\exists 10}$–proof is a succession of stages, each consisting of a sequence of lines. Adding a line to a proof means to move on to the next stage of the proof. Next, the lines of a $\mathbf{CL}^{\mathbf{s}}_{\exists 10}$–proof consist of four elements (instead of the usual three): a line number, a formula, a justification, and an *adaptive condition*. The latter is a finite subset of Ω (the set of abnormalities). As long as all elements of the adaptive condition of a line i can be considered as false, the formula on line i is considered as derivable from the premise set — remark that this is in accordance with the intuition set out in section 3. In order to indicate that not all elements of the adaptive condition of line i can be considered as false anymore, line i is marked (formally, this is done by placing the symbol $\checkmark$ next to the adaptive condition). Obviously, when a line is marked, the formula on that line is not considered as derivable anymore. Finally, markings are dynamic: at some stage of the proof, a line might be unmarked, while at a later stage, it might become marked.[14] Obviously, this proof theoretic dynamics corresponds to the dynamics involved in the use of scalar implicatures (as described in section 2).

Characterizing Proofs. Now, consider the $\mathbf{CL}^{\mathbf{s}}_{\exists 10}$–proof theory in particular. It consists of both *deduction rules* and a *marking definition*. The deduction rules determine how new lines may be added to a proof, while the marking definition determines at every stage of the proof which lines have to be marked. The deduction rules are listed in shorthand notation, with

$$A \qquad \Delta$$

---

[14] For some adaptive logics, lines that are marked might become unmarked as well. Not for the logic $\mathbf{CL}^{\mathbf{s}}_{\exists 10}$ though.

expressing that A occurs in the proof on a line with condition Δ. Consider the deduction rules below:

The Premise Rule (PREM)
If $A \in \Gamma$:

$$\frac{\cdots \qquad \cdots}{A \qquad \emptyset}$$

The Unconditional Rule (RU)
If $A_1, \ldots, A_n \vdash_{\mathbf{CL}^{\mathbf{u}}_{\exists 10}} B$:

$$
\begin{array}{ll}
A_1 & \Delta_1 \\
\vdots & \vdots \\
A_n & \Delta_n \\
\hline
B & \Delta_1 \cup \ldots \cup \Delta_n
\end{array}
$$

The Conditional Rule (RC)
If $A_1, \ldots, A_n \vdash_{\mathbf{CL}^{\mathbf{u}}_{\exists 10}} B \vee Dab(\Theta)$:

$$
\begin{array}{ll}
A_1 & \Delta_1 \\
\vdots & \vdots \\
A_n & \Delta_n \\
\hline
B & \Delta_1 \cup \ldots \cup \Delta_n \cup \Theta
\end{array}
$$

It is easily verified that the deduction rules are fully determined by the logic $\mathbf{CL}^{\mathbf{u}}_{\exists 10}$ (the **LLL** of the logic $\mathbf{CL}^{\mathbf{s}}_{\exists 10}$) and the set of abnormalities Ω. The marking definition on the other hand, strongly depends on the adaptive strategy.[15] To determine whether or not a line has to be marked at a certain stage of a proof, the adaptive strategy of the logic $\mathbf{CL}^{\mathbf{s}}_{\exists 10}$, i.e. the normal selections strategy, refers to the *Dab–consequences* of the premise set that have been derived at that stage of the proof.

Definition 17. $Dab(\Delta)$ *is a Dab–consequence of a premise set Γ at stage s of a proof iff $Dab(\Delta)$ is derived at stage s on a line with condition $\emptyset$.*

More specifically, the normal selections strategy lays down that a line i with condition Δ has to be marked at stage s in case $Dab(\Delta)$ is a Dab–consequence of the premise set at stage s.

Definition 18 (Marking for Normal Selections). *Line i is marked at stage s of the proof iff, where Δ is its condition, $Dab(\Delta)$ is a Dab–consequence of Γ at stage s.*

Defining Derivability. A formula A is derivable from a premise set Γ iff A has been derived as the second element of an unmarked line in a proof from Γ. However, defining derivability this way is rather problematic. For, markings may change at every stage, so that for every new stage, it has to be reconsidered

---

[15] In general, the marking definition constitutes the only difference between the proof theories of adaptive logics that have identical lower limit logics and sets of abnormalities (see e.g. [2]).

whether or not a formula is derivable from the premise set. Nonetheless, also a stable notion of derivability can be defined. It is called *final derivability*, which refers to the fact that for some formulas, derivability can only be decided at the final stage of a proof.

Definition 19. *A is finally derived from Γ on a line i of a proof at stage s iff (i) A is the second element of line i, (ii) line i is not marked at stage s, and (iii) every extension of the proof in which line i is marked may be further extended in such a way that line i is unmarked.*

Because of its stability, the notion of final derivability is used to define the $\mathbf{CL}^{\mathbf{s}}_{\exists 10}$–consequence relation.

Definition 20. $\Gamma \vdash_{\mathbf{CL}^{\mathbf{s}}_{\exists 10}} A$ *iff A is finally derived on a line of a proof from Γ.*

Soundness and Completeness. As $\mathbf{CL}^{\mathbf{s}}_{\exists 10}$ is a standard adaptive logic, soundness and completeness follow immediately (see corollary 2 in [3]). Hence, the soundness and completeness proofs for $\mathbf{CL}^{\mathbf{s}}_{\exists 10}$ needn't be considered here.

Theorem 3. $\Gamma \vdash_{\mathbf{CL}^{\mathbf{s}}_{\exists 10}} A$ *iff* $\Gamma \vDash_{\mathbf{CL}^{\mathbf{s}}_{\exists 10}} A$.

4.3 The Cookie Conversation

Let's return to the cookie conversation one more time. Given the conversational context at hand, the information available to John's mother is represented as follows:

(CC) $\langle \ \{(\dot{\mathsf{Some}}_\alpha)(C_\alpha, E_{j\alpha}), (\dot{\mathsf{Many}}_\alpha)(C_\alpha, E_{j\alpha}), \dot{\neg}(\dot{\mathsf{All}}_\alpha)(C_\alpha, E_{j\alpha})\} \cup \emptyset,$
$\{\langle \mathsf{All}, \mathsf{Many}, \mathsf{Some}\rangle\} \ \rangle$

In **CC**, the set $\Gamma^{\mathbf{u}}$ contains all sentences John's mother heard the nanny utter. Moreover, these utterances are placed in chronological order (actually, to represent the application of scalar implicatures in a realistic way, they should and will also enter the proof in this order). For reasons of simplicity, the set $\Gamma^{\mathbf{bk}}$ is left empty. Nonetheless, this isn't necessarily the case, for there may be a lot of background knowledge shared by John's mother and nanny. For example, they might share knowledge about John's eating habits, his likes and dislikes, etc. Finally, the set $\Gamma^{\mathbf{ls}}$ only contains one element, viz the linguistic scale $\langle \mathsf{All}, \mathsf{Many}, \mathsf{Some}\rangle$, for it is the only linguistic scale present in this conversational context.

The Cookie Conversation Formally Remastered. The $\mathbf{CL}^{\mathbf{s}}_{\exists 10}$–proof below captures the cookie conversation from the viewpoint of John's mother. Hence, the proof starts with the utterance of the nanny that John ate some cookies (see line 1 below), followed by the defeasible consequences drawn from this utterance by means of the scalar implicatures based on the scale $\langle \mathsf{All}, \mathsf{Many}, \mathsf{Some}\rangle$ (see lines 2 and 3 below).

1 $(\dot{\text{Some}}_\alpha)(C_\alpha, E_{j\alpha})$ –;PREM $\emptyset$
2 $\neg(\text{Many}_\alpha)(C_\alpha, E_{j\alpha})$ 1;RC $\{(\dot{\text{Some}}_\alpha)(C_\alpha, E_{j\alpha}) \wedge (\text{Many}_\alpha)(C_\alpha, E_{j\alpha})\}$
3 $\neg(\text{All}_\alpha)(C_\alpha, E_{j\alpha})$ 1;RC $\{(\dot{\text{Some}}_\alpha)(C_\alpha, E_{j\alpha}) \wedge (\text{All}_\alpha)(C_\alpha, E_{j\alpha})\}$

At stage 3 of the proof, no Dab–consequences of the premise set have been derived yet. Hence, no markings occur and all formulas derived on a line of the proof are considered as $\mathbf{CL}^\mathbf{s}_{\exists 10}$–derivable. However, the proof continues with the nanny's utterance that John actually ate a lot of cookies (see line 4). This obviously forces the withdrawal of one of the pragmatic conclusions drawn by John's mother (line 2 gets marked).

1 $(\dot{\text{Some}}_\alpha)(C_\alpha, E_{j\alpha})$ –;PREM $\emptyset$
2 $\neg(\text{Many}_\alpha)(C_\alpha, E_{j\alpha})$ 1;RC $\{(\dot{\text{Some}}_\alpha)(C_\alpha, E_{j\alpha}) \wedge (\text{Many}_\alpha)(C_\alpha, E_{j\alpha})\}$ ✓
3 $\neg(\text{All}_\alpha)(C_\alpha, E_{j\alpha})$ 1;RC $\{(\dot{\text{Some}}_\alpha)(C_\alpha, E_{j\alpha}) \wedge (\text{All}_\alpha)(C_\alpha, E_{j\alpha})\}$
4 $(\dot{\text{Many}}_\alpha)(C_\alpha, E_{j\alpha})$ –;PREM $\emptyset$
5 $(\dot{\text{Some}}_\alpha)(C_\alpha, E_{j\alpha}) \wedge$ 1,4;RU $\emptyset$
   $(\text{Many}_\alpha)(C_\alpha, E_{j\alpha})$

At stage 5 of the proof, a Dab–consequence has been derived on line 5. As a consequence, line 2 is marked, meaning that the formula on that line is not considered as $\mathbf{CL}^\mathbf{s}_{\exists 10}$–derivable anymore. On the other hand, line 3 is unmarked at stage 5 of the proof and it is easily verified that this will remain so, no matter how the proof is extended (no Dab–consequence yielding the marking of line 3 is derivable from the premise set). Moreover, this is also confirmed by the nanny's final utterance, viz that John didn't eat all cookies.

... ...
6 $\dot{\neg}(\dot{\text{All}}_\alpha)(C_\alpha, E_{j\alpha})$ –;PREM $\emptyset$

5   Conclusion

In this paper, I have provided a formal explication of how hearers apply scalar implicatures to get at the full intended meaning of uncooperative utterances (utterances that are not completely in accordance with the Gricean maxims). I have shown how this can be done in general, by relying on the Adaptive Logics Programme. More specifically, I have given an outline of how an adaptive logic $\mathbf{SI}^\mathbf{s}$ for scalar implicatures should look like in general. Moreover, I have characterized a particular version of $\mathbf{SI}^\mathbf{s}$, viz the logic $\mathbf{CL}^\mathbf{s}_{\exists 10}$. The latter captures applications of scalar implicatures based on the linguistic scale $\langle \text{All}, \text{Many}, \text{Some} \rangle$. The logic $\mathbf{CL}^\mathbf{s}_{\exists 10}$ does so by treating the scalar implicatures as default rules of inference. As this is completely in accordance with the characterization of scalar implicatures given by Levinson in [10], the approach characterized in this paper adequately explicates the use made of scalar implicatures in conversation.

Further Research. This paper dealt with scalar implicatures, which only constitute a (relatively small) fragment of all possible implicatures. It is still an open question whether those other implicatures can also be captured by means of adaptive logics. Moreover, all (scalar) implicatures in this paper were treated as having an equal priority. This is not always the case though. For, in certain conversational contexts, some (scalar) implicatures are given a higher priority than others. Despite the fact that the logic $\mathbf{SI^s}$ cannot cope with this phenomenon, it should be possible to construct prioritized adaptive logics that can.

References

1. Bach, K.: The Top Ten Misconceptions about Implicature. In: Birner, B., Ward, G. (eds.) Drawing the Boundaries of Meaning: Neo-Gricean Studies in Pragmatics and Semantics in Honor of Laurence R. Horn, pp. 21–30. John Benjamins, Amsterdam (2006)
2. Batens, D.: A Universal Logic Approach to Adaptive Logics. Logica Universalis 1, 221–242 (2007)
3. Batens, D., Meheus, J., Provijn, D.: An Adaptive Characterization of Signed Systems for Paraconsistent Reasoning (to appear)
4. Gazdar, G.: Pragmatics: Implicature, Presupposition and Logical Form. Academic Press, New York (1979)
5. Glanzberg, M.: Quantifiers. In: Lepore, E., Smith, B. (eds.) The Oxford Handbook of Philosophy of Language, pp. 794–821. Oxford University Press, Oxford (2006)
6. Grice, H.P.: Studies in the Way of Words. Harvard University Press, Cambridge (1989)
7. Horn, L.R.: Implicature. In: Horn, L.R., Ward, G. (eds.) Handbook of Pragmatics, pp. 3–28. Blackwell Publishing, Oxford (2004)
8. Horsten, L.: On the Quantitative Scalar Or–Implicature. Synthese 146, 111–127 (2005)
9. Jaszczolt, K.M.: Defaults in Semantics and Pragmatics. In: Zalta, E.N. (ed.) The Stanford Encyclopedia of Philosophy (2008), http://plato.stanford.edu/archives/fall2008/entries/defaults-semantics-pragmatics/
10. Levinson, S.C.: Presumptive Meanings. In: The Theory of Generalized Conversational Implicature. MIT Press, Cambridge (2000)
11. Lycke, H.: A Disjunction Is Exclusive Until Proven Otherwise. Introducing the Adaptive Logics Approach to Gricean Pragmatics (submitted)
12. Verhoeven, L.: De Disjunctie. Adaptief–Logische Formalizering van een Aantal Griceaanse Implicaturen (in Dutch). Unpublished PhD–Dissertation, Ghent University (2005)
13. Verhoeven, L., Horsten, L.: On the Exclusivity Implicature 'or' on the Meaning of Eating Strawberries. Studia Logica 81, 19–42 (2005)
14. Wainer, J.: Modeling Generalized Implicatures Using Non–Monotonic Logics. Journal of Logic, Language, and Information 16, 195–216 (2007)
15. Westerståhl, D.: Generalized Quantifiers. In: Zalta, E.N. (ed.) The Stanford Encyclopedia of Philosophy (2008), http://plato.stanford.edu/archives/win2008/entries/generalized-quantifiers/

Two Kinds of Procedural Semantics for Privative Modification

Giuseppe Primiero[1,*] and Bjørn Jespersen[2,**]

[1] Centre for Logic and Philosophy of Science,
Ghent University, Belgium
`giuseppe.primiero@ugent.be`
[2] Department of Computer Science, Technical University of Ostrava;
Institute of Philosophy, Department of Logic,
Czech Academy of Sciences, Prague, Czech Republic
`jespersen@flu.cas.cz`

Abstract. In this paper we present two kinds of procedural semantics for privative modification. We do this for three reasons. The first reason is to launch a tough test case to gauge the degree of substantial agreement between a constructivist and a realist interpretation of procedural semantics; the second is to extend Martin-Löf's Constructive Type Theory to privative modification, which is characteristic of natural language; the third reason is to sketch a positive characterization of privation.

1   Introduction

The verbal agreements between constructivist/idealist and platonist/realist semantics are so numerous and so striking that it is worth exploring the extent to which there is also substantial agreement. This paper explores some of the common ground shared by the Constructive Type Theory of Per Martin-Löf[1] and the realist Transparent Intensional Logic of Pavel Tichý.[2] We focus here on the following common features:

- a notion of construction;
- a functional language;
- a typed universe;
- an interpreted syntax.

These four features are sufficient to underpin a neutral notion of procedural semantics. Phrased in neutral terms, linguistic meaning is construed as an abstract procedure, of one or more steps, delineating what operations to apply to what operands in order to obtain a particular product as its outcome. Since

---

[*] Postdoctoral Fellow of the Research Foundation - Flanders (FWO). Affiliated Researcher IEG, Oxford University and GPI, Hertfordshire University.
[**] Project GACR 401/10/0792.
[1] See [11] and [15].
[2] See [3], [23], [24].

K. Nakakoji, Y. Murakami, and E. McCready (Eds.): JSAI-isAI, LNAI 6284, pp. 252–271, 2010.

the interpreted syntax is susceptible to type-theoretic restrictions, the range of admissible combinations of operations and operands is accordingly constrained. These procedures are structured constructions, each of whose constituents is an abstract object of a particular type.

In this paper we apply the procedural semantics sketched above to the problem of privative modification. We do this for three reasons. The first reason is to launch a tough test case to gauge the degree of substantial agreement; the second is to extend Martin-Löf's Type Theory to privative modification, which is characteristic of natural language; the third reason is to sketch a positive characterization of privation.

Property modification in the Montagovian tradition is a function from properties to properties.[3] If M is a modifier and F a property, then (MF) is the property formed by applying the function M to the argument F. Thus, $(MF)a$ is the predication of the property (MF) of the individual a. The sentential schema whose semantics we wish to study is

$$\text{``}(MF)a\text{''}.$$

The interpretation of this schema in a procedural semantics depends on the appropriate explanation of what M, F and a are, and of what logical procedures are involved in modification and predication.

A full semantic theory of modification must be able to account for the following variants:

- *Subsective*: $(M'F)a \therefore Fa$;
- *Intersective*: $(M''F)a \therefore M^*a \wedge Fa$;
- *Modal/intensional*: $(M'''F)a \therefore Fa \vee \neg Fa$;
- *Privative*: $(M''''F)a \therefore \neg Fa$.

The first variant is easily treated in a type-theoretical procedural semantics by standard subset formation, extending the language with quantifiers and λ-terms, and forming ordered pairs $\langle M, F \rangle$ where F is the functional argument of the function M whose functional value is the modified property (MF). The path from function and argument to value consists in deploying the operation of functional application. The second variant is less straightforward, as it requires a rule for replacing the modifier M by the property M^*.[4] Our conjecture, in the absence of obvious counterexamples, is that whenever "Fa" is an expression in a mathematical or logical theory, $(MF)a$ is exhausted by subsective modification, whereas for F an empirical property and a a person or an artifact, privative modification is unavoidable. In general, any semantic theory of mathematical and logical language must come with an account of modification, since the premise $(M'F)a$ contains the modifier M'.

Two examples to fix ideas:

"a is a prime number"

---

[3] See [13].

[4] See [1], §4.4, [6]. The third variant will not be considered here. See [7] for discussion.

where *prime* is a modifier of the property *number*; and

"*b* is a large elephant"

where *large* is a modifier of the property *elephant*. In the first example, we consider the least controversial kind of subsective modification, which goes along procedurally with subset formation: given a set of (natural) numbers, the modification of the property of being a number generates the subset of those numbers that have the additional property of being prime numbers.

In empirical languages, we not only have examples like "*b* is a large elephant", but also cases of privative modification, of which the following would be typical examples:

"*b* is a *forged* banknote";
"*b* is *sham* jewellery";
"*b* is a *false* friend".

According to its definition, privation merely indicates what something is *not*, namely not an *F*. We do not maintain that privation is the converse operation of subsection, and it would be too strong for the constructivist to hold that privation produces the complement of the property *F* (because there are no such types as *not being an F* or *being a non-F*). Instead our thesis is that for the constructivist privation is an extreme case of subsection. Given a set of *F*'s, privation will generate the null set of *F*'s; yet, while forming the null set of a particular property exhausts the logic of privation, its semantics is richer than that. Though both forged banknotes and railroads, say, are not banknotes, there is an intuitive sense in which forged banknotes are somehow 'closer to' banknotes than are railroads (or tea mugs or tax forms, etc.) The challenge is to make explicit what this (incomplete) approximation comes down to, which is to say something positive about what properties do define forged banknotes (etc.). Semantically, the quest is for a definition of what it is that banknotes and forged banknotes have in common. The philosophical idea which in our view ought to inform any definition of (*forged F*), say, is that being a forged *F* is as good a property as any. Hence, a procedural semantics needs to show a way of generating such a property: a constructivist semantics needs to have a way of verifying whether a particular individual has the property of being a forged banknote, and a platonist theory must be able to define the proper subset of the complement of any set of banknotes, such that the elements of that subset are forged banknotes. To do so, we characterize a privatively modified property (MF) as having some, but not all, of the properties defining *F*. So there is going to be a range of forged *F*'s, such that those sharing more of those properties are closer approximations to *F*. This idea induces a sequence of properties $G_1, \ldots, G_n$ jointly defining *F*; the more G_i are satisfied, the closer the approximation to *F*. Those forged banknotes that satisfy most G_i are virtually indistinguishable from banknotes, whereas those satisfying few are shoddy imitations (paper instead of polymer, or vice versa, wrong format, wrong colors, etc.). Still, a very poorly forged banknote will nonetheless share more defining properties with a banknote than will a railroad or a tea mug.[5]

What is wanted, overall, is a philosophically motivated and technically workable account of privative modification interpreted within a basic neutral formulation of procedural semantics. In particular, it must be shown what the type-theoretically constrained procedure for predicating a modified property of an individual looks like. In order to obtain such a technical result in the procedural semantics germane both to the constructivist and the realist approach to type theory, we have recourse to a procedure for subset formation. We then generate an appropriate procedure for privative modification by, accordingly, characterizing one form of subsective modification. However, Martin-Löf's and Tichý's respective theories will, in the final analysis, provide partially diverging accounts of such a procedure.

To sum up, this paper pursues two strands, one methodological, the other problem-oriented. The semantic problem is to provide a procedural account of privative modification in terms of subset formation. The methodological one concerns two different forms that a procedural semantics may take, namely the constructivism of Martin-Löf's Type Theory and the platonism of Tichý's Transparent Intensional Logic. The paper seeks to advance the research both on an ill-understood topic in semantics and the general debate of realism vs. anti-realism.

2   Procedural Semantics for Privative Modification

Both theories start from a notion of construction, which extends to function formation. While both operate within the confines of a typed interpreted syntax, the respective type theories work in different ways. Martin-Löf's type theory assigns a new type to each new property, laying down how to verify whether an individual has that property, whereas Tichý's type theory assigns the same type to all empirical properties of individuals. Consequently, the respective procedures for constructing a modified property are also going to differ.

---

[5] We disregard the forger's *intention* to produce forged banknotes. We realize that by disregarding the intentions of someone designing and manufacturing technological artifacts and confining ourselves to physical properties, we are guilty of a philosophical simplification. Logically, however, a property along the lines of *being intended as a forged 100-euro banknote* can be smoothly added to the list of properties jointly defining *being a forged 100-euro banknote*. Another simplification is the absence of a priority relation over the properties jointly defining the modified one. Clearly, a real-life account of modification will discriminate between the properties that are more or less relevant to the modified property. For instance, that a forged 100-euro banknote has got the watermark right may be more relevant than getting the code number wrong. Note that in a procedural semantics like Constructive Type Theory that comes with dependent types, assumptions for hypothetical judgements are normally prioritized: the present formulation is therefore a simplification where presuppositions and assumptions are all introduced at the same level of relevance.

2.1   Construction

On the constructive interpretation, predication starts by laying down all the
necessary and sufficient conditions for a judgement of the form F *set* (or equivalently F *prop*, on the *props-as-sets* identity) to be formulated: such a type
declaration is justified in terms of a judgement $f : F$ that shows a constructor
for that set, and an equality judgement $f = f' : F$, to ensure canonicity for that
element. The basic formal expressions of the theory are the standard categorical
judgements

$$f : F$$
$$f = f' : F$$

with F *set* being the appropriate type declaration. From categorical judgements
of the form $f : F$, one extends the language to hypothetical judgements as formulae of the form F' *set*$[x : F]$ which can be understood as a relation between
types, corresponding to *functional abstraction*. The justification of such a form of
judgement is given by saying that F' is a type whenever an appropriate substitution is performed by a certain canonical constructor f in the type F. Dependent
judgements are generalised to an arbitrary number of assumptions contained in
contexts (within brackets):

$$f : F[f_1 : F_1, \ldots, f_n : F_n]$$
$$f = f' : F[f_1 : F_1, \ldots, f_n : F_n]$$

where again each F_i is declared a type in an appropriate way. The theoretical
starting point of Martin-Löf's type theory is, therefore, the justification of a
typed formula in terms of its instance and the reduction of truth-conditions
to assertion-conditions. Formation – with corresponding equalities – is the first
computational rule for types; rules are completed by:[6]

– *introduction rule*, to introduce canonical elements of types with equality;
– *elimination rule*, to prove a property for a previously typed element.

As with Martin-Löf's, Tichý's theory construes procedures in a functional
fashion. Its syntax is provided by the λ-calculi, but the semantic interpretation
of it is explicitly procedural in nature.[7] The procedural aspect of Tichý's theory
is given by the fact that the λ-terms of application and abstraction do not denote,
respectively, the *result* of applying a function to an argument or arranging two
sets of entities as functional arguments and their values. Rather, in TIL, they
denote, respectively, the very *procedure* of applying a function to an argument
and of forming a function. The procedure of application is called *Composition* in
TIL and is encoded thus: $[X_0 X_1 \ldots X_n]$, where X_0 is a construction of a function,
$X_1, \ldots, X_n$ constructions of its arguments and $[\quad]$ the procedure of functional
application. The procedure of abstraction is called *Closure* in TIL and is encoded

---

[6] Cf. [11], p.24.
[7] However, especially the rules pertaining to β-conversion are susceptible to various
constraints. See [1], §2.7 for the details of TIL as a hyperintensional, partial, typed
λ-calculus.

thus: $[\lambda x_1 \ldots x_n Y]$, where $x_1, \ldots, x_n$ construct arguments, Y constructs values of a function and $[\lambda x_1 \ldots x_n Y]$ is the procedure of functional abstraction.[8]

2.2 Functional Language

The functional extension of CTT is crucial to expressing implicational and quantified formulae. If F is a type, the construction of a new type is possible by considering F' a family of sets over some $x:F$, such that $F'[x:F]$ is also a type. A function can, therefore, be construed as the judgement regarding a certain object F' *type* based on the prior judgements for a type F, possibly generalized to more types (we skip here identity on types and objects of types):[9]

$$\frac{F \ set[x:F']}{(x:F')F : set} \text{ Function Formation}$$

$$\frac{f:F[x:F']}{(x)f : (x:F')F} \text{ Function Introduction}$$

$$\frac{f:F[x:F'] \qquad f':F'}{f(f') : F[x/f]} \text{ Function Elimination}$$

The neutral formulation $(MF)a$ of an individual a instantiating the modified property (MF) is constructively expressed as a function M such that for every element x in the set F taken as argument, it returns a function $M(x)$, formally $M(x)[x:F]$. To preserve the functional aspect of M in the constructive notation, we will refer to $M(F)$ *type* as the modified type satisfied by some $f:F$; this means that the individual a from the original notation correpsonds to a *typed* element in F, expressed by a judgement of the form $f:F$, hence it will suffice to translate the modifier M into a function on f, so that $(MF)a$ will be expressed by $M(f)$. Standard modification of a property $M(F)$ is given, therefore, by functional abstraction and it produces subset formation $\{x:F \mid M(x)\}$. The case of privative modification is no exception to this general interpretation: a privative modifier will still take as arguments elements in a basic type F, hence the operation occurs at the level of extensions. It differs from a standard functional type (and standard subset formation) in that it does not define a *set of individuals* of the basic type, because its arguments no longer instantiate the original property F. Rather, the range of this modifier will consist of *functions* from the basic type F to the empty set. This shows that constructive privation represents a special case of standard subsection, specified by requiring extra conditions. That the range of the privative modifier is a set of functions of the appropriate

---

[8] Two other constructions are *Trivialization* and *Variable*. Trivializations can be dispensed with here, since we do not need to *mention* constructions; we only *use* them to obtain the entities they construct. For now, think of variables as one-step procedures for obtaining an entity relative to a sequence of assignments of entities to variables. See [1], §§1.1-1.3.2, §2.6.1.

[9] See [17], §1.7.

type – rather than individuals – can be seen as introducing a type of higher order. The bottom-up approach characteristic of the constructive philosophy is preserved, so that the Introduction Rule uses a construction $f : F$ as a premise to define a privatively modified F in terms of the empty set of F's.

The functional language of TIL is cast within a ramified type hierarchy encompassing a simple type theory, relative to which each entity of the ontology of TIL receives a type. The entities are organized into a bi-dimensional typed universe. One dimension is made up of non-constructions, the other of constructions. On the ground level of the type hierarchy there are non-constructional entities unstructured from the procedural point of view belonging to a *type of order 1*. Given a so-called *epistemic* (or, equivalently, *objectual*) *base* of *atomic types* (o-truth values, ι-individuals, τ-reals doubling as times, ω-possible worlds), the induction rule for forming *functional types* is applied: where $\alpha, \beta_1, \ldots, \beta_n$ are types of order 1, the set of partial mappings from $\beta_1 \times \ldots \times \beta_n$ to α, denoted '$(\alpha\beta_1 \ldots \beta_n)$', is a type of order 1 as well. Constructions that construct entities of order 1 are *constructions of order 1*. They belong to a *type of order 2*, denoted '$*_1$'. The type $*_1$ together with atomic types of order 1 serves as a base for the induction rule: any collection of partial mappings, of type $(\alpha\beta_1 \ldots \beta_n)$, involving $*_1$ in their domain or range, is a *type of order 2*. Constructions belonging to a type $*_2$ that construct entities of order 1 or 2, and partial mappings involving such constructions, belong to a *type of order 3*; and so on *ad infinitum*.[10]

Tichý's theory of modification proceeds, therefore, in a strictly top-down manner. First, a modified property is constructed according to the procedure of functionally applying a modifier M to a property F, and only then is the modified property (MF) predicated of an individual a.[11] What gets predicated of an individual is, strictly speaking, an *extensionalized* property, which is a function from individuals to truth-values.

An intensional entity is any function (mapping) whose domain is in the logical space of possible worlds. For most purposes, TIL takes an intension to be a function from logical space to a function from times to entities, in the manner well-known from possible-world semantics enriched with temporal parameters. Thus, an empirical property of individuals is a function from logical space to a function from times to sets of individuals, where a set of individuals is a characteristic function from individuals to truth-values. Hence, given a particular world/time pair $\langle w, t \rangle$, it is either true or false that a given individual a is a member of the set that is the extension of the property at $\langle w, t \rangle$. Formally, the type of a property is $(((o\iota)\tau)\omega)$, abbreviated '$(o\iota)_{\tau\omega}$'. The TIL abbreviation of a modified empirical property being predicated of an individual will be of the form $\lambda w \lambda t [[MF]_{wt} \, a]$.

2.3   Interpreted Syntax

The procedural way of generating privatively modified properties is based on the fact that the type-theoretical syntax is interpreted.

---

[10] See [1], §1.3.2.

[11] Note the contrast with the constructivist approach, where a modified property is obtained via application rather than abstraction.

Constructive Type Theory can be seen as one of several foundational systems for predicative constructive mathematics,[12] but its additional value is represented by a meaning theory which extends and refines the Brouwer-Heyting-Kolmogorov interpretation of intuitionistic logic.[13] CTT formalizes a proper theory of reasoning and knowledge, an interpreted system whose objects are equipped with meanings.[14] By implementing the Curry-Howard isomorphism, types are intended as polymorphic categories of predication, carrying an internal meaning that can be made explicit in terms of propositions (for which proofs are the appropriate constructors) or sets (correspondingly constructed by their elements). The fact that types represent meanings can be adapted to the interpretation of natural language semantics,[15] where reference is generally construed as the relationship between nouns or pronouns and the objects that are named by them. In a constructive procedural semantics every object comes embedded within its meaning category, by which a type gains its meaning from its constructor, and the constructor is meaningfully expressed whenever accompanied by its type ("no entity without a type").[16] As a result, any expression occurring in one of the computational rules comes embedded with types that yield meanings, and each meaning category is reduced to the corresponding syntactical construction procedure.

The syntax of TIL (its formal 'language of constructions' in which constructions are encoded) is inherently interpreted because both constructions and the entities they construct cannot be introduced without typing them first.[17] A *semantic analysis* of a piece of language executed in accordance with TIL proceeds along the following three steps.[18] First, *type-theoretic and logical analysis*: all and only logical entities (operations and their operands) being denoted by sub-expressions occurring in the overall expression under analysis receive a type, which may be drawn from the simple or ramified type hierarchy. Second, *synthesis*: the constructions of the entities mentioned are executed in accordance with the logical operations made explicit by the logical analysis in order to unveil the entity denoted by the overall expression. Third, *type checking*: by means of an annotated tree it is checked whether the type assignments check out.[19]

3   Constructive Privative Modification

Standard subsets are used in the type-theoretical setting in order to express a type that is defined by comprehension in the range of another type. Constructively, this corresponds to nothing other than a propositional function from type

---

[12] Constructive set theory, explicit mathematics and predicative topos are other examples of systems of constructive matematics.

[13] Cf. [10], [11], [17], ch.1.

[14] Cf. [17], ch.1.

[15] See [21].

[16] See also [16].

[17] See [1], §1.5.1, §2.1.2.

[18] See [1], §2.1.1.

[19] See [9] for details.

F to another type F', i.e. function formation from sec. 2.2, requiring the definition of the type in terms of the judgement $F : set[x : F']$ with an equality judgement defined on it. The appropriate introduction rule corresponds to functional abstraction $(x)f : F(x : F')$ and it is equivalent to Church's λ-abstraction. To know that the preceding rule is correct, the judgement $f(f') : F[x/f']$ must be obtained by function elimination, showing an object of the type F which satisfies also the subtype F', a typed version of β-conversion.[20]

Let us generalise and consider our subtype as M for 'modifier'; in this way one obtains the subset of elements in F satisfying M:

$$\frac{F\ set \qquad M(x)\ set\ [x : F]}{\{x : F \mid M(x)\}\ set} \qquad \text{Standard Subset Formation Rule}$$

By the side condition on canonical elements, if $f = f'$ and $M(x)$ is true for some $x : F$, one obtains equal canonical constructions of the set $\{x : F \mid M(x)\}$ when f or f' is used as input of M. That is, since every propositional function is extensional in the sense that it yields equal types when applied to equal elements, it follows from $f = f' : F$ and $M(x)\ set[x : F]$ that $M(f)$ and $M(f')$ are equal types. Consequently, from the requirement that $M(f)$ be true, we immediately get that also $M(f')$ is true.

The use of subset formation for an arbitrary property F (e.g. *banknote*) and a *privative* modifier M (e.g. *forged*) is not entirely correct, however. To preserve the constructive interpretation also for the case of privative modification, it is required that the meaning of $M(F)\ set$ be given by some (canonical) $M(f)$. By using standard subset formation, the modifier type M will yield a subset of the set of *canonical* F's. Since a *privative* modifier M is intended as a modification procedure that changes entirely the range of its input, an alteration is needed. Because a forged banknote is not a banknote in the first place, the privative modifier *forged* cannot be interpreted as a propositional function from the canonical set of banknotes to one of its (canonical) subsets. For this reason, one needs to define privative modification as an extreme version of subsection. The obvious intuition is that the basic argument $F\ set$ needs to be modified whenever used as an input of the privative modifier M in a way that allows us to turn every $x : F$ into an element of the function from F to the empty set. The first step towards obtaining such a procedure is to define appropriate constructions of the empty set and of the function from a set to the empty one, returning the empty set of elements in that set. The empty set is introduced by declaring the following constants:[21]

$$\{\} : set;$$
$$case_{\{\}} : El(Z(x))\ [Z : (\{\})\ set, x : El(\{\})].$$

The first constant simply declares the collection with no elements to be a set; the case step gives the empty set of Z's elements, by applying a set Z to any element x

---

[20] See [17], §1.8. For an analysis of functions and types and the reference of abstract terms, see [18].

[21] Cf. [15], p.21.

on condition that Z be an element of the collection of empty sets, and x an element of any set in that collection. Both of these constructions are crucial to the formulation of the privative modifier M. The idea is to use a canonical type declaration F *set* and to apply a modifier M to any $x : F$, under conditions that $x : El(\{\})$ and $F : El(\{\})$ *set*. By this, we do not mean to construct (in the standard way) an arbitrary empty set, nor to show a (constructively inadmissible) canonical element for not-F. For the canonical constructive empty set does not allow distinguishing among different empty sets (which is what we need, if we interpret privative modification as a construction of the empty set); and there is constructively no way to give a definitional procedure for a negative type such as *the set of non-banknotes*, because its conditions cannot be canonically specified, in case such set should include everything that does *not* satisfy the conditions for being a banknote. Instead, we give the appropriate assertion conditions for a function that takes any element in the set of banknotes to the complement of such a set, because in this case it is completely specified what the conditions for its input are, and the function only requires that those conditions remain (entirely or partially) unsatisfied. Formally:

Privative Subset Formation Rule

$$\frac{F \; set \qquad M(x) \; set[F : (El(\{\})) \; set; x : El(F(f))]}{\{x : F \mid M(x)\} \; set}$$

This construction defines a function M over the set F; the result is not a canonical subset of F, for given any x in F as its input, $M(x)$ returns a set of functions to the empty set. The apparent mismatch between F in the first premise and its occurrence in the context of the second premise is easily explained: the first premise declares a type which, by the given case formation rule for the empty set, is taken as valid input for the type of elements in the empty set and used as a condition for the second premise. In the latter, F (a function) is employed as the argument of a function application rule: namely, M is the function and $x : El(F(f))$ gives the input. Nothing needs to be said explicitly about M, provided the needed information is contained in the context under which M is a valid construction.[22]

---

[22] We consider $El(\{\})$ in the second premise as not entirely arbitrary, instead it contains an object of the type F defined by the first premise: hence one might require that f be not only an arbitrary object in $\{\}$, but, more specifically, an object in the set $\{x : \{\} \mid F(x)\}$. This makes any (standard) restriction over F impossible. The second premise needs to be taken conditionally, where its conditions are not meant to be interpreted in terms of subsection. The context in which the modifer M is applied to f requires that F be an element in the collection of empty sets; then, f is declared one arbitrary element in this empty set, and finally the set obtained by functional application $F(f)$ is considered. This gives the empty set of F's, restricting all arbitrary elements of the empty set to those obtained by only considering functions from fs to the definitional set, in turn declared empty. The formulation of the second premise is therefore conditional on the requirement of an empty set F, and that whenever we consider an $M(f)$ we know it leads to an empty set of F's. The crucial point is precisely not to introduce a subset of F's, but a set of functions satisfied by an empty argument.

When the Privative Subset Formation Rule is applied to the example of *forged banknote*, one starts from the set of banknotes and, by applying the appropriate conditions on that set, one wishes to obtain the empty set of banknotes:

$$\frac{banknote\ set \qquad forged(x)\ set[banknote:(El\{\})\ set;\ b:El(banknote(x))]}{\{x:banknote\ |\ forged(x)\}\ set}$$

It is essential, therefore, to operate with *typed* empty sets.

Privative modification treated as output of the empty-set function lays down the distinction between the output of $M(F)$ – for M some privative modifier like *forged* and F an argument, e.g. *banknote* – and any other empty set: what is the difference between constructing the empty set of banknotes in terms of the set of forged banknotes and any other way of constructing a set none of whose elements is a banknote? This problem is constructively solved by putting forward an appropriate equality rule governing $M(F)$ with respect to the set F:

Equality Rule on Sets

$$\frac{F\ set \qquad F = F'\ set \qquad M(x)\ set[F = F':(El\{\})\ set;\ f:El(F = F'(x))]}{\{x:F = F'\ |\ M(x)\}\ set}$$

By this rule, for any equivalent set taken as argument of the modifier, the same empty set is obtained. For any set G with its own constructor $g \neq f : F$ the modifier $M(x)\ set[x:G]$ shall return a different empty set (namely, the empty set of G's, different from the empty set of F's). This obviously allows defining the difference between $M(F)$ (*forged banknotes*) and G (*railroads*, say) as empty sets of banknotes in a different sense: the former will, strictly speaking, be the set of function constructors from the set of banknotes to the empty set; the second set will contain no constructor of the set of banknotes at all, hence being empty with respect to any such individual.

The introduction rule instantiates the procedure which, starting from a typed object, returns a privatively modified one:

Introduction Rule

$$\frac{f:F \qquad m:M(f)[F:(El\{\})\ set;\ f:El(F(f))]}{f:\{x:F\ |\ M(x)\}}$$

where F can be taken to be the set of banknotes and f an instance of that set, and M the modifier *forged*.

In the introduction rule one starts from the premise that a canonical element f in the set F is given; provided $M(f)$ is true, i.e. there is a canonical element m of the set of functions from F to the empty set, we know that f will yield a canonical element in the set of modified F's when taken as the argument of the empty-set function of $M(F)$. By the associated equality rule, if $f = f'$ are elements in F, and if there is an m such that $M(f)$ is true, f and f' will yield canonical elements in the set of modified F's; and from $f = f':F$ and $m:M(f)$ it follows that $m:M(f')$. Notice that according to the constructive requirement on the introduction rule, in order to form the set of modified F's, one needs to know at least one instance $m : M(f)$, and because the latter relies on a function applied to f, it is a further presupposition that f be known. For example, in the case of *forged banknote*, in order to display or recognise a forged banknote one needs to be able to lay down the conditions for knowing what a banknote is.

The set of rules is rounded off by an appropriate elimination rule, which makes one able to specify how to extract a modified property from its corresponding set. Formulating an elimination rule for the subset theory is a notoriously difficult matter. It is impossible to give in constructive type theory an elimination rule that captures the way one has introduced elements in a subset, because there is no explicit construction of the element $m : M(f)$ for a standard subset $\{x : F \mid M(x)\}$.[23] In the case of privative modification, the elimination rule is supposed to formalise the procedure which, starting from an element of a privatively modified property (*forged banknote*, say), will return another modified element defined over the former; this means that variables will occur bound in the second construction. The informal meaning of the elimination rule is to enable positive predication for privatively modified entities. Saying that a banknote can be identified by ascertaining that it reacts to ultra-violet lamps emitting light at around 365 nanometres[24] can be rephrased by saying that a forged banknote will fail to react to uv-lamps emitting light at around 365 nanometres; similarly, one may want to state of a false friend that he or she is a seasoned liar, or that sham jewellery is an "abomination [. . .], a lie, a pretension".[25] In the following, let Δ abbreviate the conditions on a privatively modified set as given by the second premise in its introduction rule. In the corresponding elimination rule, one starts from an instance of a privatively modified property $M(f)$ satisfying $x : \{f : F \mid M(f)[\Delta]\}$; then, another function $f'(x)$ of type $M'(x)$; by substituting f in the free occurrences of x in $M'(x)$, one concludes that $f'(f)$ is an element of the newly modified type $M'(f)$:

Elimination Rule

$$\frac{f : \{x : F \mid M(x)[\Delta]\} \qquad f'(x) : M'(x)[x : F, m : M(x)]}{f'(f) : M'(f)}$$

3.1   Degrees of Modification

Standard typing rules do not as yet say anything relevant about the sense in which modification comes in degrees, given that there are different sorts of forged banknotes. For example, in the light of a description of a banknote as a green piece of polymer with a hologram printed on it, there are different ways in which a forged banknote may be forged: it may be a piece of polymer which is either not

---

[23] Cf. [22] for a full explanation, the solution proposed and the consequences for the deductive power of the theory.

[24] Pamphlet of the Bank of England, downloadable at
    `http://www.bankofengland.co.uk/banknotes/kyb_lo_res.pdf`.

[25] From the *Routledge Manual of Etiquette*, 2007, p. 175.

green or lacks the appropriate hologram, or it can be a green piece of something other than polymer with or without a hologram printed on it. All in all, an individual that lacks all three properties fails to qualify as a forged banknote. We shall explain these differences by introducing a formal notion of degrees of modification.

The use of *dependent* types has been shown to be crucial to the definition of the subset formation rule, both in its standard format and its privative variant. We want now to make a dependency relation explicit also for the argument of the modifier function, which will make it possible to differentiate among privatively modified F's. Take

$$F \ set[x_1 : F_1, \ldots, x_n : F_n]$$

to be the formal way of saying that F is a canonical set whenever each $x_i : F_i$ is a type-theoretical expression satisfied by an appropriate element $[x_i/f_i]$, where each F_i is a definitional property of F.[26] The rule for defining the privative modifier can be analytically formulated with respect to its application to the definitional properties F_i of F:

Dependent Privative Subset Formation

$$\frac{F \ set[x_1 : F_1, \ldots, x_n : F_n] \qquad M(x) \ set[F_i : (El\{\}) \ set; x : El(F_i(x))]}{\{x : F \mid M(x)\} \ set}$$

where $1 \leq i \leq n$. This new rule says that $M(x)$ is a modified F in view of the empty set of F_i, for every $\bigvee F_i \in F$ up to $\bigwedge_{-1} F_i$ defining F, that is by privation with respect to some – up to all bar one – of its definitional properties.

Depending on the selection and combination of F_i, one obtains different *degrees of modification*. A standard recursive definition of the factorial of the integer n[27] is used in the following for the standard combinatorial result of d elements extracted from n.[28] In the following we shall use n to indicate the number of F_i occurring in the dependency context of definitional properties of F, so that we shall call the degree d of modification M of a property F the number of n definitional properties of F with respect to which a privative modifier is applied. By the combinatorial result, the following can be easily stated:

– there will be n distinct modifications of degree $d = 1$, corresponding to the privation of $x : F$ with respect to F_i for some $i \in n$ in the set of conditions for $F \ set$;

---

[26] In the present treatment of type-theoretical predications, we are referring to standard types requiring a finitistic formulation of a dependency relation from a context of assumptions. In [12], a non-standard extension of intuitionistic type theory with infinite objects was introduced, which represents a generalization of the finitistic frame, relying on the latter for justification. The negation of predicates at one stage or more in the infinite dependent structure of contexts can be formulated in that frame in a way that resembles the notion of unsatisfied conditions introduced here. As mentioned in the Introduction, we are relying on the simplification that elements in the dependency context come without any priority relation.

[27] $n! = 1$, if $n = 0$ and $n(n-1)$, if $n \geq 1$.

[28] $C_n^d = n!/d! \cdot (n-d)!$.

– there will be a combinatorial number of distinct modifications of degree $d = i < n$ in view of the rule for C_n^i, corresponding to the privation of $x : F$ with respect to the union $\bigcup\{F_1, \ldots, F_i\}, 2 \leq i < n-1$ in the set of conditions for F set.

Following this rule, an individual determined by 10 properties will accommodate a total of 198,720 possible combinations of modification, counting all the modifications of one property, those of two properties and so on, up to counting 10 possible combinations of modification involving 9 properties (obtained by the calculation $3, 628, 800/362, 880 = 10$). For a simple example, consider the definitional presentation of the set of banknotes introduced above, for which three different modifications of degree 1 are possible, making forged banknotes forged due to their being deprived of just one defining property:

$$banknote\ set[polymer, green, hologram]$$
$$\frac{forged(x)\ set[F_i : (El\{\})\ set; x : El(F_i(x))]}{\{x : banknote \mid forged(x)\}\ set}$$

where F_i is a variable for any of the properties of being made of polymer, of being green or of having a hologram. A modification of degree 2 would take into account two defining properties; as a result, an instance of the following constructor would be a forged banknote by failing to be made of green polymer (or any other combination):

$$banknote\ set[polymer, green, hologram]$$
$$\frac{forged(x)\ set[F_{i,j} : El(\{\})\ set; x : El(F_{i,j}(x))]}{\{x : banknote \mid forged(x)\}\ set}$$

where again $F_{i,j}$ instantiate two defining properties.

3.2   Iteration of Modifiers

The formulation of degrees of modification enables us to make comparisons among different instances of the same modified type. In particular, it enables us to express, in the metatheory, that a particular modified set is at a certain degree of approximation to its original counterpart. In the case of *forged banknote*, a privative modification of degree 1 will be a closer approximation to *banknote* than will a privative modification of degree 2. This squares with natural-language predicates like 'is a well-made forged banknote', whose use presupposes various degrees to which a forged banknote may succeed in passing for what it is a forgery of.

This remark leads directly to the next case we want to analyse, namely the iteration of modifiers. The modifier *well-made* needs to qualify *forged banknote*, otherwise one ends up with the infelicitous *((well-made forged) banknote).*[29]

---

[29] Brackets are used as scope indicators. Note that if *well-made* is to modify *forged*, then because the latter is a first-order modifier (modifying, as it does, a non-modifier), the former must be a higher-order modifier like, e.g., *very*. See [6] for discussion of higher-order modification.

Whether *well-made* modifies *forged banknote* or *forged, well-made* is a subsective modifier, and we do not want to extract well-made forged banknotes from a set of banknotes. For the iteration to be such that, given a set of forged banknotes, one extracts only the well-made ones, one has to be sure that the construction of *(well-made (forged banknote))* uses a correct application of different subset formation rules.

Consider the by now well-known construction of *forged banknote* and let us abbreviate again the additional conditions on the privative subsection as Δ. Now the construction of *(well-made (forged banknote))* is of the following form:

$$\frac{banknote\ set \qquad forged(x)\ set[\Delta]}{\{x:banknote \mid forged(x)\}\ set \qquad well\text{-}made(x)\ set[x:banknote \mid forged(x)]}$$
$$\frac{}{\{x:banknote \mid well\text{-}made \times forged(x)\}\ set}$$

This construction applies first the privative subset formation rule and then the standard subset formation rule to the resulting set of functions, thus obtaining the cartesian product of two families of functions over correctly defined sets.

On the other hand, the construction of *((well-made forged) banknote)* is an illegitimate one. The predicate 'is a (well-made forged) banknote' does not split the application of the modifiers into two steps, instead the formal construction combines via the cartesian product the standard subsective modifier and the privative subsective modifier. The resulting construction is ill-defined because the subsective modifier *well-made* has the categorical set *banknote* as its arguement, whereas the privative modifier *forged* applies to functions defined over an empty set:

$$\frac{banknote\ set \qquad well\text{-}made(x)\ set[x:banknote] \times forged(x)\ set[\Delta]}{\{x:banknote \mid well\text{-}made(x) \times forged(x)[\Delta]\}\ set}$$

A specific case of iteration of modifiers is the iteration of privative modifiers. This kind of iteration avoids the problem of the previous case, because in both cases the modifiers are privative, hence they share the same conditions. The iteration will give the cartesian product of the sets of functions that are arguments of the modifier. The following construction is an example of a formation rule regulating *burned forged banknote*:

$$\frac{banknote\ set \qquad forged(x)\ set[\Delta]}{\{x:banknote \mid forged(x)\}\ set \qquad burned(x)\ set[forged(x)[\Delta]]}$$
$$\frac{}{\{x:banknote \mid forged \times burned(x)[\Delta]\}\ set}$$

Burned is privative because a burned F is not an F, though it originally was an F. Not all pairs of privative modifiers cancel each other out, such that a burned forged banknote would be a banknote. Furthermore, though both *forged* and *burned* are privative, their logical behaviour does not overlap entirely. In particular, "*a* is a burned banknote" is an example of resultative predication[30]

---

[30] See [2], p. 226ff.

while "a is a forged banknote" is not. From a being a burned banknote, it follows that a is not a banknote (because a pile of ashes does not make a banknote), but it is presupposed that a started out as a banknote (otherwise there would have been no banknote to burn). So *burned* comes with a dynamic dimension that *forged* lacks: a forged banknote was never a banknote and only remains an approximation to one.[31]

4   Realist Privative Modification

4.1   Predication of Modified Properties

A property is an intensional entity of type $(((o\iota)\tau)\omega)$, abbreviated '$(o\iota)_{\tau\omega}$', which is a function from worlds (ω) to functions from times (τ) to sets of individuals $((o\iota))$. A property modifier, by contrast, is an extensional entity, because it is not indexed to possible worlds. Instead it is a function-in-extension between two intensions. Since a property modifier is a function that takes one property to another, its type is $((o\iota)_{\tau\omega}(o\iota)_{\tau\omega})$. So in order to construct a modified property, the procedure of functional application (Composition) is called for:

$$[modifier\ property]$$

The predication of a property of an individual goes via two instances of functional application. First, the relevant property is extensionalized so as to obtain a set from a property. Second, the set is applied to the individual to obtain a truth-value. The philosophical motivation is that individuals exemplify empirical properties only relative to worlds and times.[32] Schematically, predication is this Closure:

$$\lambda w\lambda t\ [property_{wt}\ a]$$

This Closure, which constructs a possible-world proposition (a function from worlds to functions from times to truth-values), would be the logical form of the sense of a sentence like, "a is a banknote".

The schema of the predication of a modified property of a is this Closure:

$$\lambda w\lambda t\ [[modifier\ property]_{wt}\ a]$$

This Closure would be the logical form of the sense of a sentence like, "a is a forged banknote" or "a is a burned banknote".

If the property constructed by $[modifier\ property]$ is itself modified, the resulting predication looks like this:

$$\lambda w\lambda t\ [[modifier'\ [modifier\ property]]_{wt}\ a]$$

This would be the form of, say, "a is a burned forged banknote" or "a is a well-made forged banknote". In all three cases the semantic analysis culminates in the assignment of a propositional construction to a sentence as its sense.

---

[31] As for a being a well-made forged banknote, the degree to which a qualifies as being well-made is a reflection of the quality of the craftsmanship of the forgery.

[32] See [5] for details.

4.2   The Requisites of Privation

True to its top-down approach, TIL accounts for a property like *being a forged banknote* in terms of other properties being 'stacked upon it', to wit, the set of properties that are individually necessary and jointly sufficient for an individual to have that property. Such a set is called the *essence* of the property in question, and each element is called a *requisite*.[33] The type of a requisite, when a relation-in-extension between two properties, is $(o(o\iota)_{\tau\omega}(o\iota)_{\tau\omega})$, while the type of the essence of a property is $((o(o\iota)_{\tau\omega})(o\iota)_{\tau\omega})$: the *essence* function takes a property to the set of properties that are its requisites. Formally, F being of type $(o\iota)_{\tau\omega}$ and p ranging over the same type, these two constructions converge in the same set of properties:

$$[essence\ F] = \lambda p\ [Req\ p\ F]$$

The requisite relation is defined in the following manner. Let X, Y be intensional constructions such that X, Y are first-order constructions ranging over the type $(o\iota)_{\tau\omega}$ (i.e. X, Y are property variables) and let x range over ι.[34] Then:

$$[Req\ YX] = \forall w \forall t\ [\forall x\ [[True_{wt}\lambda w \lambda t\ [X_{wt}x]] \rightarrow [True_{wt}\lambda w \lambda t\ [Y_{wt}x]]]]$$

Gloss *definiendum* as, "Y is a requisite of X", and *definiens* as, "Necessarily, at every $\langle w, t \rangle$, whatever x instantiates X at $\langle w, t \rangle$ also instantiates Y at $\langle w, t \rangle$."

Logically, privation comes down to, say, *being a banknote* and *being a forged banknote* having an empty intersection at every $\langle w, t \rangle$. This is obtained thus:

$$[Req\ \lambda w \lambda t\ \neg[banknote_{wt}\ x][forged\ banknote]]$$

We say that the property constructed by *[forged banknote]* has, *inter alia*, the requisite property constructed by $\lambda w \lambda t\ \neg[banknote_{wt}\ x]$. This is to say that if, at some $\langle w, t \rangle$ or other, an individual x is in the extension of *[forged banknote]* then x is in the extension of the property constructed by $\lambda w \lambda t\ \neg[banknote_{wt}\ x]$.

Hence, the proposition that *not being a banknote* is a requisite of *being a forged banknote* is equivalent to the proposition constructed thus:[35]

$$\forall w \forall t\ [\forall x\ [[forged\ banknote]_{wt}\ x] \rightarrow [\neg[banknote_{wt}\ x]]]$$

What is special about the sort of non-banknote that is not a tea mug, a railroad or a tax form, but a forged banknote? Given a $\langle w, t \rangle$, the set constructed by $[banknote_{wt}]$ will have a complement in which we find tea mugs and all the rest, including forged banknotes, but the set constructed by $[[forged\ banknote]_{wt}]$ will be a well-defined proper subset of that complement.[36] To define the notion

---

[33] See [1], §4.4. Requisites play pretty much the same role as do *presuppositions* in constructivism.

[34] See [1] §4.1, def. 4.1. See also §4.1 for *True*, which is the propositional property of being true at $\langle w, t \rangle$.

[35] For the record, '$\forall y$' abbreviates '$[^0\forall[\lambda y]]$', y ranging over an arbitrary type α, $\forall$ a function of type $(o(o\alpha))$, and $^0\forall$ being the Trivialization of this function.

[36] See [7] for a positive characterization of the proper subset of the complement of any set of F's containing forged F's.

of the subset of forged banknotes within the set of non-banknotes, we need to express that no forged banknote is a banknote and that some non-banknotes are forged banknotes:

$$\forall w \forall t \; [[[All \; [forged \; banknote]_{wt}][\lambda x \; \neg[banknote_{wt} \; x]]] \; \wedge$$
$$[[Some \; [\lambda x \; \neg[banknote_{wt} \; x]]] \; \lambda x[[forged \; banknote]_{wt} \; x]]]$$

We invoke the quantifiers *All, Some*, here of type $((o(o\iota))(o\iota))$.[37] *All* is the function from the set constructed by $[F_{wt}]$ to the set of all those sets that contain the set constructed by $[F_{wt}]$ as a subset. *Some* is the function from the set constructed by $[F_{wt}]$ to the set of all those sets that share a non-empty intersection with the set constructed by $[F_{wt}]$.

In the Introduction we argued that a forged banknote is an (intended) approximation to a banknote. We also made the (simplistic) assumption that *being green* and *being made of polymer* exhaust *being a banknote*. Thus, one reason why a may be a forged banknote is because a, though being made of polymer, fails to be green. Therefore, at some $\langle w, t \rangle$, a may have some, though not all, of the properties making up the essence of being a banknote, q ranging over $(o\iota)_{\tau\omega}$:

$$\lambda w \lambda t \; [[[Some \; \lambda q \; [q_{wt} \; a]] \; [essence \; banknote]] \; \wedge \; \neg[[All \; \lambda q \; [q_{wt} \; a]] \; [essence \; banknote]]$$

A forged banknote is any individual that is not a banknote and which is either made of polymer but fails to be green, or is green but fails to be made of polymer. If we add a third property, e.g. *having a hologram*, it becomes an option that a non-banknote may have either one or two of those three properties and, therefore, qualifies as a forged banknote to a lower or higher degree. Degrees of modification would be captured in TIL by spelling out which of the requisite properties of being a banknote a given forged banknote possessed.

5   Conclusion and Further Research

Above we set out the philosophical and technical features of two different conceptions – one constructive, the other platonist – of what a procedural semantics for privative modification may look like. These two conceptions of privative modification are, however, only the first step toward dealing with modification in general within these two frameworks. Subsective and privative modification are the easiest two of the altogether four forms of modification mentioned in the Introduction. The modal/intensional variant, on the other hand, represents the most challenging case both philosophically and technically. For one thing, its very logic is far from being clear, since it is not sufficient to simply infer the classical tautology $Fa \vee \neg Fa$ from $(MF)a$. Future research will be devoted to extending both Martin-Löf's Constructive Type Theory and Tichý's Transparent Intensional Logic so as to include a worked-out semantics for intensional modifiers. In particular, the application of CTT to intensional modification will take its lead

---

[37] See [1] §1.4.3.

from [19] and [20], which present a modal type theory including syntactic rules for defining possibility judgements made from *open* assumptions.[38]

References

1. Duží, M., Jespersen, B., Materna, P.: Procedural Semantics for Hyperintensional Logic: Foundations and Applications of Transparent Intensional Logic. Logic, Epistemology and the Unity of Science, vol.17. Springer, Berlin (2010)
2. Jackendoff, R.: Semantic Structures. MIT Press, Cambridge (1990)
3. Jespersen, B.: Significant sententialism in Transparent Intensional Logic and Martin-Löf's Type Theory. In: Childers, T., Majer, O. (eds.) The Logica Yearbook 2002, pp. 117–131 (2003)
4. Jespersen, B.: Explicit intensionalization, anti-actualism, and how Smith's murderer might not have murdered Smith. Dialectica 59, 285–314 (2005)
5. Jespersen, B.: Predication and extensionalization. Journal of Philosophical Logic 37, 479–499 (2008)
6. Jespersen, B.: Property modification and the rule of pseudo-detachment (in submission)
7. Jespersen, B., Carrara, M.: Two conceptions of technical malfunction (in submission)
8. Jespersen, B.: How hyper are hyperpropositions? Language and Linguistics Compass 4, 96–106 (2010)
9. Jespersen, B., Duží, M., Materna, P.: The *logos* of semantic structure. In: Stalmaszczyk, P. (ed.) Philosophy of Language and Linguistics. The Formal Turn, vol. I, pp. 85–101. Ontos-Verlag, Farnkfurt (2010)
10. Martin-Löf, P.: Constructive mathematics and computer programming. In: Cohen, J.J., et al. (eds.) Sixth International Congress for Logic, Methodology and Philosophy of Science, pp. 153–175. North-Holland, Amsterdam (1982)
11. Martin-Löf, P.: Intuitionistic Type Theory. Bibliopolis, Naples (1984)
12. Martin-Löf, P.: Mathematics of infinity. In: Martin-Löf, P., Mints, G. (eds.) COLOG 1988. LNCS, vol. 417, pp. 146–197. Springer, Heidelberg (1990)
13. Montague, R.: English as a formal language. In: Visentini, B., et al. (eds.) Linguaggi nella societá e nella tecnica, Milan, pp. 189–224 (1970); Reprinted in R.H. Thomasson (ed.). Formal Philosophy. Yale University Press, London (1974)
14. Nordström, B., Petersson, K., Smith, J.: Programming in Martin-Löf's Type Theory. Oxford University Press, Oxford (1990)
15. Nordström, B., Petersson, K., Smith, J.: Martin-Löf's Type Theory. In: Abramsky, S., Gabbay, D., Maibaum, T.S.E. (eds.) Handbook of Logic in Computer Science, pp. 1–38. Oxford University Press, Oxford (2000)
16. Primiero, G.: The determination of reference in a constructive setting. Giornale di Metafisica 26, 483–502 (2004)

---

[38] The authors wish to thank the referee for helpful comments pertaining especially to §2. Bjørn Jespersen is indebted to Marie Duží for very helpful suggestions regarding §4.2. Giuseppe Primiero wishes to thank Bartosz Wieckowski for very helpful comments leading to clarifications in §3. The research reported herein was conducted while Bjørn Jespersen was affiliated with the Section of Philosophy, Delft University of Technology, The Netherlands. He gratefully acknowledges its financial support enabling him to participate at the LENLS VI conference.

17. Primiero, G.: Information and Knowledge. A Constructive Type-Theoretical Approach. Logic Epistemology and the Unity of Science, vol.10. Springer, Berlin (2008)
18. Primiero, G.: Proceeding in abstraction. From concepts to types and the recent perspective on information. History and Philosophy of Logic 30, 257–282 (2009)
19. Primiero, G.: Epistemic modalities. In: Primiero, G., Rahman, S. (eds.) Acts of Knowledge: History, Philosophy and Logic, pp. 207–231. College Publications, London (2009)
20. Primiero, G.: Constructive contextual modal judgements for reasoning from open assumptions. In: Proceddings of the Computability in Europe Conference (to appear, 2010)
21. Ranta, A.: Type-Theoretical Grammar. Oxford University Press, Oxford (1990)
22. Salvesen, A., Smith, J.: The strength of the subset type in Martin-Löf's Type Theory. In: Proceedings of LICS 1988. IEEE, Edinburgh (1988)
23. Tichý, P.: The Foundations of Frege's Logic. De Gruyter, Berlin (1988)
24. Tichý, P.: Collected Papers in Logic and Philosophy. In: Svoboda, V., Jespersen, B., Cheyne, C. (eds.) Filosofia, Czech Academy of Sciences, Prague. University of Otago Press, Dunedin (2004)

On the Nature and Formal Analysis of Indexical Presuppositions*

Igor Yanovich

Massachusetts Institute of Technology
yanovich@mit.edu

Abstract. This paper is divided into an empirical part and a theoretical part. The empirical part rediscovers a peculiar creature which was found by [Cooper, 1983] and then plunged back into the abyss of oblivion — indexical presuppositions, a special kind of "presuppositions" which can only, according to Cooper, be satisfied in the actual world. Cooper claimed that presuppositions of the gender features of free anaphoric pronouns induce such non-projecting "presuppositions". I enlarge the empirical scope of the discovery by showing that, first, indexical presuppositions are induced by gender features of all anaphoric pronouns, not just the free ones; and second, that their satisfaction requirements are more complicated than simple "satisfaction in the actual world only." On the theoretical side, I sketch three possible analyses of indexical presuppositions. The first one relies on direct copying of features from the antecedent, but meets certain serious problems. More work is needed in order to figure out if that theory can actually be worked out. The second theory employs a very complex constraint on choosing the name for the world variable of the pronoun, crucially using both syntax of the sentence and the model against which it is interpreted. The third one, in a sense the least conventional of all three, moves all the work to semantics, and analyzes indexical presuppositions as conditions on a rich structure of worlds, somewhat doubling the structure of embedding in the formula, which replaces the standard single possible world evaluation parameter.

1 Data: What Indexical Presuppositions Are

1.1 How Indexical Presuppositions Were Found

In the Kaplanian family of approaches to indexicality, there is a distinction made between "regular" expressions and indexical expressions, where the latter depend

---

* This paper has benefitted greatly from the discussions with and comments of Simon Charlow, Patrick Grosz, Kai von Fintel, Irene Heim, Pauline Jacobson, Salvador Mascarenhas, Reinhard Muskens, Chris Potts, Philippe Schlenker, Anna Szabolcsi, and Hedde Zeijlstra, as well as with and of the audiences at the LF Reading Group at MIT in the Spring of 2009, the 2009 Southern New England Workshop in Semantics (SNEWS) at UMass, the Semantics Seminar at NYU, and LENLS VI workshop in Tokyo. All errors in the paper are, of course, mine. I also want to thank half a dozen of patient and generous English native speakers whose judgements I cite in the paper.

K. Nakakoji, Y. Murakami, and E. McCready (Eds.): JSAI-isAI, LNAI 6284, pp. 272–291, 2010.
© Springer-Verlag Berlin Heidelberg 2010

on the context of utterance, as opposed to evaluation parameters[1]. Unlike non-indexical expressions, such expressions as *I* are to be strictly identified with the speaker of the actual context, without the possibility to be shifted by any operator. Kaplanian indexicals include words such as *I, she, that, now, actual,* an d so forth. Not all of the uses of some of those words are indexical, though: pronouns like "she" or "he" are indexical in their unbound uses (when they are accompanied by pointing of some sort), but are not indexical when bound.

[Kaplan, 1989] argues that the minimal adequate system which will be able to make good predictions about indexicals must have *double indexing* of a sort: it must distinguish between the actual contexts (*contexts*, in Kaplan's own terms) and the evaluation parameters which can be shifted (Kaplan's *circumstances*). The argument for that comes from the following consideration: for actual contexts, we want to use only those ones where, say, the speaker is located in the place and time when the utterance is taking place, so that our theory will recognize the sentence *I am here now* as true in every context. On the other hand, the set of possible circumstances must include "improper" contexts, e.g., where the speaker is at some different place, so that the sentence *It is necessary that I am here now* would not be predicted to be true (it should actually be false most of the time, because it is usually not necessary that I stay where I am at the moment — I could have been in thousands of different places instead.) Of course, the degree to which this argument is convincing depends on how much you want to be always able to say that *I am here now* cannot be false; if you do not at all, the argument fails.

But there are purely linguistic arguments in favor of using at least as much power as double indexing can give us for the interpretation of natural language, which come from [Kamp, 1971] and [Vlach, 1973]. Consider an example *Dick once thought that he would now be the president.* It is essential to be able to "remember" that we can refer to the actual time of the utterance from inside of the *would* embedded clause. If we have double indexing to the evaluation parameters and to the global context, we can do that for this example — *now* will be able to pick up the context time, not the evaluation time.

Under Kaplan's theory, the only thing about an indexical which relies on the actual context is its reference as such, the object it designates. [Cooper, 1983], in his analysis of anaphoric pronouns, has replicated the same distinction between non-indexicals and indexicals in his analysis of the interpretation of gender features. All gender features introduce, according to Cooper, a presupposition. Bound pronouns (that is, non-indexicals) introduce normal presuppositions, which interact with plugs, filters and holes as other presuppositions do. Unbound pronouns (that is, indexicals) also introduce presuppositions, but they must be

---

[1] To be more precise, [Kaplan, 1989] makes a distinction between pure indexicals, like *I*, which do not require anything extra-linguistic to determine their reference, and true demonstratives, like *those* or (unbound) *she*, which are incomplete without something else, for instance, a pointing gesture. I will ignore this distinction here, and call the unbound anaphoric pronouns with which we will be concerned simply "indexicals."

satisfied by the actual context, not the current evaluation context (Kaplan's circumstance). Consider the following data:

(1) Cooper's (13c-e), p. 180:

    a. Bill said that she talked.

    b. Bill hopes that she talked.

    c. Bill regrets that she talked.

(2) a. Bill said that Mary's wife works for Google.

    b. Bill hopes that Mary's wife works for Google.

    c. Bill regrets that Mary's wife works for Google.

[Cooper, 1983] notes that in 1, featuring free pronouns, the *speaker* has to believe that the individual referred to by the pronoun is female, even though the pronoun itself is within a complement to a propositional attitude verb. For factive verbs like *regret* it is unsurprising, but for such non-factive verbs as *hope* it is unexpected: presuppositions in their complements do not have to project to the actual world, cf. the presupposition of existence of Mary's wife in 2.

Cooper's point is that in 2, the presupposition of Mary's wife's existence interacts with the verbs in an expected way: 2a does not inherit the presupposition of existence, because Bill's words might have been wrong or misleading; 2b does not presuppose that Mary's wife exists either, but presupposes that Bill believes so; and 2c does presuppose that Mary has a wife because of the factive properties of *regret*. But in all examples in 1, the whole sentence inherits the presupposition that the referent of *she* is female in the actual world.

Cooper dubbed this special kind of presuppositions induced by free pronouns as in 1 *indexical presuppositions*, as opposed to normal presuppositions as in 2.

Bound pronouns, on the other hand, do not exhibit this special presuppositional behavior, according to Cooper (we will see shortly that contrary to his claim they do.)

(3) Cooper's (16), p. 182:

    a. Bill said that a neighbour$_1$ thinks that she$_1$ saw John.

    b. Bill hopes that a neighbour$_1$ thinks that she$_1$ saw John.

    c. Bill regrets that a neighbour$_1$ thinks that she$_1$ saw John.

Cooper argues that the presupposition that the referent of *a neighbor* is female in 3 interacts with the matrix verbs in the same expected way as the presupposition of existence in 2:

> "In the case of (16a) (= our 3a) with the verb of assertion *say* there is no presupposition that a neighbor is female (Bill might have been wrong) nor that Bill assumes that a neighbor is female (he might have been deliberately misleading). In the case of (16b) (=3b) with the verb of psychological attitude *hope* there is again no presupposition that a neighbor is female but it is presupposed that Bill assumes that a neighbor

is female. In the case of (16c) (=3c) where we have the factive verb *regret* we do get the presupposition that a neighbor is female." ([Cooper, 1983], p. 182)

This behavior is parallel to the one Cooper assumes for normal presuppositions. His conclusion is that free and bound pronouns induce different kinds of presuppositions: free pronouns contribute indexical presuppositions, and bound pronouns, normal presuppositions:

> "Our general claim is, then, that the gender consequences associated with a pronoun are presuppositions when the pronoun is bound but that something different than the normal case of presupposition is going on when the pronoun is free." ([Cooper, 1983], p. 182).

Cooper's analysis certainly makes sense when viewed from the perspective of a Kaplan-style theory of indexicality. Non-indexicals induce normal presuppositions, while indexicals, being rigid designators, can only introduce presuppositions about the actual context of the utterance.

There are two possible ways to modify this view. One was taken by the mainstream formal semantics following Cooper. The idea that the interpretational import of pronominal features is (normally) presuppositional was widely accepted, and has made its way into the textbooks. But possibly because the general focus of linguistic research has shifted away from the philosophical problems with indexicals, Cooper's actual analysis that divides pronouns into two classes carrying different sorts of presuppositions was largely forgotten. Even though the focus has somewhat shifted back now, with a flourishing literature on monsters, to our knowledge the nature of presuppositions of pronominal gender features has not received any new attention yet, as if the matter were closed. It is remarkable that [Schlenker, 2003] actually credits Cooper for a *unified*, or at least unifiable, account of presuppositions of indexical and non-indexical pronouns. So the first direction is to fo rget about the difference between normal and indexical presuppositions altogether.

In the rest of this paper, I pursue the second direction: after we examine actual data more carefully, it appears that, first, the distinction between the two kinds of presuppositions does exist, and second, that indexical presuppositions are much more widespread than Cooper suggested. Namely, both indexical and non-indexical pronouns carry them, as well as, it seems, the descriptive content of definites in general.

1.2 Both Free and Bound Pronouns Induce Indexical Presuppositions

Cooper restricts the emergence of indexical presuppositions to indexical pronouns, but uses only a very limited set of examples to argue for that. As it has happened with the famously prohibited, yet apparently existing monsters, the wrong choice of the small set of examples is to blame. Under a closer examination, all pronouns bear what may be called indexical presuppositions, though their satisfaction properties are more complex than the simple rule of "satisfaction in the actual context only."

I will not provide the full range of pronominal evidence for the reasons of space, restricting myself to several representative examples.

Scenario 1. My friend Richard and I went to a bar yesterday, and I talked to an acquaintance of mine who I know to be a woman. But Richard for some reason thinks it is a man, and I am not able to convince him that he is wrong.

When speaking about Richard's thoughts, we can only use a feminine pronoun, not a masculine pronoun:

(4)   *I*: Richard thought that the person I talked to yesterday in the bar lost ***his/^{OK}her** keys.

(5)   a.   *I*: After talking to Richard$_1$ today about [that person I talked to in the bar last night]$_3$, I figured out that Richard thought he$_1$ should help her$_3$ out when she$_3$ realized she$_3$ lost her$_3$ car keys.

      b.   *I*: After talking to Richard$_1$ today about [that person I talked to in the bar last night]$_3$, I figured out that Richard thought he$_1$ should help him$_3$ out when he$_3$ realized he$_3$ lost his$_3$ car keys.

There are two sets of worlds in these examples: W, the set of all of the speaker's epistemic alternatives (that is, worlds compatible with what the speaker considers to be actual), with members w, and W', with members w', which are worlds compatible with Richard's beliefs as they are according to the speaker. In all w worlds, the person Richard and the speaker saw at the bar is female (the speaker thinks it is a woman); in all w' worlds, this person is male (the speaker thinks that Richard is sure it is a man.)

If the gender constraint induced by the gender feature of the pronoun is a normal presupposition, as it should be according to Cooper, then we expect that we should be able to satisfy the presupposition either in the w worlds or in the w' worlds. Yet the latter option is actually out, irrespectively of whether we have a c-commanding antecedent, as in 4, or a non-c-commanding one as in 5. We can only use *she*, the pronoun whose feature corresponds to the gender of the referent in w-worlds, not *he*, which corresponds to its gender in w'-worlds.

That is exactly the behavior of Cooper's indexical presuppositions: the presupposition must be satisfied in the actual world. The novel empirical point of Scenario 1 is that indexical presuppositions are induced by pronouns with antecedents, not by indexicals.

The satisfaction pattern of indexical presuppositions contrasts both with the behavior of normal presuppositions, such as the existence presuppositions induced by definite descriptions, and that of regular DP-internal predicate constituents, such as DP-internal adjectives within indefinite DPs, a standard test case for *de re/de dicto* readings.

Scenario 2. I know that Ann does not have a dog, and I also know that my friend Richard is sure that she has one. Moreover, he told me that he thinks Ann's dog is fed well.

Certainly I can describe that state of affairs (as I've already did in the previous sentence) with the following example:

(6)   Richard thinks that **Ann's dog** should be fed well.

Uttering 6 does not commit me to the belief that Ann has a dog. Yet uttering the versions of 4 and 5 in Scenario 1 with masculine pronouns is not consistent with my belief that the person Richard and I met in the bar is female.

Turning to DP-internal predicates of indefinite DPs, when embedded under an intensional operator, such predicates can be interpreted with respect to the actual world (*de re*), or some set of non-actual worlds (*de dicto*). For instance, adjectives like *male* or *female*, supposedly denoting the same gender properties that the gender features of pronouns presuppose of the pronoun's referent, do not have to be always evaluated against the actual world:

Scenario 3. Richard works in the HR department and they are now looking for a driver. Richard is in charge of the selection process, and he already conducted all the interviews. There is a clear top candidate, whom Richard believes to be a man. But I know that he is wrong about that, because this person is my friend and I know that it is a woman, not a man.

(7) I: Richard wants to hire a ^{OK}male/^{OK}female driver.

In the speaker's epistemic alternatives w, the person Richard wants to hire is a woman. In Richard's epistemic alternatives w', the same person is a male. In 7, we can use either a predicate true of this person in w, which is the adjective *female*, making a *de re* statement, or a predicate true of it in Richard's belief worlds w', which is *male*, thus making a *de dicto* statement. That contrasts with the behavior of the gender restriction invoked by the pronouns in 4 and 5 which has to be satisfied in the actual world.

Thus the properties of being a male or a female as such are not special in any way, and are not responsible for the indexical nature of pronominal gender presuppositions; what matters is not the choice of properties, but the way they are introduced. When those properties are introduced by the gender features of anaphoric pronouns, then they are indexically presupposed.

We conclude that [Cooper, 1983] was wrong when he argued that the behavior of presuppositions induced by bound and free pronouns is different. In fact, bound pronouns show the same indexical presupposition behavior as free pronouns: their presupposition can be satisfied only in the actual world.

Another conclusion to draw from Scenario 1 and 4-5 is that it does not matter whether we have a pronoun syntactically bound by a c-commanding antecedent, or a pronoun which is coreferent to some non-c-commanding DP. In what follows I will assume there is no significant distinction between the two kinds of antecedents, and will not include the examples testing this distinction into the paradigms.

Cooper's own characterization of what indexical presuppositions do is simple: he argues they must be satisfied in the actual world. However, Cooper reports only a very limited set of examples probing the behavior of indexical presuppositions. Let's take a look at what happens when we embed pronouns.

The simplest possible case which shows that satisfaction conditions are more complex is the case of talking about an imaginary individual:

Scenario 4. Beth knows that Andrew, a music teacher, does not have any students at the moment, but Andrew himself mistakenly thinks that he has one girl student. It is then possible to refer to this imaginary female student of Andrew's with a feminine pronoun:

(8)   *Beth*: According to Andrew, [his student]$_1$ always tries hard to reach her$_1$ / *his$_1$ goals.

Scenario 4 shows that if the referent does not exist in the actual world, we are allowed to use a pronoun with a gender feature as long as that feature matches the gender property of the individual in the non-actual set of worlds W'. It may seem to be quite a natural state of affairs, but the facts might have been different: it might have been that for fictional individuals we would be able to use pronouns of either gender, or that we would not be able to use a pronoun with a gender feature at all.

In fact, one of those two alternative way of how things might have been is predicted by Cooper's account of indexical presuppositions, which says they can be satisfied only by an actual-world property. But either no property can hold of an individual in a world where that individual does not exist, or all properties are true there.

So the observation just reported is not trivial: the fact that we must use in Scenario 4 a gendered pronoun with a gender feature matching the gender property in Andrew's belief worlds means that Cooper's characterization of the behavior of indexical presuppositions is wrong. If the referent does not exist in the actual world, we have to choose a feature agreeing with its gender in some other set of worlds.

We have seen in Scenario 1 what happens when the referent exists in the actual world, but someone mistakenly believes it is of different gender than it actually is: the actual-world gender wins. The same holds if we compare the gender of some individual in two sets of non-actual worlds embedded one under another:

Scenario 5. 1. Just as in 4, Andrew the music teacher does not have any students at the moment, and he knows that.
2. Beth's beliefs about how the world is (w'): Andrew has a boy student.
3. Beth's beliefs about Andrew's beliefs (w''): Andrew has a girl student.

So according to Beth, Andrew mistakenly thinks that his student is a girl, though it is a boy in Beth's own beliefs.

(9)   a. ^{OK} Beth believes that Andrew trusts [his student]$_7$ so much that he allows him$_7$ to play his violin.

   b. * Beth believes that Andrew trusts [his student]$_7$ so much that he
       allows her$_7$ to play his violin[2].

The winner is the set of worlds which is as close in the chain of accessibility
relations to the actual world as possible (in simplest, and most frequent, cases
this role is given to the actual world itself.) That generalization is more far-
reaching than our data directly support: we have not checked what will happen
in cases of deeper embedding. But first, it is already very hard to get the relevant
judgements even for the cases like those in Scenario 5 above, with just two
levels of embedding; the deeper we go, the harder it becomes. Secondly, it is
hardly possible that the grammar could treat the cases of one-level and two-level
embedding in one way, while cases of, say, four-level embedding in a completely
different manner. That is why it is fairly safe to assume the most general form of
the rule: only the gender property in the least embedded set of worlds matters.

   What happens if the individual is actual, but its gender is unknown (that is,
is different in different epistemic alternatives of the speaker)?

Scenario 6. The Russian name Sasha can belong both to a girl and to a boy.
Suppose that I plan to visit some old friends of mine who I have not met for
years. I know that they have a kid, and for some reason I want to bring some
present for that kid with me. I know the name of the kid, which is Sasha, but I
do not know whether it is a girl or a boy.

   (10)   Before I visit Sasha$_5$'s parents, I want to buy a present for *him$_5$ / *her$_5$/
          /$^{?/OK}$[her or him]$_5$.

However, when we explicitly restrict Sasha's gender by an *if*-clause, we can
successfully use pronouns with specific gender features:

   (11)   I am at the end of my wits. If Sasha$_1$ is a boy, I should buy him$_1$ a doll.
          But if Sasha$_2$ is a girl, I'd rather buy her$_2$ a toy car.

Under the standard Kratzerian view on what conditional clauses contribute,
the *if*-clause "if Sasha is a boy" here explicitly restricts the current set of the
speaker's epistemic alternatives W, producing $W_1 \subset W$, the set of those epistemic alternatives where Sasha is male. The matrix clause following that *if*-
clause is evaluated in this subset of W, and we can use the pronoun *he*. The
second *if*-clause restricts the evaluation domain to $W_2 \subset W$ where Sasha is a
girl, in a similar manner. It is worth noting that *if*-clauses do not simply make
possible free use of whatever pronouns:

   (12)   *If Sasha$_1$ is a boy, I should buy her$_1$ a doll. *But if Sasha$_2$ is a girl, I'd
          rather buy him$_2$ a toy car.

So we can use a corresponding gendered pronoun only when we restrict the
set of epistemic alternatives to a subset in which the gender is uniform.

---

[2] Judgements for such sentences are very hard to get. But as long as English speakers
can process the sentences at all, the contrast clearly goes this way.

The following scenario shows that a counterfactual if-clause cannot override the real-world gender property[3]; such cases fall into the same category with the attitude verbs, as in Scenario 1.

Scenario 7. There is a kid named Sasha, and I know that Sasha is a girl.

(13)   I will buy Sasha a toy train. Well, if Sasha were a boy, I would buy *him/^{OK}her a doll.

What distinguishes the use of if-clauses in 11 and 13 is that in 11 the if-clauses select a subset of the set of worlds one of which may turn out to be actual, while in 13 the if-clause selects a counterfactual set of worlds. Even though we do not know in 11 which gender the kid actually belongs to when we use the pronoun, the pronoun's gender feature nevertheless has to correspond to the gender property as it is in some of the possibly actual worlds under consideration. But in 13, the kid is a girl in all possibly actual worlds, so in line with our generalization that the least embedded set of worlds in which the referent exists dictates the choice of the gender feature, the pronoun has to be feminine.

The behavior of indexical presuppositions of pronouns bound by quantifiers is essentially the same. However, they provide us with a new test case: we can check what happens when the individuals a pronoun ranges over are of different gender in some set of worlds (say, in the actual world).

Scenario 8. There are both female and male students at UMass Amherst.

(14)   Every UMass student is content with {*her / *his / $^{OK/?}$ his or her / $^{OK/?}$ their} grades.

When there are both male and female individuals in the domain of the quantifier, it is not possible to use a pronoun like his or her. Only gender-neutral

---

[3] In fact, the real picture is more complicated — though not because of the counterfactual if-clause. Some speakers are consistently more likely to be sympathetic to the use of a pronoun matching a non-actual gender in the presence of an overt obviously gendered DP — even if the DP is non-referential, as in 13.

More generally, some people find acceptable certain uses of gendered pronouns which do not match the actual world gender. My favorite theory about it at the moment is that it is possible to construct counterparts in different world sets as being crucially distinct individuals on a certain criterium. The usual rules of gender presupposition satisfaction will not apply simply because there would be no real-world counterpart of the imaginary individual in one sense. Cf. the following sentence, to be interpreted against Scenario 1:

(1)   [That man that Richard saw in [that woman I talked to yesterday]$_s$]$_3$ would not, according to Richard, tell me what [he]$_3$ really thinks.

pronouns may be possible (which may include for some speakers *his* or *her* used as gender-neutral, default pronouns[4].)

Of course, it is possible to restrict the domain of quantification to just male or just female individuals (which is in a sense parallel to restricting the relevant worlds with *if*-clauses), and use pronouns with specific gender features, as in 15. After that the domain of quantification will no longer be of mixed gender.

(15)   Every male UMass student is content with his grades.

Scenario 9. Smith College, one of the Five Colleges of Western Massachusetts, is a women's college. Imagine that Smith has recently gone coed, but not everyone knows about it yet, and Beth reads a letter to some newspaper by a Smith alumna who thinks that Smith is still a women's college. At the same time, Beth already knows that Smith is coed now.

(16)   *Beth*: *This alumna strongly believes it should be made an absolute principle that every Smith College student meet **her** adviser at least twice a week.

Beth cannot utter 16 in this situation (unless she consciously uses "she" as a gender-neutral pronoun, cf. fn. 4.) The sentence carries a presupposition that all Smith students are female according to Beth's own beliefs, and since in her beliefs Smith is coed, this presupposition can not be satisfied.

Thus cases when the quantification domain includes individuals of both genders behave similarly to the cases when the gender of a single referent is unknown: only gender-neutral pronouns may be used without further restriction, and it is not possible to override that by how things are in some embedded set of worlds.

The full set of generalizations about the occurrence and behavior of indexical presuppositions is as follows:

---

[4]   How do we know that a pronoun is used as a "gender-neutral"? It is helpful that the current range of acceptable English default gender patterns is varied: all variants of the following sentence contain a default pronoun (obviously there are both female and male students in the world, so the pronoun has to be default for the sentence to make sense in our world.)

(1)   [Every   student]$_6$ in   the   world   wants   to   become   better   than $\{$his$_6$/her$_6$/their$_6$/[her or his]$_6\}$ teachers.

Not all speakers find all variants shown in 1 acceptable, but each of the four is used by some. This variability allows us to distinguish default gender use from normal, or "proper" gender use: only in examples with such variability we have default usage, as the same variants are not acceptable for cases like the following:

(2)   a.   [John's son]$_6$ wants to become better than $\{$*her$_6$ / * their$_6$ / *[her or his]$_6\}$ teachers.

   b.   [Every girl]$_6$ in our school wants to become better than $\{$*his$_6$ / */?? their$_6$ / *[her or his]$_6\}$ teachers.

(17) a. Free and bound pronouns behave alike: both induce indexical presuppositions. Whether the antecedent c-commands the pronoun coreferring with it does not matter.

b. It does not matter whether the antecedent is quantificational.

c. Indexical presuppositions may only be satisfied in the highest (=closest to the actual world in the chain of embedding) set of worlds where the individual of which the presupposition is predicated exists.

d. If the gender property is non-uniform across such a set, the presupposition fails.

e. If it is possible to look only at a gender-uniform subset of a non-uniform set of worlds (e.g., it is possible when we restrict the relevant set of worlds with a *realis* if-clause), the presupposition is evaluated only against that subset, and thus may be satisfied, though it would be failed in the whole set.

As a final empirical remark, indexical presuppositions seems to be invoked not only by pronominal gender features, but by other kinds of descriptive predicates inside of definites:

(18) Beth believes that my dog is a unicorn. She also believes that {*the unicorn / ^{OK}it / ^{OK}my dog} is playing in the garden now.

(19) Beth believes there is a unicorn in the garden. ^{OK}She also believes the unicorn is grazing now.

Thus there might be nothing special about gender features: anaphoric pronouns may be analyzed as concealed definite descriptions ([Elbourne, 2005], a.o.), with their features being the predicative content, which is subject to the general indexical presupposition rule.

2   Analysis: Three Alternative Theories

2.1   A Very Brief History of Indexicality

Cooper built his system on the Kaplanian distinction between indexicals, including unbound anaphoric pronouns, and non-indexicals, including bound pronouns, using a relatively straightforward double-indexing logic. It is important to note that the double indexing was not the final point of the studies of time and world dependencies in natural language (the two domains are largely parallel.) In the tense domain, [Gabbay, 1974] argued that semantics has to keep track of all time points introduced during evaluation of a sentence. [Saarinen, 1978] has proven that natural language must have the power to have infinitely many time points available for back reference. [Enç, 1981] has shown that nouns behave differently from verbs: roughly, they are evaluated at a time, but this time seems to be contextually supplied, in a similar way assignment functions give values to free pronouns. [Cresswell, 1990] concludes that natural language is as expressive as a language wit h explicit quantification over worlds, with the argument transferrable to times.

It is thus quite obvious by now that a lot more expressive power is needed than just keeping track of two points — the current evaluation point and the actual point, be they times or worlds. To name just a few recent contributions to the debate, [Schlenker, 2005] argues for complete symmetry between the domains of individuals, times and worlds — and thus for explicit quantification over times and worlds. [Kusumoto, 2005] argues for a relatively restricted tense system where all predicates have time arguments, but not all of them have those arguments represented in syntax. Only the ones present in syntax may be used for subsequent reference, so the resulting system is somewhat less expressive than a fully articulated explicit quantification system: the power of the logic is essentially the same, but not all language expressions are allowed to use that power. [Keshet, 2008] tries to find ways to restrict the system with free use of world and time variables in a way which would prevent overgeneration. Thus the main question of the debate, in a simplified form, is how we should restrict the very expressive system which postulates covert world/time variables for every predicate.

Note that the original result driving the Kaplanian distinction is lost in this perspective. After it has become obvious that we need much more than just two points to interpret our sentences at, linguists have become preoccupied with how exactly those additional points are to be introduced and handled rather than in the distinction between Kaplan's contexts and circumstances. That is probably a part of the sociological reason why Cooper's treatment of free pronouns was largely forgotten.

On the other hand, the empirical foundation of Kaplan's two-way distinction has been shattered as well when it was discovered that monsters actually exist, both in understudied languages like Amharic and in less studied corners of English, [Schlenker, 2003], though non-shiftable indexicals, like the English *I*, are also possible.

How does our data on gender in anaphoric pronouns fit into that general picture? We see that both bound and unbound pronouns introduce non-trivial restrictions, which can be called indexical, as they pertain to how the least embedded relevant set of worlds is — which is close enough to the classical indexicality, with the difference that not only the actual world matters. But this indexicality of the presuppositions is not straightforwardly related to the Kaplanian indexicality of reference for special lexical elements. In a world where Kaplanian indexicality is actually not quite hard-wired into the interpretation of language, which is close to our world, as the monster data suggest, this finding is not surprising.

The question then is what kind of expressive power we need to capture the indexical presuppositions. The next section sketches several possible alternatives.

2.2 Three Stabs at the Account

Theory 1: Copy the antecedent's world argument. Imagine a theory of predicates within pronouns following the spirit of [Percus, 2000] for DP-internal predicates. Suppose we have world indices on the predicates of the antecedents and on pronouns themselves, too. Suppose also the pronoun copies its world argument from the antecedent — that would be one of the principles of the binding theory for indices.

There is no complex procedure of searching the right index here, and thus the main question this theory immediately raises is whether it can be made adequate empirically.

Theory 2: Constraining the resolution of the world variable. Imagine a different kind of system where pronouns do not copy anything from their antecedents, but instead it is required that the world arguments of indexically presupposed predicates be resolved to the highest possible of world variables, where "possible" means that the resulting extensional predicate will not be trivially false or true. (In other words, if an index is "possible", it means that the referent must exist at that index.)

This looks nice, but requires a lot of expressive power and inter-modular interaction. Here is how the constraining mechanism will work:

We take the individual the pronoun refers to and check if it exists at the actual world. If yes, we use the actual world as the world argument. If not, we go one step of embedding lower and check if that individual exists at all worlds of that level of embedding. If yes, we use the variable name bound by the quantifier over those worlds; if not, we repeat the same step, going yet further down.

To be able to do that, the module which executes this procedure must be able to access both world variable names, the syntactic structure of the sentence (and read off it the embedding structure between sets of worlds), and, crucially, the model against which it needs to check the existence of that individual.

It is useful to compare Theory 2 with the procedure of presupposition resolution used in the Binding Theory of presuppositions in DRT, [van der Sandt, 1992], [Geurts, 1999]. The DRT procedure takes a condition or a referent marked as being presupposed and pushes it towards the matrix DRS one step at a time. If at some point the presupposition can be identified with some conditions and referents already present in the structure, we resolve it to them, and stop: the presupposition is bound now. If we do not find such a binder, we need to accommodate the presupposition into the DRS, and we simply change its status from presupposed to asserted. We do that in the top-to-bottom manner, since after we checked for all possible binders, we find ourself in the highest DRS.

There are certain similarities between the two procedures: both privilege the top DRS, and involve, at least in a certain sense, movement from the bottom to the top, and then possibly back again. But differences are serious: first, indexical presuppositions can never be bound; second, for regular presuppositions global accommodation is just a preference, but the principle guiding the resolution of indexical presuppositions is absolute; finally, the presupposition resolution procedure does not need to check the model — it remains strictly syntactic, on the level of the syntax of DRS-s, but the indexical presupposition procedure does need to check the actual model to determine where the presupposition should end up.

Theory 3: A rich structure of worlds under discussion. An alternative is to shift all the work to the semantics. Indexical presuppositions will be simple uniform operators, but our translation logic will be flexible enough to allow them to be interpreted in the right way.

To build such a system, called Enriched Indexicality (EI), we replace the world evaluation parameter with a new, richer evaluation parameter containing a complex structure $\mathcal{f}$, which can be thought of as the "map" of worlds under discussion. This "map" is, formally speaking, a forest with the epistemic alternatives of the speaker as the roots, plus a pointer to some node(s) which is (are) "current". $\mathcal{f}$ is essentially the point-generated rooted submodel of the propositional modal logic underlying our translation logic.

Note that the same world w_1 may appear many times in $\mathcal{f}$. So it is more appropriate to use *pointers* to worlds in $\mathcal{f}$ — "addresses" of nodes in $\mathcal{f}$ — rather than of worlds. Given a pointer, we can easily find the world which occupies that address, but the reverse is not true. The contents of a pointer — the world which is stored at the node — is denoted by $^*p$ for a pointer p.

The world corresponding to the current world in the standard semantics is given by an internal pointer $\mathcal{f}.cr$ (cr from "current"), which is always defined. We can think about it as if one of the nodes in $\mathcal{f}$ is red color, and all other nodes are black. This red node corresponds to the current world in the usual semantic systems. If we only use the world sitting in $\mathcal{f}.cr$, and make no use of the rest of the structure, that would be the same as if we used usual semantics. So usual semantic systems are special cases of our new one.

Starting from some world inside of $\mathcal{f}$, we can always reach the root of the tree in which that world resides simply going in steps each of which takes us from some world to its parent. We can define a natural preorder relation on worlds in $\mathcal{f}$: $w_1 \succ_{\mathcal{f}} w_2$ iff we pass through w_1 when we go from w_2 up to the root in $\mathcal{f}$.

The purpose of $\mathcal{f}$ is to represent the sets of worlds introduced by intensional operators in such a way that all members of some set introduced by a more embedded operator would be situated on the same level, and also farther from the roots of their trees than members of other sets introduced by operators on top.

To ensure that $\mathcal{f}$ is such an "accurate map", we need to have meaning postulates guaranteeing a proper building procedure. While only intensional operators themselves know enough to add the relevant worlds to the structure, we should constrain their freedom of modifying $\mathcal{f}$. E.g., we do not want some operators erasing all the content of $\mathcal{f}$, or replacing it with some totally unrelated stuff.

We crucially distinguish between new quantifiers over worlds (like attitude verbs) and restrictors (like realis if-clauses). The first type adds a new layer of worlds to the map. The second type erases some part of the lowest level of worlds from the structure, leaving just a subset of them in it.

For expository reasons, we present "translation" instructions for converting usual semantics into our semantics rather than define the new semantics from the scratch.

First we define some notational sugar to make our life simpler:

(20) *Augmenting $\mathbf{f}$:*
Let $\mathbf{f} + W$, where W is a set of worlds, denote the result of adding all worlds in W as daughter nodes to the node $\mathbf{f}.cr$.

(21) *Changing the current world in $\mathbf{f}$:*
Let $\mathbf{f}[p]$, where p is a pointer to some node in $\mathbf{f}$, be exactly the same as $\mathbf{f}$ except that $\mathbf{f}[p].cr = p$.

(22) *Taking the daughter set of the current world in $\mathbf{f}$:*
If p is a pointer to a node in some $\mathbf{f}$, let $dtr(p)$ denote a set of pointers to all daughter nodes of p.
More formally, $dtr(p) = {}^{def}\{w|\mathbf{f}.cr \succ w \land \neg\exists w' : (\mathbf{f}.cr \succ w' \land w' \succ w)\}$

(23) *Redefining world-dependence:*
If there is a meaning $Q(w)$ in (your favorite) usual semantics which denotes some expression dependent on world w, then in the corresponding Enriched Indexicality theory $Q(\mathbf{f})$ is defined and is true whenever Q is true in $^*(\mathbf{f}.cr)$.

(24) *Functional Application normally passes the $\mathbf{f}$ argument intact:*
The default interpretation rule for $[\![\alpha(\beta)]\!]^{\mathbf{f}}$ is: $[\![\alpha(\beta)]\!]^{\mathbf{f}} = [\![\alpha]\!]^{\mathbf{f}}([\![\beta]\!]^{\mathbf{f}})$

Now consider some/your favorite "standard" meaning of a first-kind intensional operator Op of type $\langle st,st \rangle$, having the following general form[5]:

(25) $\lambda Q.\lambda w.\{w'|w' \in acc_{Op}(w)\} \, \xi \, \{w''|w'' \in acc_{Op}(w) \land Q(w'')\}$, where acc_{Op} is the accessibility relation returning a set of worlds accessible by some measure of accessibility from the world w; Q is the propositional argument of Op; ξ denotes the relation between the set of worlds $acc_{Op}(w)$ and its intersection with the set of worlds where the proposition P is true.

This is a general format for expressing a relation ξ between two sets of possible worlds: the whole set of worlds accessible from some w and its intersection with the set of worlds in which some proposition Q is true. In the case of a universal modal, ξ would be identity; in the case of an existential model, ξ would be the relation of having a non-empty intersection; and so forth, if we need more relations.

---

[5] This is an oversimplified picture of what such operators may be; e.g., a good case can be made that some attitude verbs take into account an ordering on the worlds they quantify over, not only the set as such, like in Heim's analysis of *want*. However, once we have dealt with an oversimplified version in the text, it is easy to treat more realistic cases in the same way.

We make two changes to this entry in 25 to adapt it to our EI system: first, its argument will no longer be w, it will be $\mathbf{f}$ instead. The world $*(\mathbf{f}.cr)$ will be used instead of w where we need contingency upon what the current world is. In effect, $acc_{Op}(\mathbf{f})$ is the same thing as $acc_{Op}(w)$ was: both essentially return the set of worlds accessible by some measure from the current evaluation world. It is just that $acc_{Op}(\mathbf{f})$ contains much more information than needed for this simple task.

Secondly, we will not simply evaluate Q in the individual worlds from $acc_{Op}(\mathbf{f})$, the set of worlds accessible from w. We will supply Q with an augmented version of $\mathbf{f}$ obtained by adding all worlds in $acc_{Op}(\mathbf{f})$ to $\mathbf{f}$. So unlike in standard intentional semantics when we simply pass an individual world to an expression in the scope of an intentional operator, in the EI system we pass down the information about the whole chain of possible world embedding recorded in $\mathbf{f}$ as well. The difference is thus that, again, we pass down a lot of information unnecessary for the evaluation under normal circumstances.

This is how we do that: $(\mathbf{f} + acc_{Op}(\mathbf{f}))$ is such $\mathbf{f}'$ where all the members of the set $acc_{Op}(\mathbf{f})$ — the set of worlds accessible from the current evaluation world in $\mathbf{f}$ — have been added as daughters of the node $\mathbf{f}.cr$, the node pointing to the current evaluation world.

To determine which worlds will be specific evaluation parameters, we restrict our attention to such pointers p that point to one of the daughter nodes we have just added — we single them out with $dtr(\mathbf{f}'.cr)$. Then we check whether Q is true in those worlds, by supplying it with $(\mathbf{f} + acc_{Op}(\mathbf{f}))[p]$ — the augmented structure we just built, but with the current world in it shifted to p, where p points to one of the newly added nodes.

To make the result more readable, let's define $\mathbf{f}' = (\mathbf{f} + acc_{Op}(\mathbf{f}))$ (note that the current world of $\mathbf{f}$ and $\mathbf{f}'$ is the same). The new meaning then is:

(26) $\lambda Q.\lambda \mathbf{f}.\{w'|w' \in acc_{Op}(\mathbf{f})\}\ \xi$
      $\{w''|w'' \in acc_{Op}(\mathbf{f}) \wedge \exists p \in \mathbf{f} :\ p \in dtr(\mathbf{f}'.cr) \wedge {}^*p = w'' \wedge Q(\mathbf{f}'[p])\}$

The second set of worlds w'' contains such worlds which are pointed to by the daughter nodes of the current world in $\mathbf{f}'$ — that is, it contains only the worlds from $acc_{Op}(\mathbf{f})$; furthermore, w'' is containing only those such worlds in which Q is true — this is what $Q(\mathbf{f}'[p])$ says.

How to apply similar transformations to standard meanings in order to get EI meanings for other kinds of intensional operators is straightforward, and we omit them (with the exception of if-clauses, to which we return later.)

Despite the fact that the meaning in 26 looks more complex, it is actually equivalent to the old meaning in 25 in cases when Q only pays attention to what the current world of $\mathbf{f}'[p]$ is and ignores the rest of the structure. Our innovation, thus, allows us to preserve as much of the usual semantics as we want. The new part here is that we can also utilize the extra information contained in $\mathbf{f}$ when evaluating Q. And that is exactly what we will do to account for pronominal indexical presuppositions.

Note that our system for augmenting $\mathbf{f}$ is static, not dynamic. We manipulate the parameter we pass down to an embedded constituent, but the result of that manipulation is accessible only in that constituent. Whatever we do to $\pmb{f}$, that does not change the contexts of other constituents. Of course, it is possible to build a dynamic theory where that will not be so, but indexical presuppositions can be accounted for in a completely static setting.

It is useful to show how the evaluation structure $\pmb{f}$ will look like for a constituent under several levels of embedding. Suppose we have two levels of embedding, one that introduces w' worlds, and another one which introduces a set of w'' worlds for each w' world (so for w'_1 there will be a set W''_1, and so forth.) Consider a proposition evaluated in the most local, w'' context, such as $Q =$ *the parrot talked to Mary* in *Ann said that her sister thinks the parrot talked to Mary*. For the whole sentence to be true, Q must be true in any w'' world, under each w' branch. At any w''-world where we check the truth of that proposition, through the current $\pmb{f}$ we have access to: 1) this w'' world itself (it will be given by $\pmb{f}.cr$); 2) to the whole set of W'' worlds, as they are already in $\pmb{f}$; 3) to the w'-world creating the current branch — it will be the parent of $\pmb{f}.cr$; 4) to the wh ole W' set — it is all worlds of the same embedding level as the w'-parent; 5) to the root world w (the actual world). There is, however, no access to other w'' sets, branching from different w' worlds.

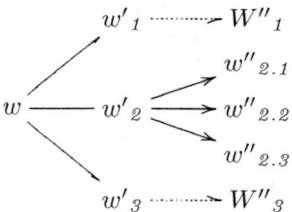

Let a special predicate $lives(a)(w)$ which is true iff individual a is in the domain of individuals existing in world w. Let $level(p)(\pmb{f})$ be the set of all nodes in $\pmb{f}$ which have the same number of nodes between them and the roots of their trees as p has.

Having prepared those simple devices, we add a new operator of the EI logic $\odot$ which will introduce indexical presuppositions. $\odot$ takes a predicate P and an individual b. b is returned if the check $\odot$ performs is successful. P is the indexically presupposed predicate. $\odot$ checks if the presupposition corresponding to P is satisfied in the current $\pmb{f}$, and if it is not, returns undefinedness.

(27) $\odot(P_{\langle e,t\rangle})(b_e)$ is defined as b_e w.r.t. $\pmb{\mathfrak{M}},\pmb{f}$ iff
$\exists p_1 : \forall p_2 \in level(p_1)(\pmb{f}) : lives(b)(^*p_2) \wedge$
$\wedge \neg\exists p_3 : (p_3 \succ_{\pmb{f}} p_2 \wedge lives(b)(^*p_3)) \wedge$
$\wedge P(b)(^*p_2),$
undefined otherwise

27 looks for a set of worlds on the same level of embedding $level(w_1)(\pmb{f})$ such that b lives at all of those worlds $(lives(b)(^*p_2))$, and does not live in any of the parent worlds of those worlds $(\neg\exists p_3 : (p_3 \succ_{\pmb{f}} p_2 \wedge lives(b)(^*p_3)))$.

The existential quantification over p_1 is essentially existential quantification over levels. If there is no such a level at all of the world of which b lives, the whole formula is false, and the operator is undefined. If there is such a set, then $\odot$ checks whether at each of those worlds P holds of b $(P(b)(*p_2))$. If yes, the indexical presupposition is satisfied. If not, we get the presupposition failure.

Thus there are two ways to fail an indexical presupposition: first, there may be no such a set of worlds of the same embedding that an individual lives at all of them; second, if that individual does not satisfy the indexically presupposed property in the highest of such sets of worlds, the presupposition also fails.

The meanings for gender features will be simple, as the real presuppositional work is done by the satisfaction properties of $\odot$. If we believe in featural decomposition, we may use the semantics for the gender feature as in 28; the result is supposed to look very much like 29 (except that we omit the semantics of the other features from it.)

(28) $[\![\text{FEM}]\!] = \lambda x_e.(\odot(\lambda y.female(y))(x))_e$

(29) $[\![\text{SHE}]\!] = (\odot(\lambda y.female(y))(x))_e$

$\odot$ straightforwardly captures our generalizations for the intensional operators of the first kind, like attitude verbs; for *if*-clauses, another operation of $\mathfrak{f}$ modification, a deletion operation should apply:

(30) Let $\mathfrak{f} \cap W$ be exactly like $\mathfrak{f}$ except that all the worlds on the leaves of $\mathfrak{f}$ which are not in the set W are deleted from the structure. In other words, $\mathfrak{f} \cap W$ purges all non-W worlds from the lowest level of $\mathfrak{f}$.

(31) $[\![\mathbf{if}(P_{\langle s,t\rangle})(Q_{\langle s,t\rangle})]\!]^{\mathfrak{f}} = Q(\mathfrak{f} \cap (\lambda w.P(w)))$

The prediction this meaning makes, taken together with $\odot$, matches the data: the worlds remaining after the deletion are at the same level as the whole set was, and the purged worlds are simply not seen. So Q will not even know that the gender was non-uniform in the first place. Of course, it remains to be seen if this analysis works well for the semantics of if-clauses as such or not.

EI is cast in a static setting, but at the same time uses a very rich notion of context of evaluation; instead of the traditional use of a single world as an evaluation parameter we use a very complex structure over possible worlds. The crucial difference between that kind of rich context and the dynamic notion of context is that in the static EI the changes to evaluation parameters are not seen outside the constituent where they are made. In other words, there is no transparency: e.g., the interpretation of the first conjunct does not (have to) affect the interpretation of the second. All richness of context is passed from the top to the bottom, but not sideways.

If there are no indexical presuppositions present, the EI theory is equivalent to the standard semantics that was the base to which EI was added: the definitions for EI we have given were formulated as translation instructions for modification of an existing semantic system. We can think of the EI theory as of an extra module added to an otherwise standard theory in order to account for a part of the data.

2.3   Testing the Three Theories

We consider two crucial test cases. First, recall 16. For Theories 2 and 3, the result is straightforward: they directly say that the actual world gender should win. For Theory 1, it is more complicated: since it says that pronouns copy the world argument of their antecedents, to get the pronoun's feature interpreted at the actual world we need to assume that the antecedent is interpreted de re.

At first sight that seems to be a disastrous prediction: why would a quantified DP have to be interpreted de re? But a closer examination reveals that Theories 2 and 3 in fact also need the antecedent to be read de re:

Suppose that the alumna does not have wrong beliefs about anyone's gender, she only does not necessarily know who are current Smith students. Since she thinks Smith is a woman college, there will only be female Smith students in her belief worlds, and by our assumption, they are female also in the actual world.

Then what Theories 2 and 3 predict is that the presupposition requires each person who is a Smith College student in the alumna's worlds to be female in the actual world. But all such persons are female, by our assumption. Then the presupposition is satisfied even if Smith is actually coed: all that matters is which gender the individuals who may be students according to the alumna are.

So if we allow *every Smith College student* to be read *de dicto* in 16, Theories 2 and 3 make wrong predictions. They would make the right ones if we require the *de re* interpretation of the antecedent, but then Theory 1 will do equally well. The requirement itself remains surprising and unusual, and should be investigated further, but it does not tease different theories apart.

The second test case involves if-clauses. Recall 11. Theory 1 says that we must simply copy the world argument from the antecedent to the presupposed gender feature. The antecedent in both cases is *Sasha*, and in order to get the pronouns right, we must suppose that the if-clauses must introduce a new world variable, and that *Sasha* has to be interpreted *de dicto* in both if-clauses. Then we will get the results right. This looks somewhat suspicious: for the previous test case, Theory 1 had to assume obligatory *de re* interpretation, and here with if-clauses it has to have *de dicto*.

For Theory 3, on the other hand, this case does not present any substantial difficulty — we have sketched above in section 2.2 what the account would look like. For Theory 2, a parallel account treating if-clauses in a different way from other intensional operators should probably be possible, though it has to be spelled out first in order to judge it.

2.4   Relative Merits of the Three Theories

The conclusion is that it is now hard to say with certainty which of the three ways to go outlined above is better. Theory 1 seems to be somewhat more restrictive, but also requires non-trivial assumptions about what should be interpreted *de re* and *de dicto*. While its requirements in the case of 16 are essential for the two other theories, the obligatory *de dicto* reading in 11 seems to be a genuine problem, as well as the conflict between the two requirements.

Theories 2 and 3 are both more complex, and are equivalent empirically as far as our data is concerned. However, while Theory 3 uses only semantics, Theory 2 crucially requires non-trivial interaction between different modules of grammar, and thus is less restrictive. As for Theory 3, the operation we need to account for indexical presuppositions uses only a small part of the possibilities which the use of evaluation trees offers. It remains to be investigated whether there are other linguistic phenomena that can receive natural explanation in the EI framework, and/or whether the power of EI can be successfully restricted; but so far this theory seems to be the best bet.

References

[Cooper, 1983] Cooper, R.: Quantification and Syntactic Theory. Studies in Linguistics and Philosophy, vol. 21. Reidel, Dordrecht (1983)

[Cresswell, 1990] Cresswell, M.: Entities and Indices. Kluwer, Dordrecht (1990)

[Elbourne, 2005] Elbourne, P.: Situations and individuals. MIT Press, Cambridge (2005)

[Enç, 1981] Enç, M.: Tense without Scope: An Analysis of Nouns as Indexicals. PhD thesis, University of Wisconsin (1981)

[Gabbay, 1974] Gabbay, D.M.: Tense logics and the tenses in english. In: Moravcsik, J. (ed.) Logic and Philosophy for Linguists, Mouton, The Hague (1974)

[Geurts, 1999] Geurts, B.: Presuppositions and Pronouns. Elsevier, Oxford (1999)

[Kamp, 1971] Kamp, H.: Formal properties of "now". Theoria 37, 227–273 (1971)

[Kaplan, 1989] Kaplan, D.: Demonstratives. In: Almog, J., Perry, J., Wettstein, H. (eds.) Themes from Kaplan. Oxford University Press, Oxford (1989)

[Keshet, 2008] Keshet, E.: Good Intensions: Paving Two Roads to a Theory of the De re/De dicto Distinction. PhD thesis, MIT (2008)

[Kusumoto, 2005] Kusumoto, K.: On the quantification over times in natural language. Natural Language Semantics 13, 317–357 (2005)

[Percus, 2000] Percus, O.: Constraints on some other variables in syntax. Natural Language Semantics 8, 173–229 (2000)

[Saarinen, 1978] Saarinen, E.: Backward-looking operators in tense logic and in natural language. In: Hintikka, J., Niiniluoto, I., Saarinen, E. (eds.) Essays on Mathematical and Philosophical Logic, pp. 341–367. Reidel, Dordrecht (1978)

[Schlenker, 2003] Schlenker, P.: A plea for monsters. Linguistics and Philosophy 26, 29–120 (2003)

[Schlenker, 2005] Schlenker, P.: Ontological symmetry in language: A brief manifesto. Mind and Language 21(4), 504–539 (2005)

[van der Sandt, 1992] van der Sandt, R.: Presupposition projection as anaphora resolution. Journal of Semantics 9, 333–377 (1992)

[Vlach, 1973] Vlach, F.: 'Now' and 'Then': A Formal Study in the Logic of Tense Anaphora. PhD thesis, University of California, Los Angeles (1973)

Non-standard Uses of German 1st Person Singular Pronouns⋆

Sarah Zobel

Georg-August Universität Göttingen, 37073 Göttingen, Germany
szobel@gwdg.de

Abstract. The purpose of this paper is to shed light on a phenomenon concerning the German first person singular pronoun *ich* which challenges the standard view on the semantics of first person singular pronouns, i.e. that they are always speaker-referential. The presented data shows a non-standard use of first person singular *ich* which I analyze to have a similar semantics to the German impersonal (generic) pronoun *man*. The analysis for non-standard *ich* is shown to be modifiable to also model the deictic use of *ich*. Finally, I bring up some related problems that merit further investigation.

1 Introduction

The purpose of this paper is to shed light on a phenomenon concerning the German first person singular pronoun *ich* which challenges the standard view that the semantics of first person singular pronouns is as in (1).

(1) $[\![\text{pronoun}_{1PSg}]\!]^c = c_S$ where c_S is the speaker of the context c^1

The core of the standard semantics is - in short - that a first person singular pronoun expresses speaker-referentiality. Keeping this in mind, consider the colloquial German data in (2) and (3).

(2)   Wenn **ich** als Mannschaft gewinnen will, dann muss ich motiviert auf
        if   I   as team   win.INF   want then must I   motivated on
        den Platz gehen.
        the field   go.INF

(3)   **Ich** muss als Fußballnation eine solche Mannschaft dominieren können.
        I   must as soccer-nation a   such team   dominate   can

---

⋆ I thank Dirk Buschbom, Eva Csipak, Ilaria Frana, Magdalena Schwager and Arnim von Stechow for their comments on various versions of this paper. Thanks also go to Thomas Graf for discussions on the data. All mistakes are of course my own.

1 Malamud [17] explicitly assumes the standard semantics in (1). Kratzer [10] proposes that first person singular pronouns are composed from the two meaningful features [1st] and [singular], which also amounts to saying that first person singular pronouns refer to a unique speaker of a context.

Even though both sentences[2] include the German first person singular pronoun *ich*, neither of them can be understood as talking about the speakers directly (obviously neither of the two speakers thinks of himself as a soccer team or soccer nation). The only sensible interpretations possible for (2) and (3) are paraphrased in (4) and (5) respectively. I will call these paraphrases the *non-standard readings* of the sentences. An occurrence of *ich* in a sentence with a non-standard reading is said to be in its *non-standard use*.

(4)   If (one as) a team wants to win, then one/they has/have to enter the field motivated.

(5)   (The national team of) a "soccer nation" (i.e. a nation known for being good at soccer) has to be able to have the upper hand over a (contextually salient) team (*eine solche Mannschaft*).

Interestingly, neither (4) nor (5) talks about the speaker even though (2) and (3) contain tokens of the first person singular pronoun *ich*.

This paper attempts to answer the following questions: What is the semantic contribution of *ich* in the non-standard readings? Does this new data provide enough reasons to discard the standard semantics in (1)?

The paper is organised as follows. In Sect. 2, I present a possible analysis of the data and show why it should not be pursued. Section 3 offers a detailed data discussion. In the fourth section, I give a purely semantic formalization of the non-standard use for *ich*. I show that the proposed meaning for the non-standard use can be easily modified to capture the semantics of the standard speaker-referential use of *ich*. Section 5 concludes.

2   A Counterfactual Analysis?

Before I provide an in-depth data discussion of the non-standard uses of German *ich* (cf. Sect. 3), I argue against a treatment in terms of what I will call the *counterfactual hypothesis*. The starting point of the counterfactual hypothesis is to assume that (2) and (3) are in fact hidden counterfactuals with a meaning similar to the English counterfactuals given in (6-a) and (6-b). The rationale for this assumption is that by analyzing (2) and (3) as counterfactuals one could retain the standard semantics for *ich*.

(6)   a.   If I were a team and wanted to win, I would have to enter the field motivated.
       b.   If I were a soccer nation, I would have to be able to have the upper hand over a team (of a contextually salient kind).

---

[2] Both pieces of data were taken with small modifications from discussions on the internet. (2) was taken from http://www.welt.de/print-welt/article532778/ Schlechte_Argumente_fuer_den_Aufnahmeantrag_an_die_G_14.html and (3) from http://www.rp-online.de/public/article/sport/fussball/nationalelf/wm/744344/ Die-deutsche-Mannschaft-muss-sich-steigern.html

However, there are at least three reasons why such an analysis should not be pursued.

First, consider (3). In contrast to example (2), the sentence in (3) is not an overt conditional. One would have to stipulate a covert *if*-clause that includes parts of the matrix clause, namely the *als*-phrase. In addition to that, the *als*-phrase would have to be analysed in both sentences as expressing non-factive predication. This is at odds with Jäger's [8] argument that *als*-phrases in this use contribute a factive, presupposition-like meaning.[3]

Second, example (2) is an indicative anankastic conditional (cf. [4]). Indicative mood in conditionals, as Stalnaker [27] observed, implies non-counterfactuality. Therefore, the mood of the verb can not contribute counterfactuality. In fact, there is no obvious part that could.

Third, the non-standard reading is actually unavailable for overt German counterfactuals.

(7)   ?Wenn **ich** als Mannschaft gewinnen wollen würde,   dann
        if   I   as team   win   want would.KONJ, then
        müsste   ich motiviert auf den Platz gehen.
        must.KONJ I   motivated on the field go.INF

The German counterfactual in (7), which is the explicitly counterfactual version of the indicative conditional in (2), can only be understood in the marked context where the speaker alone constitutes a team: *If I as a team would want to win, then I would have to enter the field motivated.* So, overt counterfactuality actually blocks the non-standard reading that the counterfactual analysis is trying to draw on.[4]

However, counterfactuals with *als*-phrases containing semantically plural nouns like *Mannschaft* (team) do not always require marked contexts, as is illustrated by the following example taken from another soccer fanpage.

(8)   Wenn **ich** das als Mannschaft von PSG   mitbekommen
        if   I   that as team   of Paris-St-Germain noticed
        hätte,   wäre   ich, glaube ich, aus   Protest nicht angetreten.
        had.KONJ, would.KONJ I   guess I   out-of protest not   played

---

[3] Apart from this fact, the *als*-phrase can not be reanalysed to contribute counterfactuality since it is optional and not required for a sentence to get a non-standard reading (see Sect. 3.4).

[4] It was suggested to me that this is not a good argument against the *counterfactual hypothesis* if one assumes that the counterfactuality is brought in by the pronoun itself. If that is the case, an overt counterfactual would contain an embedded counterfactual element. This stacking of counterfactuality could then be seen as the reason for the unavailability of the *non-standard reading*. Putting the counterfactuality inside the first person pronoun, however, takes away the initial motivation to pursue the *counterfactual hypothesis* in the first place, i.e. to retain the standard meaning for the indexical.

'If I had been the team of Paris St. Germain and had noticed what was going on, then, I guess, I wouldn't have played out of protest.'[5]

An in-depth exploration of why sentence (7) seems more marked than example (8) is out of the scope of this paper. As a first idea, the reason for (7) to be less acceptable is that it is impossible for the speaker as a single individual to win a soccer game. In example (8), on the other hand, the predicate *mitbekommen* (*'notice'*) is also compatible with the speaker as a semantically singular subject.[6]

3 Data Discussion

3.1 Putting Things into Context

In order to get a feeling for the meaning and the use of sentences with a non-standard reading, it is important to take a close look at the contexts in which such examples surface. Out of the blue, the sentences (2) and (3) are very odd. It seems that, without the appropriate context, one prefers to interpret *ich* speaker-referentially.

The following two scenarios are constructed contexts that "trigger" the non-standard readings of (2) and (3).

Scenario 1: Imagine you are a soccer expert who is often consulted to evaluate games on TV. In the match you have just seen, the team that lost played weakly from the beginning since the players were obviously not motivated. During the evaluation of the game, the interviewer asks for your opinion as to why this team lost. You consider it entirely obvious what went wrong during that match. So, you answer the interviewer's question by uttering (9) (repeats (2)).

(9)   Naja, wenn **ich** als Mannschaft gewinnen will, dann muss ich
        well   if   I   as team   win.INF   want then must I
        motiviert auf den Platz gehen.
        motivated on the field go.INF
        'Well, if (one as) a team wants to win, then one/they has/have to enter
        the field motivated.'

Scenario 2: Imagine again that you are a soccer expert. This time you are asked to evaluate an international match: Germany against Faroe Islands. Embarrassingly, Germany only won by 1-0. Right before the evaluation starts, you hear an interview with the coach of the German national team, who says that he is quite content with the his team's performance. You can not believe that the coach could be content with such a weak performance in light of the expectations that are usually placed on a national team of a country known to be good at soccer. So, when the interviewer asks you during the evaluation whether you share the coach's opinion on the match, you answer with (10) (repeats (3)).

---

[5] http://www.roteteufel.de/archive/index.php/t-24375.html

[6] I thank Magdalena Schwager (p.c.) for suggesting this line of argument to me.

(10)   Nein, **ich** muss als Fußballnation eine solche Mannschaft dominieren
         no,   I   must as soccer-nation a   such   team   dominate
         können.
         can
         'No, (the national team of) a "soccer nation" has to be able to have
         the upper hand over a team (of a contextually salient kind; *eine solche
         Mannschaft*).'

The two scenarios highlight the stance the speaker takes with respect to the
interviewer's questions about the matches. In both scenarios, the use of *ich* in
the answer signals that the opinion expressed is something that the speaker
thinks is (or should be) unobjectionable.

What further corroborates the idea that the speaker considers the expressed
proposition unobjectionable is the possible co-occurrence of the discourse particle
doch. In the literature, this discourse particle has been analyzed as signaling a
contradiction or inconsistency between two propositions (cf. for example [5], [7]
and [16]).

Gast [5] argues that *doch* has the following two characteristic features: First,
a proposition p containing *doch* is taken for granted by the speaker who also expects the addressee to take it for granted. And second, the speaker assumes that
the addressee takes $\neg p$ for granted. Adopting a dynamic system in the tradition
of Heim [6], he proposes that a sentence containing *doch* is used to eliminate
contradictions from an input context to give a consistent output context.

In the data sample I collected for non-standard *ich*, *doch* occurs frequently
and is in principle compatible with all of the collected examples. The data in
(11), (12) and (14) in the following section show the compatibility of *doch* with
non-standard *ich*.

(11)   Ich find   das ist ein total   doofes Argument! Ich kann **doch**   als
         I   think that is a   absolutely dumb   argument   I   can   DOCH as
         Brautpaar   nicht von meinen Gästen erwarten dass sie   mir
         bridal-couple not   from my   guests   expect   that they me
         quasi   die Feier finanzieren!
         more-or-less the party pay
         'I think this is an absolutely dumb argument! The bridal couple can't
         expect their guests to more or less pay the party!'[7]

(12)   Ich kann **doch**   als Schiedsrichter in so   einer Situation wie vor
         I   can   DOCH as referee   in such a   situation like before
         der Halbzeit keinen Elfmeter für Siegen geben.
         the half-time no   penalty   for Siegen give
         'A referee can't give a penalty to Siegen in a situation as it occurred
         right before the half time.'[8]

---

[7] http://www.urbia.de/archiv/forum/th-2142726/Wieviel-Geld-zur-Hochzeit-
schenken.html

[8] http://www.sportfreunde-siegen.de/content/view/893/16/

The second point demonstrated by the two scenarios is the "relation" between the context and the utterance. One can distinguish between two types of context-utterance pairs.

The first type is exemplified in (10). Scenario 2 provides a context for example (10) in which the prejacent of the highest scoping modal ('*muss*') is not true, i.e. the national team could not dominate the other team. This observation can be generalized for all analogous examples.

Consider also example (12). This sentence was uttered after a soccer match in which the referee gave a penalty shot to Siegen right before the half time. The sentence in (12) contains a negated possibility modal which is logically equivalent to a necessity modal taking a negated argument ($\neg\Diamond\phi \Leftrightarrow \Box\neg\phi$). To give an analogous description of the relation between context and utterance to the one given for exmaple (10), one needs to consider the equivalent formulation with the necessity modal. In this case, the prejacent of the necessity modal ($\neg\phi$) is not true in the context. The same reasoning applies to (11).

For (9) and analogous examples the situation is even more complicated. Most of the conditionals in the collected data contain a modal in the consequent. There are, however, some examples where the consequent contains only an indicative finite verb. This means that these two cases have to be differentiated[9]. Thus for conditionals like (9) containing a modal, it has to be said that the prejacent of the modal in the consequent is false in the context, whereas for conditionals without a modal the consequent - as it is - is not true in the context. In the case of (9), the team in scenario 1 was unmotivated from the beginning which contradicts the prejacent, i.e. that the team enters the field motivated.

What has been said up until this point suggests that sentences containing non-standard *ich* express an opinion or expectation of the speaker's that has not been met in the context. The criteria for "not having been met in the context" are unfortunately hard to grasp. For the examples of the second type, it seems to suffice that the speaker thinks that someone does not share his or her opinion.

Example (13), illustrating the second type, is taken from a forum discussion about the political correctness of a funny/offensive mother's day poem that was intended as a joke (this is a sample of Austrian colloquial German). In the course of the discussion, some commenters said they know that the poem is offensive but they still find it funny, while others told their reasons for not finding it funny at all, eg. example (13).

(13)   Wenn ich als Familie die Frau/Mutter so   auslauge, bis sie
        if   I   as family the wife/mother so-much wear-out until she
        -vorzeitig-   alt und schiach ist, und dann lach ich drüber,
        before-her-time old and ugly   is and then laugh I   about-that
        dann ist das einfach niveaulos und wäh.
        then is that simply dumb   and disgusting

---

[9] If one assumes like Kratzer [12] that all conditionals contain a modal in the consequent, i.e. that conditionals with no overt modal contain a covert modal element, the two cases collapse.

'If a family wears the wife/mother out until she is old and ugly before her time and then they laugh about that, then that's simply dumb and disgusting.'[10]

In the above case, the context permitting the non-standard use of *ich* is a sequence of matching and non-matching opinions given by other users in the thread. The speaker expects that there should be no one who thinks that laughing about a worn out, overworked mother is not dumb and disgusting. Since there are some people that have no problem laughing about the poem, the speaker assumes that they do not agree with her about this point. Therefore, it can be said that her expectations are not met in the context.

The characterization that the speaker voices an expectation that is not met in the context provides a clue for determining the kind of modal found in sentences with non-standard *ich*. In all of the examples the speaker's expectations have a normative flavour. The soccer teams talked about in example (2) and (3) are soccer teams that conform to a certain standard for soccer teams held by the speaker. In the case of example (11), the speaker holds it morally objectionable to ask for money from one's guest. Thus, her utterance expresses something she thinks is impossible behaviour for a bridal couple conforming to her moral standard. In sum I propose that the modals found with non-standard readings have (possibly among others) a *stereotypical* or *moral* flavour. Kratzer [13] defines a *stereotypical conversational background* for a modal as a function f which returns for a world w the set of propositions that gives the expectations concerning what w is like, i.e. "the normal course of events" for w. An analogous definition is given for morally accessible worlds from a world w.

To summarize, it was shown that the non-standard reading of *ich* requires a certain kind of context in which the speaker's expectations - the opinion expressed by the non-standard meaning which the speaker takes as unobjectionable - is not met.

3.2   Emotional Involvement

In the given examples, the use of non-standard *ich* also signals *emotional involvement* on the part of the speaker about the matter at hand. I suggest that the non-standard readings should be listed among those constructions that are grouped under the term *emotive language* (Potts and Schwarz to appear). Potts and Schwarz characterize *emotive language* as "words and constructions that are more or less dedicated to the task of conveying information about our attitudes and emotions" [23, 2]. The next piece of data is a clear cut example that shows *emotional involvement* on the part of the speaker.

(14)   sie nimmt nie was für ihre tochter zu essen für unterwegs mit [...] in meinen augen ist das eine rabenmutter ich muss doch als mutter mit einem kleinkind dafür sorgen dass es immer zu essen bekommt egal wo es gerade ist ich muss einfach immer was dabei haben...

---

[10] http://www.parents.at/forum/archive/index.php/t-253616.html

'She never brings something to eat for her daughter when they are out.
[...] In my opinion she is a bad mother. A mother with a toddler has
to see to it that the toddler gets food, no matter where the child is. A
mother really has to have something to eat with her...'[11]

Example (14) was taken from a women-oriented forum with the central topic
of in-family relationships. The author of this passage is a pregnant woman (not
yet a mother). In the given passage she talks about a woman (a mother) in
her family that does not take good enough care of her daughter. The post has
the title "I'm pregnant and I feel screwed by my partner" (*'Ich bin schwanger
und fühl mich von meinem Partner verarscht'*), which clearly suggests strong
feelings for the topic at hand on the part of the author. Further expressivity is
achieved by her use of expressive language (*'Rabenmutter'* - Engl. *'bad mother'*)
and clusters of exclamation marks near the end of each paragraph.

A somewhat less explicit example that nicely shows the emotional involvement
by the speaker is example (11), where the author also uses expressive language
(*'total doof'* - Engl. *'totally dumb'*) and exclamation marks.

Since the non-standard use of *ich* is emotive, it is not surprising that the
native speakers I consulted judge this way of giving one's opinion as decidedly
unobjective. They evaluate the non-standard use of *ich* as unsuitable for serious,
objective argumentation.

3.3   Genericity

Even though, as was shown in the previous section, the utterances in (9), (10) and
(14) express the speaker's expectations with respect to a certain situation, the
paraphrases suggest that the speaker informs the addressee about an expectation
he holds in general and not only with respect to this particular utterance context,
i.e. regarding all entities in the set denoted by the argument of *als* and not only
the salient entity from the context of utterance. That the non-standard readings
express genericity in the domain of individuals is shown by the possibility to
extend example (10) by (15).

(15)   ...egal   ob   **ich** Deutschland, Italien oder Brasilien bin.
         ...no-matter whether I   Germany,   Italy   or   Brazil   am

When (15) follows the example (10) above, it has roughly the following meaning:
*...and it does not matter, whether one considers Germany, Italy or Brazil (or
any other team).* To give such an exemplary list of teams (or nations) is a natural
extension of (10). In contrast, the extension in (16), which restricts the opinion
given in (10) to only the German national team, is incoherent.

(16)   ?...aber nur wenn **ich** die deutsche Nationalmannschaft bin.
         ...but   only if   I   the German   national-team   am

---

[11] http://forum.gofeminin.de/forum/relationsfamille/__f1465_relationsfamille-Ich-bin-
schwanger-und-fuhl-mich-von-meinem-partner-verarscht.html

So one can say, that the opinion is expressed about every entity that is an element of the set denoted by the argument of *als*.

Sentences like (9) and (10) also express genericity in the spatio-temporal domain. The genericity is easier to see when the number of elements in the set denoted by the complement of *als* is restricted to just one element restraining genericity over individuals, as in example (17).

(17)   Wenn **ich** als deutsche Nationalmannschaft gewinnen will, dann muss
        if   I   as german   national-team   win.INF   want then must
        ich motiviert auf den Platz gehen.
        I   motivated on the field go.INF
        'If the German national team wants to win, then the team has to enter the field motivated.'

Sentence (17) expresses that in any usual match situation involving the German national team (i.e. no matter against which opponent they are playing) it is the case that if the German national team wants to win, the team has to be motivated from the start.

Given that restricting the set denoted by the complement of *als* restricts genericity over individuals, one also expects that leaving out the *als*-phrase altogether results in total genericity in that domain. Example (18) shows that this is the case.

(18)   Wenn **ich** gewinnen will, dann muss ich motiviert auf den Platz
        if   I   win.INF   want then must I   motivated on the field
        gehen.
        go.INF
        'If one wants to win, then one has to enter the field motivated.'[12]

The paraphrase using the impersonal pronoun *one* suggests that example (18) is freely exchangable with example (19), in which *ich* has been replaced by the German impersonal pronoun *man*.

(19)   Wenn **man** gewinnen will, dann muss man motiviert auf den Platz
        if   one   win.INF   want then must one motivated on the field
        gehen.
        go.INF
        'If one wants to win, then one has to enter the field motivated.'

The seeming interchangability of (18) and (19) is not accurate, however, since the use of *ich* suggests a subjective opinion (cf. Sect. 3.1), whereas *man* is usually used for objective arguments. Thus, the only difference between (18) and (19) lies in the respective subjectivity and objectivity on the speaker's side.

---

[12] This sentence has of course also a speaker-referential reading: *if I want to win, then I have to enter the field motivated.*

3.4   Summary and a First Analysis

What can be concluded from the data discussion? What does the first person singular pronoun *ich* contribute meaningwise to the sentence?

There is a certain parallel between the non-standard use of *ich* and a use of demonstratives that Lakoff [15] calls *emotional deixis*. Lakoff states that *emotional deixis* covers "a host of problematic uses, generally linked to the speaker's emotional involvement in the subject-matter of his utterance" [15, p.347]. An example she provides is given in (20).

(20)   I see there's going to be peace in the mideast. This Henry Kissinger really is something!
         [15, p.347]

The core of the discussion is that the effect of *emotional deixis* is to achieve camaraderie between the speaker and the hearer, which makes these forms colloquial. The speaker tries to create emotional closeness and a sense of participation in the hearer by giving the utterance more vividness.

In a recent paper, Davis and Potts [1] argue that *affective demonstratives* (demonstratives used for *emotional deixis*) are semantically marked elements in competition with the unmarked definite article *the*. They follow Horn in assuming *division of pragmatic labor* - unmarked forms are used to express unmarked meanings and marked forms are used to express marked meanings - and argue that *affective demonstratives* "generate an exclamative profile", thus expressing a more marked meaning than the unmarked definite article.

As we have seen, the non-standard occurrences of *ich* are substitutable by the impersonal pronoun *man* (cf. Sect. 3.3). The substitution apparently does not change anything on the truth conditional level of the sentence, but the emotional flavour of the non-standard reading is lost. Applying the same reasoning as Davis and Potts, the impersonal pronoun *man* could be seen as the unmarked form the marked non-standard *ich* is competing against. If this is indeed the case, one would expect non-standard *ich* to have a similar semantic make up as *man*.

One question suggested by (18), where the *als*-phrase has been removed, is which elements are actually needed to obtain the non-standard reading. As far as I can tell at this point, the non-standard reading always involves a modal or generic sentential context: one finds the non-standard reading with universal and existential modals and indicatives under a generic interpretation, although universal modals seem to be prevalent. The *als*-phrase and discourse particles are optional, although the particle *doch* often improves and enforces the non-standard reading where the sentence can also be understood as the speaker talking about himself.

4   Formalization

In this section I formulate a purely semantic account that tries to unify the non-standard and the standard indexical use of *ich*. I argue that *ich* contributes a (more or less semantically adorned) variable in the non-standard as well as the standard deictic use.

4.1   A Theory of the Structure of Indexicals

For a unifying, purely semantic account of the facts, one needs to discard the standard semantics for *ich* in (21) (repeats (1)) since it always forces speaker-referentiality.

(21) $[\![\text{pronoun}_{1PSg}]\!]^c = c_S$ where c_S is the speaker of the context c

To discard (21) and to allow *ich* to refer to other individuals besides the speaker means that one partly departs from Kaplan's [9] direct-referentialist view that *ich* is a "pure indexical", i.e. that *ich* automatically picks out the speaker of the context.

One of the works that criticize the direct-referentialist picture is [21]. In this paper Nunberg specifically argues against the assumption that indexicals give rise to singular propositions. He presents data where indexicals do not contribute a single individual (or group) but a property. A sentence with a property contributing indexical is non-singular (as long as there are no other singular terms) and thus, Nunberg argues, the assumption that indexicals give rise to singular propositions has to be discarded. He consequently proposes, in contrast to what direct-referentialists assume, that the referent of an indexical is determined not directly, but in two stages. He suggests that indexicals are more complex than what is usually assumed and posits that an indexical has three components[13]: a *deictic component*, a *classificatory component* and a *relational component*.

The deictic component picks out an individual from the context. Nunberg calls this individual the *index*. The index connects the final semantic value of the indexical to the context. Nunberg [21, p.20] notes that this component corresponds to the standard semantics, e.g. as for *ich* in (21).

The relational component specifies the relation in which the index stands to the final semantic value.

The classificatory component consists of features that restrict the final semantic value (e.g. animacy, singularity . . .).

To summarize, an indexical denotes an individual (or individual concept), whose features match the classificatory component, and which stands in a certain relation to the contextually chosen index.

Coming back to the original puzzle of the non-standard use of *ich*, Nunberg's three component account sounds very promising in light of the needed flexibility required for the meaning of non-standard *ich*; the account provides the necessary amount of freedom in the choice of referent.

---

[13] Nunberg relativizes the three-component account again by assuming that non-participant terms, i.e. demonstratives and demonstratively used third person pronouns, lack a relational component [21, p.23]. Since this paper looks only at participant terms, I will gloss over this fact.

The sentences in (22) are two of the examples Nunberg gives to motivate the three-component analysis.

(22)   a.   *Condemned prisoner*: I am traditionally allowed to order whatever I like for my last meal.

       b.   *President*: The Founders invested me with sole responsibility for appointing Supreme Court justices.

       [21, p.20f]

Nunberg argues that *I* in (22-a) can not pick out the speaker since it is not a tradition for the speaker that he is allowed whatever he wants for his last meal. Rather, it is a tradition for anyone with the property of being a condemned prisoner. Thus Nunberg concludes that *I* picks out the property of being a condemned prisoner. The argumentation for example (22-b) runs analogously. Since the Founders did not actually invest the sole responsibility for appointing Supreme Court justices in the current president himself, *I* in (22-b) picks out the property of being the president of the United States.

     Elbourne [3] formalizes Nunberg's account. He straightforwardly implements the idea of the three components by explicitly putting the deictic component and the relational component into the syntax. The requirements posed by the classificatory component are added as presuppositions to the meaning of the indexical. Thus syntactically, an indexical has the complex structure in (23).

(23)   [indexical [R_1 i_2]]   [3, p.421]

The two variables R_1 and i_2 constitute the relational and the deictic component, respectively. i_2 is a variable of type e and R_1 is a variable of type $\langle e, \langle se, st \rangle \rangle$, i.e. a variable for intensional relations between individuals and individual concepts. The values for both variables are determined from the context. On a technical note, this means that they are left unbound and are determined by the variable assignment, which constitutes a parameter of the interpretation function.

     Regarding the meaning of the overt lexical item of the complex indexical, Elbourne generalizes Nunberg's observation that, in certain contexts, indexicals can contribute properties. He proposes that the meaning of an indexical is in fact always a definite description[14]formed from R and i. Example (24) is Elbourne's proposed meaning for English third person singular *it* (he does not explicitly formalize the *classificatory component*).

(24) $[\![\text{it}]\!] = \lambda f_{\langle se,st \rangle}.\lambda s.\iota x(f(\lambda s'.x)(s) = 1)$

       [3, p.421]

     The first argument of *it*, $f_{\langle se,st \rangle}$, is the result of applying the contextually supplied relation R to the given index i. The informal paraphrase of the final

---

[14] Even though Nunberg talks about properties, Elbourne [3, p. 420] argues that the properties in Nunberg's examples always denote singleton sets in the relevant minimal situations. He takes this observation as the starting point of his analysis and implements them as definite descriptions.

meaning of the complex pronoun [it $[R_1 \, i_2]$] in the situation of evaluation is: *the unique individual x such that x stands in relation R to i.*

4.2 Emotional Involvement and Speaker Empathy

It is usually assumed in the literature that *emotional involvement* signalled by *emotive language* needs to be modelled on a different level than the truth conditional meaning of a sentences (cf. for example [22] and [24]). As has been alluded to in Sect. 3.1, the speaker expresses certain expectations with the use of non-standard *ich*, which seem to constrain the set of individuals to those that conform to a certain normative standard held by the speaker. Since one usually considers one's normative standards to be applicable to oneself, one tries to conform to them and one will tend to identify with the group of people conforming to them. This identification can be observed in example (25), which contains both an occurrence of non-standard *ich* and the objective impersonal *man*.

(25)   **Ich** kann **als Kunde** wohl erwarten, dass für den Preis das Paket oder der Brief auch korrekt zugestellt wird. Und wenn **man** das eben für einen so niedrigen Preis nicht kann, dann darf man so einen niedrigen Preis auch nicht anbieten.
'A **client** should be allowed to expect that a package or letter will be delivered correctly for the price that is charged. And if **one** can't do that for such a low price, then one just shouldn't offer such a low price.'

The speaker of (25) clearly sympathizes with the clients rather than the mail service providers. This kind of perspective-taking of the speaker (*speaker empathy*) is observable also with other impersonal pronouns.

I consider two previous proposals for modelling empathy. Malamud [17,18] looks at the impersonal use of the English second person singular pronoun *you* and proposes that *you* involves *hearer empathy*. Moltmann [19,20] looks at English generic *one* and suggests that *one* involves a special kind of speaker empathy.

Both proposals share that empathy is modelled by means of a special relation that is required to hold between the speaker/hearer and the values of a variable that is contributed by the impersonal pronoun.[15] Example (26) is the meaning Malamud proposes for impersonal *you*.

(26) $\llbracket \text{you} \rrbracket^c = \lambda s.\lambda P.\exists y[persona(y, addressee(c), s) \,\&\, P(y, s)]$
[17, p.25]

Hearer empathy is modelled by the *persona*-relation, $\lambda y.\lambda x.\lambda s.persona(y, x, s)$, that relates the later existenially quantified variable y to the addressee of the context.

---

[15] I will not get into details here, since the exact characterisation of the relations is secondary to formalizing the meaning of non-standard *ich*. For the details, please see [17] and [19].

Moltmann [19, p.24] proposes that generic *one* ranges only over such individuals that the speaker identifies with. She models this semantic restriction by letting *one* introduce a *qua*-predicate that takes two arguments, an individual variable and the property $\lambda y[Izy]$ (*I* is the *identification relation*), that holds for any y that z identifies with.[16] The notion of identification relation is conceptually further specified as a notion of pretence: the speaker applies a predicate to a value of generic *one* on the basis of "projecting himself" onto that value.

Technically, both of the above accounts are very well combinable with the account for pronouns presented in the previous section. Both Malamud's *persona* relation and Moltmann's *identification* relation could be fitted into the *relational component*.

Thus I follow Malamud and Moltmann in modelling speaker empathy by assuming an *identification relation* (different from Malamud's and Moltmann's) that restricts the set of entities to those that the speaker identifies with. A precise technical account for the *identification relation* is still to be proposed. A possible starting point is to restrict the individuals to those that conform to the stereotypical, moral or otherwise normative standards held by the speaker, parallel to the various possible flavours found with the modals contained in non-standard readings.

4.3   Adding Up the Parts

In this section, I bring together Malamud's [17,18] and Moltmann's [19,20] work on "empathy pronouns" with Elbourne's [3] formalization of Nunberg's [21] three-component account of "ordinary indexicals". I also reconsider the speculations made at the end of Sect. 3.4 that - given *division of pragmatic labor* - the semantic make up of non-standard *ich* is similar to the semantic make up of impersonal pronoun *man*.

With Nunberg's three-component analysis laid out, only the values for the three components need to be determined. Nunberg himself has a short section on the English first person singular pronoun *I*. There he briefly states that, like other indexicals, *I* has all three components of meaning and he specifies the values of the three components. For the index the deictic component always picks the speaker of the utterance, the relational component requires the index to instantiate[17] the final interpretation and the classificatory component restricts the interpretations to an animate syntactically singular individual (or individual property).

---

[16] Moltmann [20] proposes that *one* introduces a complex variable that contains an individual variable and the property, $\lambda y.Izy$, that should hold of any value assigned to the individual variable.

[17] Nunberg (1993:20) talks about *instantiation of the interpretation* since he allows for properties as final interpretations. Concretely this means, that when the interpretation is an entity, the index has to be identical to the interpretation, and when the interpretation is a property, the index as to be a member of the set denoted by the property.

Following Nunberg, I assume for non-standard *ich* that the deictic component picks out the speaker. Even though *ich* in the non-standard reading does not refer to the speaker, the speaker is crucial for modelling speaker empathy since the individuals that are ultimately considered vary with respect to the speaker. Thus, the index is the common core of the non-standard and the standard use.

The classificatory component I will also adopt without change because a sentence is plainly ungrammatical if *als* takes a plural and/or inanimate entity as its complement, (27).

(27)   a. *Ich muss   als Mütter   meinen Kindern etwas   zu essen
           I   have-to as mothers.pl for-my children something to eat
           mitnehmen.
           take-along
        b. *Wenn ich als Schraubenzieher eine Schraube festschraube...
           if   I   as screwdriver   a   screw   fix...
           (the sentence is fine if the screwdriver is humanized)

For the relational component, however, I use the identification relation specified in Sect. 4.2.

As for the syntactic structure of the indexical, I adopt Elbourne's proposal [3]. Thus, *ich* has the complex structure in (28).

(28)   [ich [R_1 i]]

Regarding the interpretation, I assume that $[\![R_1]\!]^g \in D_{\langle e,\langle e,st\rangle\rangle\rangle}$ and $[\![i]\!]^g \in D_e$. Semantically, I depart from Elbourne in that I do not assume that the *ich* forms a definite description from R and i. As was shown in Sect. 3.3 the data under discussion expresses a maximally general subjective opinion of the speaker (maximally with respect to the *als*-phrase). In particular this means that *ich* can not be analysed to refer to one unique individual. Therefore, an analysis as definite description can not capture this basic characteristic. I propose that non-standard *ich* is an indefinite, as it shows indefinite like behaviour with respect to quantificational variablity and binding through a generic operator (see [14]).

At least three possibilities to model indefinites are discussed in the literature. The first is to analyze indefinites as properties that are existentially closed at a higher point in the structure, see [6]. The second possibility is to model them in a dynamic system, eg. [2], or thirdly, to use choice functions [25], [26]. I choose the third option since overt existential quantification leads to technical complications when I consider the standard indexical use of *ich* later on.

The definition of intensional choice function, which I adopt, is taken from Romero [26, p.7] who attributes this definition to Irene Heim.

(29)   Intensional Choice Function: A function $f_{\langle\langle e,st\rangle,\langle se\rangle\rangle}$ is an intensional choice function $(\mathrm{ICH}(f))$ iff for all P in the domain of f and for all w in the domain of $f(P)$: $P(f(P)(w))(w) = 1$

An intensional choice function f in (30) is existentially bound at the highest level in the structure and constrained by the predicate ICH which ensures that f is a choice function.

Putting it all together, the meaning in (30) formalizes the semantic contribution of *ich* to the non-standard readings.

(30) $[\![\text{ich}]\!]^{w,c,g} = \lambda Q_{\langle e,st \rangle}.\lambda P_{\langle e,st \rangle}.\lambda s.[P(f(Q)(s))(s)]$

Like in Elbourne's proposed meaning for *it* in (24), the first argument, $Q_{\langle e,st \rangle}$, is filled by the result of applying $[\![R_1]\!]^g$ to $[\![i]\!]^g$. Q, a property, is the argument of an intensional choice function f that returns an individual concept whose value in the situation of evaluation is an element of the set denoted by Q in the situation of evaluation. Specifically for non-standard *ich*, the variable assignment returns $\lambda x.\lambda y.\lambda s.$identifies-with$(y)(x)(s)$ (the identification relation modelling speaker empathy) for R_1. Since the deictic component always picks out c_S (the speaker of context c), one can fix i to be c_S. Consequently, $Q = \lambda y.\lambda s.$identifies-with$(y)(c_S)(s)$.

The interpretation of the complex structure underlying the pronoun is given in (31).

(31) $[\![[\text{ich } [R_1 \ c_S]]]\!]^{w,c,g} = \lambda P.\lambda s.[P(f(\lambda y.\lambda s'.\text{identifies-with}(y)(c_S))(s))(s)]$

The proposed semantics in (31) formalizes speaker empathy and gives *ich* an indefinite semantics that is compatible with genericity. It also creates the desired parallel to the meaning of impersonal *man*. Malamud [17,18] proposes (32) for the meaning of *man*, partially based on Kratzer's work [11]. She assumes that *man* has the complex syntactic structure in (32-a) which consists of a determiner *Det* and an element *SE*. The semantics of the two lexical items is given in (32-b) and (32-c).

(32)   Slightly adapted from Malamud [18]
    a.   man $= [Det \ SE]$
    b. $[\![Det]\!]^{c,w} = \lambda x.\lambda P.\exists y[y \in \text{HUMANS} \ \& \ P(y,w)]$
    c. $[\![SE]\!]^c = c_S$

In sum, *man* is a generalized quantifier with existential force, which is parallel to the meaning proposed in (30). The only difference is that for non-standard *ich* I make use of choice functions.

To give an exemplary truth condition for example (33), I provide the meanings for the other parts of the sentence.

(33)   Ich muss als Nationalspieler   motiviert spielen.
        I   must as national-team-player motivated play.
        'A player of the national team has to play with motivation.'

The semantics of the modal *müssen* is the same as for English *must*, for which I adopt the meaning proposed by Kratzer [12] in (34). A modal in this proposal has two parameters, f and g which are assigned a *conversational background* and an *ordering source* respectively. The functions f and g together pick out the optimal worlds accessible from w, $O(w,\text{f},\text{g})$. Concretely, g induces an ordering on the worlds picked by f for which a set of optimal worlds can be determined.

(34) $[\![\text{müssen}]\!]^{w,c,g} = \lambda\phi.\lambda s.\forall w' \in O(w,\text{f},\text{g})[w' \in \phi]$

Drawing on Jäger [8], I let the *als*-phrase contribute a presupposition for the individual picked out by the choice function. Jäger analyses English *as*-phrases in the framework of Discourse Representation Theory (DRT) as inducing a presupposition that has to be resolved (i.e. successfully added to the discourse representation structure) either by simple resolution or via accomodation. Both variants amount to identifying the argument of the presupposed predicate with the argument of the predicate the *as*-phrase attaches to. For a Montague-style system, as I am using, I propose the following semantics for *als* which modifies a predicate by adding a presupposition to its argument.

(35) $[\![\text{als}]\!]^{w,c,g} = \lambda P_{\langle e,st\rangle}.\lambda Q_{\langle e,st\rangle}.\lambda x.\lambda s : P(x)(s) = 1.Q(x)(s)$

Consequently, (33) has the truth condition in (36).

(36) $[\![(33)]\!]^{w,c,g}$ is defined if
national-team-player($f(\lambda y.\text{identifies-with}(y)(c_S))(s))(w) = 1$
and if defined $[\![(33)]\!]^{c,g} = 1$ iff $\exists f \forall w' \in O(w,\text{f},\text{g})[\text{ICH}(f)$
& play-with-motivation($f(\lambda y.\text{identifies-with}(c_S)(y)))(w')]$

4.4   Standard Deictic *ich*

In this section, I show that the proposed meaning for the non-standard use of *ich* can be modified to model also the speaker-referential use of *ich*.

As was already suggested in the last section, the value of R essentially determines the set from which the final semantic value is picked. For deictic *ich* according to Nunberg, one needs a relation that the speaker instantiates. To capture speaker-referentiality it suffices to assign the identity relation ($\lambda x.\lambda y.\lambda s.x{=}y$ *in* s) to R. Since for *ich* Elbourne's i component has the fixed value c_S, a sentence containing deictic *ich* is only true if the choice function f picks out the speaker.

(37)   a.   Ich bin müde.
              I   am tired
        b. $[\![\text{Ich bin müde}]\!]^{w,c,g} = \lambda s.\text{tired}(f(\lambda y.\lambda s'.c_S = y \text{ in } s')(s))(s)$

This shows that the proposed meaning in (30) in fact covers the non-standard and the deictic use of *ich*.[18] The strong point of this unified treatment is that it

---

[18] In Sect. 4.3 I mentioned that I use choice functions to model indefiniteness since using an existential quantifier leads to technical problems for standard deictic *ich*. Consider (i).

accounts for the fact that the non-standard use of *ich* shares a semantic core of speaker-relatedness with the deictic use.

As was already noted in Fn. 12, some of the data with a non-standard use of *ich* also have a sensible speaker-referential interpretation, as in (38-b).

(38)　　Wenn ich als Spieler gewinnen will, 　dann muss ich motiviert 　auf den
　　　　if 　I 　as player 　win 　　　　want, then must I 　motivated 　on the
　　　　Platz gehen.
　　　　court go.

　　　a.　　If (one as) a player wants to win, then he/she has to enter the field
　　　　　　motivated.
　　　b.　　If I being (in my role as) a player want to win, then I have to enter
　　　　　　the field motivated.

Regarding the two interpretations (38-a) and (38-b), the unified treatment says the difference in the interpretation lies only in the difference of the value assigned to the relational component.

5　Conclusion and Outlook

In this paper, I have shown that *ich* has an unexpected non-standard use that challenges the standard view that first person singular pronouns are always speaker referential.

The data discussion has demonstrated that the non-standard use of *ich* signals that the speaker informs the hearer about a rule he believes to hold in the actual world, but which is violated in the context of utterance. It was also shown that the non-standard reading signals emotional involvement on the part of the speaker. One aspect of the speaker's involvement I identified as speaker empathy.

I discussed and adopted the theory of indexicals given in Nunberg [21] and parts of its formalization by Elbourne [3]. To model speaker empathy I looked at the analyses of English impersonal *you* [17,18] and generic *one* [19,20] which, as the authors argue, also involve forms of empathy. For the meaning proposed in the end for non-standard *ich*, I parted from Elbourne's technical proposal that pronouns are definite descriptions and reanalysed non-standard *ich* to form an indefinite parallel to the meaning proposed for german impersonal *man* [17]. I showed that the meaning given for the non-standard use could be modified to model (albeit unconventionally) the normal deictic use of *ich*.

One possible point for criticism is the complete freedom regarding the relational component. As the proposal stands right now, there are no restrictions that would block any two-place relation to be picked for the relational component. This problem is already present in Nunberg's proposal [21] where the only

---

(i)　　　$[\![\text{ich}]\!]^{w,c,g} = \lambda Q.\lambda P.\lambda s.\exists y[Q(y)(s) \wedge P(y)(s)]$

Given this semantics the sentence in (37-a) would have the meaning $\exists y[c_S = y \wedge \text{tired}(y)(s)]$, which can be paraphrased as 'I exist and am tired' which is not the desired result. I thank an anonymous reviewer for pointing this out to me.

restriction on the relation is that the speaker *instantiates* it, i.e. that the speaker stands in this relation to himself (see Fn. 17).

An alternative that might be preferable to the unifying account I presented is to see the non-standard and the standard use of *ich* as an instance of true polysemy rather than context dependence. If one pursues a formalization based on this assumption, the analysis and the technical parts up to Sect. 4.3 could be adapted without change, since unifying the non-standard and the standard deictic use has not been the core motivation for the analysis I proposed. The application to standard deictic *ich* shown in Sect. 4.4 has been an automatic result of adapting Elbourne's account [3]. Thus, if the proposed semantics for *ich* in (30) is restricted to the non-standard use, the relational component can be fixed to the identification relation. This would eliminate the problem of the unrestricted relational component. The only context dependence would be brought in by the choice function and c_S, which would remain the common core of the non-standard and the standard deictic use. Therefore, if one does not object to the assumption of two distinct lexical items ich_1 and ich_2, the new data does not force one to discard the standard semantics for deictic German *ich*.

As always, there are still open issues that need to be looked at.

The first question is how to capture the second occurence of *ich* in the consequent of the conditional in data such as (2). This second occurrence of *ich* seems to be donkey-bound by the *ich* in the *if*-clause. In sentences such as (2), the other occurrences of first person singular pronouns can be analyzed as *fake indexicals* as proposed in [10]. Fake indexicals is the term for bound occurrences of first and second person pronouns that are not independently speaker- or hearer-referential, eg. *my* in *'Only I did my homework'*, which implies that nobody else did their homework rather than that nobody else did the speaker's homework. As far as I know, nobody has proposed a treatment for fake donkey indexicals, yet.

Second, one would wish for a comparison with German impersonal second person singular *du* and a cross-linguistic search for first person singular pronouns in other languages with similar non-standard readings as *ich*.

Last but not least, also the speaker's *emotional involvement* has not yet been treated satisfactorily.

However, even though there are still remaining open questions, I have offered a first analysis for non-standard *ich* which can be taken as basis for further investigation on this topic.

References

1. Davis, C., Potts, C.: Affective demonstratives and the division of pragmatic labor. In: Aloni, M., Bastiaanse, H., de Jager, T., van Ormondt, P., Schulz, K. (eds.) Preproceedings of the 17th Amsterdam Colloquium, pp. 32–41 (2009)
2. Dekker, P.: Existential Disclosure. Linguistics and Philosophy 16, 561–587 (2008)
3. Elbourne, P.: Demonstratives as individual concepts. Linguistics and Philosophy 31, 409–466 (2008)
4. von Fintel, K., Iatridou, S.: What to Do If You Want to Go to Harlem: Anankastic Conditionals and Related Matters. Ms. MIT Press, Cambridge (2005)

5. Gast, V.: Modal particles and context updating - the functions of German *ja, doch, wohl* and *etwa*. In: Vater, H., Letnes, O. (eds.) Modalverben und Grammatikalisierung, pp. 153–177. Wissenschaftlicher Verlag (2008)
6. Heim, I.: The Semantics of Definite and Indefinite Noun Phrases. Doctoral Dissertation: University of Massachusetts (1982)
7. Hentschel, E.: Funktion und Geschichte deutscher Partikeln. Niemeyer (1986)
8. Jäger, G.: On the semantics of "as" and "be". a neo-carlsonian acount. In: Kim, M., Strauss, U. (eds.) Proceedings of NELS, vol. 31 (2001)
9. Kaplan, D.: Demonstratives. In: Almog, Perry, Wettstein (eds.) Themes from Kaplan. Oxford University Press, Oxford (1977/1989)
10. Kratzer, A.: Making a Pronoun: Fake Indexicals as Windows into the Properties of Pronouns. Linguistic Inquiry 40, 187–237 (2009)
11. Kratzer, A.: German impersonal pronouns and logophoricity. Presentation at Sinn und Bedeutung II, Berlin, Germany (1997)
12. Kratzer, A.: Modality and Conditionals. In: Semantik: ein internationales Handbuch. de Gruyter, Berlin (1991).
13. Kratzer, A.: The Notional Category of Modality. In: Eikmeyer, H., Rieser, H. (eds.) Words, worlds, and contexts: new approaches in word semantics, pp. 38–73 (1981)
14. Krifka, M., Pelletier, F.J., Carlson, G.N., ter Meulen, A., Chierchia, G., Link, G.: Genericity: An Introduction. In: The Generic Book, pp. 1–124. University of Chicago Press, Chicago (1995)
15. Lakoff, R.: Remarks on This and That. In: Proceedings of the Chicago Linguistics Society, Chicago, vol. 10, pp. 345–356 (1974)
16. Lindner, K.: Wir sind ja doch alte Bekannte - The use of German *ja* and *doch* as modal particles. In: Abraham, W. (ed.), Discourse Particles, Amsterdam, pp. 163–202 (1991)
17. Malamud, S.: Impersonal indexicals: you, man and si. Draft (2007), http://people.brandeis.edu/~smalamud/iiss.pdf
18. Malamud, S.: Semantics and pragmatics of arbitrariness. Doctoral Dissertation: University of Pennsylvania (2006)
19. Moltmann, F.: Generalising Detached Self-Reference and the Semantics of Generic One. Mind and Language (to appear)
20. Moltmann, F.: Generic one, arbitrary PRO, and the first person. NLS 14, 257–281 (2006)
21. Nunberg, G.: Indexicality and Deixis. Linguistics and Philosophy 16, 1–43 (1993)
22. Potts, C.: Conventional implicature and expressive content. In: Maienborn, C., von Heusinger, K., Portner, P. (eds.) To appear in Semantics: An International Handbook of Natural Language Meaning. Mouton de Gruyter, Berlin (2008)
23. Potts, C., Schwarz, F.: Affective 'this'. Linguistic Issues in Language Technology 3(5), 1–30 (2010)
24. Potts, C., Alonso-Ovalle, L., Asudeh, A., Bhatt, R., Cable, S., Davis, C., Hara, Y., Kratzer, A., McCready, E., Roeper, T., Walkow, M.: Expressives and identity conditions. Linguistic Inquiry 40, 356–366 (2009)
25. Reinhart, T.: Quantifier Scope: how labor is divided between QR and choice functions. Linguistics and Philosophy 20, 335–397 (1997)
26. Romero, M.: Intensional Choice Functions for *Which* Phrases. In: Proceedings of SALT IX, pp. 255–272. CLC Publications, Ithaca (1999)
27. Stalnaker, R.: Indicative Conditionals. Reprint in Stalnaker, R. (1999); Context and Content (1975)

Part IV
Learning with Logics and Logics for Learning

The Sixth Workshop on Learning with Logics and Logics for Learning (LLLL2009)

Akihiro Yamamoto[1], Kouichi Hirata[2], and Shin-ichi Minato[3]

[1] Graduate School of Informatics, Kyoto University, Japan
akihiro@i.kyoto-u.ac.jp
[2] Department of Artificial Intelligence, Kyushu Institute of Technology, Japan
hirata@ai.kyutech.ac.jp
[3] Graduate School of Information Science and Technology
Hokkaido University, Japan
minato@ist.hokudai.ac.jp

1   The Workshop

Nowadays the theory of Machine Learning attracts much attention not only as a subject in Artificial Intelligence, but also for developing new techniques for mining useful knowledge from various types of data. Originally Machine Learning means methods of developing learning machines for generating appropriate rules with which training data can be explained. Computational Logic is originally from mechanizing the activity of explaining why propositions and rules hold. The two areas are connected on the points representing and explaining rules. Based on the connection, the workshop of Learning with Logics and Logics for Learning (LLLL) has been organized to bring together researchers who are interested in both of the areas of Machine Learning and Computational Logic, and to have intensive discussions on various relations between them with making the interchange more active. More precisely the LLLL workshop is aiming at clarifying both how logic is useful for learning and how learning contributes new types of logic.

The LLLL workshop was started as a domestic workshop in January and December in 2002. In every year from 2005 to 2007, the LLLL workshop was held as an international workshop collocated with the Annual Conferences of Japan Artificial Intelligence Society (JSAI). The sixth workshop, LLLL2009, was held in Kyoto from July 6th to 7th in 2009, as a satellite event of the JSAI International Symposia on AI (JSAI-isAI). The workshop was collocated with the Fourth International Workshop on Data-Mining and Statistical Science (DMSS2009), which was held from July 7th to 8th. All of the international LLLL workshops were supported by the Special Interest Group of Fundamental Problems in Artificial Intelligence (SIG-FPAI) in JSAI. The post-workshop proceedings of LLLL 2009 was included in this volume with other international workshops held in JSAI-isAI.

To the programming committee of LLLL2009 we invited 9 researchers working on the relations between logic and learning. We invited Prof. Thomas Zeugmann

K. Nakakoji, Y. Murakami, and E. McCready (Eds.): JSAI-isAI, LNAI 6284, pp. 315–316, 2010.

and Prof. Marta Arias for their special talks in the workshop. The papers submitted to the call for papers were reviewed by the PC members, and 12 papers were selected for the contributed talks. More information about the workshop and past workshops are available at the LLLL homepage:

http://www.iip.ist.i.kyoto-u.ac.jp/LLLL/LLLL.html

2   The Post-workshop Proceedings

For the post-workshop proceedings, we requested the contributors to submit revised versions of their workshop papers. The revised papers were reviewed by the PC members again. Finally, we have selected the following 4 papers for this volume.

The paper by Kameda and Tokunaga investigates inferability of unbounded unions of languages from positive data in the Gold-style learning. Their work is under the assumption that every language is a closed set system, which is based on the analogy between learning from positive data and computational algebra.

The paper by Katoh *et al.* proposes a method of mining time-series patterns from input event sequences. They introduce a new class of patterns consisting of k-partite episodes and provide efficient algorithms deriving patterns in the class which frequently appears in the input.

The paper by Ouchi and Yamamoto also treats inferability of languages from positive data in the Gold-style learning. They investigate it by designing learning procedures using refinement operators and the MINL strategy.

The paper by Satoh investigates the method for computing minimal models in propositional logic, which is useful for learning algorithms when data are represented as models and rules are in the form of formulae. He introduces the concept of positively minimal disjuncts in Disjunctive Normal Forms for the investigation.

Acknowledgments

We believe that the success of the LLLL2009 workshop was brought by the support of the invited speakers, all of the program committee members, all speakers of contributed papers, and all audiences who attended the workshop. We would like to express our thankfulness to the program committee members by listing their names:

Yoji Akama, Marta Arias, Tamás Horváth, Katsumi Inoue,
Tetsuhiro Miyahara, Taisuke Sato, Takayoshi Shoudai, Hiroo Tokunaga,
and György Turán,

We would express our great thankfulness to Prof. Takashi Washio for his proposal and elaboration to the collocation of LLLL and DMSS, and the co-chairs of DMSS, Prof. Akihiro Inokuchi and Prof. Masashi Sugiyama, for their great support to the collocation. At last we would like to express our thankfulness to Dr. Kumiyo Nakakoji, Dr. Yohei Murakami, and Prof. Elin McCready for inviting us to JSAI-isAI and publishing the post-workshop proceedings.

Inferability of Unbounded Unions of Certain Closed Set Systems

Yuichi Kameda and Hiroo Tokunaga

Department of Mathematics and Information Sciences
Graduate School of Science and Engineering,
Tokyo Metropolitan University
1-1 Minami-Ohsawa, Hachioji-shi 192-0397, Japan
tokunaga@tmu.ac.jp, kameda-yuuiti@ed.tmu.ac.jp

Abstract. In this article, we study inferability from positive data for the unbounded union of certain class of languages. In order to show inferability, we put an emphasis on a characteristic set of a given language. We consider a class of closed set systems such that there exists an algorithm for generating a characteristic set consisting of one element. Two concrete examples of closed set systems with such algorithms are given. Furthermore, we consider applications of these examples to the study of transaction databases.

1 Introduction

In this article, we continue to study inferability from positive data for the set union of languages. In the following, "inferability" always means "inferability from positive data" [4,1]. The class of languages we study here is so called a *closed set system* (see §2 for its definition). In the previous article [5], we studied a learning procedure for the class of bounded unions of a Noetherian closed set system by using characteristic set. The aim of this article is to consider the class of unbounded unions of certain closed set systems. As we have seen in [2], any Noetherian closed set system has finite elasticity and this implies that the class of bounded unions of a Noetherian closed set system has also finite elasticity by the result of Wright [12] and it is inferable by [12,9]. The class of unbounded unions of a Noetherian closed set system, however, is not inferable in general. In fact, in [2, Theorem 9], a necessary and sufficient condition was given (see Theorem 5). For example, by Theorem 5, we see that the class of unbounded unions of ideals of a polynomial ring of $\mathbb{Q}$-coefficient is inferable if and only if its number of variable is one. Since there seems to be few results on the class of languages satisfying the condition of Theorem 5, it is worthwhile to consider explicit classes of such languages, which is our purpose of this article.

Likewise our previous article [5], the notion of *characteristic set* (see §2 for its definition) plays an important role. By [2], any element of a Noetherian closed set system has its characteristic set. In this article, we study two explicit closed set systems such that there exist an algorithm (or a recursive mapping) to "unify" a characteristic set into a single element for each language.

K. Nakakoji, Y. Murakami, and E. McCready (Eds.): JSAI-isAI, LNAI 6284, pp. 317–330, 2010.
© Springer-Verlag Berlin Heidelberg 2010

This article goes as follows: In §2 we review definitions and theorems about the inferability from positive data and closed set systems. In §3 we introduce the notion of the unbounded union of languages and examine the mapping δ to "unify" a characteristic set into one element. We actually construct δ under two particular circumstances in §§4 and 5. In §6 we consider applications of our results in §§4-5 to the study of transaction databases. This is one of goals for our study: an application of algebra to knowledge discovery in databases, and vice versa.

2   Preliminaries

2.1   Inferability from Positive Data

In this article, a *language* L is a subset of some countable set U such that L is expressed $L(G)$ by some finite expression G. We call this finite expression a *hypothesis*. A set of all hypotheses $\mathcal{H}$ is called *hypothesis space*. Let $\mathcal{L}$ be the set of all languages $\{L(G) \mid G \in \mathcal{H}\}$. We assume that $\mathcal{L}$ is *uniformly recursive*, that is, there is a recursive function $f(w, G)$ such that $f(w, G) = 1$ iff $w \in L(G)$ for every $w \in U$ and $G \in \mathcal{H}$.

A *positive data* (or *positive presentation*) of $L \in \mathcal{L}$ is an infinite sequence $\sigma : s_1, s_2, \ldots$ of elements of L such that $L = \{s_1, s_2, \ldots\}$. An *inference algorithm* M is that:
- M receives incrementally elements of a positive data σ of a language,
- M outputs a hypothesis $G_n \in \mathcal{H}$ when M receives n-th element of σ.

$\mathcal{L}$ is *inferable in the limit from positive data* if there exists an inference algorithm M satisfies that for all $L \in \mathcal{L}$ and an arbitrary positive data of L, the output sequence of M converges to a hypothesis G such that $L(G) = L$.

A *finite tell-tale* of $L \in \mathcal{L}$ is a finite subset S of L such that L is minimal in the class $\{L' \in \mathcal{L} \mid S \subseteq L'\}$ with respect to set inclusion. If L is minimum, S is called a *characteristic set* of L. Note that the idea of characteristic set is essentially the same as that of *test set* in [6].

Theorem 1. ([1]) *$\mathcal{L}$ is inferable in the limit from positive data if and only if there exists a procedure to enumerate elements of a finite tell-tale of every $L \in \mathcal{L}$.*

Theorem 2. ([7]) *If every $L \in \mathcal{L}$ has a characteristic set, then $\mathcal{L}$ is inferable from positive data.*

We say that (*i*) $\mathcal{L}$ has *finite thickness* if the set $\{L \in \mathcal{L} \mid w \in L\}$ is finite for any $w \in U$ and (*ii*) $\mathcal{L}$ has *infinite elasticity* if there exists an infinite sequence $w_0, w_1, \ldots$ of elements of U and infinite sequence $L_1, L_2, \ldots$ of languages such that $\{w_0, \ldots, w_{n-1}\} \subseteq L_n$ but $w_n \notin L_n$. We say that $\mathcal{L}$ has *finite elasticity* if it does not have infinite elasticity.

Theorem 3. ([7],[8]) *1. If $\mathcal{L}$ has finite elasticity, then every L in $\mathcal{L}$ has a characteristic set.*
2. If $\mathcal{L}$ has finite thickness, then $\mathcal{L}$ has finite elasticity.

Let $\mathcal{L}'$ be a subclass of $\mathcal{L}$. Then, it holds clearly that:

Proposition 1. *1. If $\mathcal{L}$ is inferable from positive data, then $\mathcal{L}'$ is. 2. If every $L \in \mathcal{L}$ has a characteristic set in L, then every $L' \in \mathcal{L}'$ has a characteristic set in L'.*

2.2   Closed Set System

Let 2^U be the power set of U. A mapping $C : 2^U \to 2^U$ is called a *closure operator* if C satisfies:
(CO1) $X \subseteq C(X)$,
(CO2) $C(C(X)) = C(X)$, and
(CO3) $X \subseteq Y \Rightarrow C(X) \subseteq C(Y)$
for each $X, Y \in 2^U$. A set $X \subseteq U$ is called *closed* if $X = C(X)$. A *closed set system* $\mathcal{C}$ is the class of all closed sets of a closure operator.

Remark 1. In a closed set system, the intersection of arbitrary number of closed sets is closed, but the union of closed sets is not necessarily closed.

Remark 2. There is another definition of closure operator that is somewhat different from ours as follows. Let $\mathcal{F} \subseteq 2^U$. A closure operator is a mapping $C : \mathcal{F} \to \mathcal{F}$ that satisfies (CO1)-(CO3) as above. Here we call this a closure operator on $\mathcal{F}$.

In the following, we regard $\mathcal{C}$ as a class of languages and assume that it is recursive. If a closed set $X \in \mathcal{C}$ is represented $X = C(Y)$ for some finite set $Y \subseteq U$, X is called a *finitely generated closed set*.

Lemma 1. ([2]) *Let $X = C(Y)$ be a closed set. The followings are equivalent:*
1. Y is finite,
2. Y is a finite tell-tale of X, and
3. Y is a characteristic set of X.

An immediate consequence of Lemma 1 and Theorem 1 is as follows:

Corollary 1. *$\mathcal{C}$ is inferable from positive data if and only if every closed set is finitely generated.*

A closed set system $\mathcal{C}$ is *Noetherian* if it contains no infinite strictly ascending chain of closed sets. This condition is equivalent to finite elasticity [2, Theorem 7]. Hence it follows that:

Corollary 2. *A Noetherian closed set system is inferable from positive data.*

From Proposition 1, a subclass of a closed set system inherits the properties such as inferability. Henceforth, we regard a subclass of closed set system as a closed set system.

Remark 3. According to the definition in Remark 2, a subclass of a closed set system becomes a closed set system. Let $\mathcal{C}$ be a closed set system and C be a closure operator associated with $\mathcal{C}$. Let $\mathcal{C}'$ be a subclass of $\mathcal{C}$. Then a closure operator C' associated with $\mathcal{C}'$ can be constructed as follows: $C' = C|_{\mathcal{F}}$, where $\mathcal{F} = \{X \subseteq U \mid C(X) \in \mathcal{C}'\}$ and $|_{\mathcal{F}}$ is restriction to $\mathcal{F}$. So a subclass of a closed set system is a closed set system.

3   Unbounded Unions of Closed Set Systems

We start this section by defining the bounded union of languages.

Definition 1. *Let $\mathcal{L}$ be a class of languages. The bounded union $\cup^{\leq k}\mathcal{L}$ of $\mathcal{L}$ is the class defined by*

$$\cup^{\leq k}\mathcal{L} = \{L_1 \cup \ldots \cup L_m \mid m \leq k, L_i \in \mathcal{L} \ (i = 1, \ldots, m)\}.$$

In [12], Wright showed that:

Theorem 4. *([12]) If $\mathcal{L}$ has finite elasticity, then $\cup^{\leq k}\mathcal{L}$ also has finite elasticity. In particular, $\cup^{\leq k}\mathcal{L}$ is inferable from positive data.*

The unbounded union of languages is defined as follows:

Definition 2. *([10]) Let $\mathcal{L}$ be a language. The unbounded union $\mathcal{L}^*$ of $\mathcal{L}$ is the class*

$$\mathcal{L}^* = \{L_1 \cup \ldots \cup L_m \mid \forall m \in \mathbb{N}, L_i \in \mathcal{L} \ (i = 1, \ldots, m)\}.$$

where $\mathbb{N}$ denotes the set of all positive integers $\{1, 2, \ldots\}$.

In [2], de Brecht et al. gave a necessary and sufficient condition for unbounded unions of closed set systems to be inferable.

Theorem 5. *([2]) Let $\mathcal{L}$ be a closed set system. $\mathcal{L}^*$ is inferable from positive data if and only if every closed set $L \in \mathcal{L}$ is equal to a union of finitely many closed sets generated from a single element.*

To study the inferability of $\mathcal{L}^*$, let us start with the following proposition.

Proposition 2. *Let $\mathcal{L}$ be a closed set system such that every $L \in \mathcal{L}$ has a characteristic set and let $\mathcal{U}$ be the family of all finite subset of U. If there exists a map $\delta : \mathcal{U} \to U$ satisfying the condition*

$$(\star) \qquad \delta(S) \in L \Leftrightarrow S \subseteq L$$

for all $S \in \mathcal{U}$ and $L \in \mathcal{L}$, then every $L \in \mathcal{L}$ has a characteristic set consisting of one element.

Proof. Let S be a characteristic set of an arbitrary $L \in \mathcal{L}$. We show that $\{\delta(S)\}$ is a characteristic set of L. From the condition $(\star)$, $\{\delta(S)\} \subseteq L$. Assume that there exists a $L' \in \mathcal{L}$ such that $\delta(S) \in L'$. By applying $(\star)$ for L', we have $S \subseteq L'$. Since S is a characteristic set of L, L must be a subset of L'.

Combining Proposition 2 and Theorem 5, we have:

Corollary 3. *Let $\mathcal{L}$ be the same as in the previous proposition. If δ satisfies the condition $(\star)$, then $\mathcal{L}^*$ is inferable from positive data.*

In the following sections, we present two examples of $\mathcal{L}$ and δ.

4   Invariant Subspaces of a Linear Transformation of a Vector Space

Let V be an infinite dimensional vector space with countable basis over the set of rational numbers $\mathbb{Q}$. More precisely,

(a) V is a vector space over $\mathbb{Q}$, and
(b) V has a countable subset $\mathcal{B}$ called a basis such that (b1) each $v \in V$ can be uniquely written by a linear combination of finite number of elements of $\mathcal{B}$ and, (b2) each $g \in \mathcal{B}$ can not be written by any linear combination of finite number of elements of $\mathcal{B} \setminus \{g\}$.

Note that V is enumerable. We fix one basis $\mathcal{B} = \{g_1, g_2, \ldots\}$ of V. First we consider the class of all finite dimensional subspaces, $\mathcal{V}$, of V.

Lemma 2. *1. $\mathcal{V}$ is a closed set system. 2. Every $W \in \mathcal{V}$ has a characteristic set.*

Proof. (1) The mapping $\langle \cdot \rangle : 2^V \to 2^V$, $S \mapsto \langle S \rangle$, where $\langle S \rangle$ is the subspace spanned by S, is a closure operator associated with $\mathcal{V}$. (2) Since W is finite dimensional, there is a finite subset $F \subset W$ such that $\langle F \rangle = W$. By Lemma 1 and (1), F is a characteristic set of W.

Remark 4. $\mathcal{V}$ is not Noetherian. In fact, there is an infinite strictly ascending chain

$$\langle g_1 \rangle \subset \langle g_1, g_2 \rangle \subset \langle g_1, g_2, g_3 \rangle \subset \cdots .$$

Nevertheless one can show that the bounded union $\cup^{\leq k} \mathcal{V}$ is also inferable. Note that we can not apply Theorem 4 to show the inferability of $\cup^{\leq k} \mathcal{V}$. Instead, we make use of the argument given in Takamatsu et al. [11]. Choose $W = \langle w_1, \ldots, w_r \rangle \in \mathcal{V}$ and let M be a $(r \times (rk - 1))$-matrix in $\mathbb{Q}$-entries such that any k column vectors are linearly independent. Put

$$\left(w_1', \ldots, w_{rk-1}' \right) = (w_1, \ldots, w_r) M.$$

Then by the same argument in the proof of [11, Theorem 9], we can show that $\{w_1', \ldots, w_{rk-1}'\}$ is a characteristic set of W in $\cup^{\leq k} \mathcal{V}$.

On the other hand, $\mathcal{V}^*$ is not inferable. In fact, put $W = \langle g_1, g_2 \rangle$. Choose any finite subset $\{v_1, \ldots, v_r\}$ of W. Then, the union $\langle v_1 \rangle \cup \ldots \cup \langle v_r \rangle$ of 1-dimensional vector spaces $\langle v_i \rangle$ $(i = 1, \ldots, r)$ is proper subset of W. This means that any finite subset of W can not be a finite tell-tale. Thus $\mathcal{V}^*$ is not inferable.

Let $T : V \to V$ be a fixed linear transformation such that $T(g_i) = a_i g_i$ $(a_i \in \mathbb{Q})$ for each i. Here we assume that a_i's are distinct. A subspace W of V is called T-invariant if $T(W) \subseteq W$. Let $\mathcal{V}_T$ be the class of all finite dimensional T-invariant subspaces of V. From Lemma 2 (1) and Remark 3, $\mathcal{V}_T$ is a closed set system.

Lemma 3. *Every $W \in \mathcal{V}_T$ has a characteristic set of the form $\{g_{i_1}, \ldots, g_{i_n}\}$ in $\mathcal{V}_T$.*

Proof. The next lemma implies that we can write $W = \langle g_{i_1}, \ldots, g_{i_n} \rangle$. It is easy to verify that $\{g_{i_1}, \ldots, g_{i_n}\}$ is a characteristic set of W.

Lemma 4. *Let $W \neq \{0\}$ be a finite dimensional subspace of V. W is T-invariant if and only if there exists $g_{i_1}, \ldots, g_{i_n}$ such that $W = \langle g_{i_1}, \ldots, g_{i_n} \rangle$.*

Proof. If there exists $g_{i_1}, \ldots, g_{i_n}$ such that $W = \langle g_{i_1}, \ldots, g_{i_n} \rangle$, then clearly W is T-invariant. Suppose that W is T-invariant. Let w be a nonzero vector of W. We can write $w = c_1 g_{k_1} + \ldots + c_m g_{k_m}$, where every $c_i \neq 0$. Since W is T-invariant, $T(w) = c_1 a_{k_1} g_{k_1} + \ldots + c_m a_{k_m} g_{k_m} \in W$. Similarly, $T^2(w), \ldots, T^{m-1}(w) \in W$. Now, let A be the $(m \times m)$-matrix on the right in the equation

$$
(w, T(w), \ldots, T^{m-1}(w)) = (g_{k_1}, \ldots, g_{k_m}) \begin{pmatrix} c_1 & c_1 a_{k_1} & \cdots & c_1 a_{k_1}^{m-1} \\ c_2 & c_2 a_{k_2} & \cdots & c_2 a_{k_2}^{m-1} \\ \vdots & \vdots & & \vdots \\ c_m & c_m a_{k_m} & \cdots & c_m a_{k_m}^{m-1} \end{pmatrix}.
$$

A is a matrix obtained by multiplying each column of a Vandermonde's matrix by nonzero scalars. Since a_i's are distinct,

$$
\det A = \left(\prod_{j=1}^m c_j \right) \cdot \left(\prod_{p<q} (a_{k_p} - a_{k_q}) \right) \neq 0,
$$

so there is the inverse matrix A^{-1}. This means that g_{k_j}'s can be written by linear combinations of $w, T(w), \ldots, T^{m-1}(w)$. Thus, $G = \{g_{k_1} \cdots g_{k_m}\} \subseteq W$. If $\langle G \rangle \neq W$, repeat the method above for $w^{(1)} \in W \setminus \langle G \rangle$, and add the yielding $\{g_{k_1}^{(1)}, \ldots, g_{k_{m_1}}^{(1)}\}$ to G. This procedure ends in finite number of steps since W is finite dimensional.

Theorem 6. *$\mathcal{V}_T^*$ is inferable from positive data.*

Proof. Let $S = \{v_1, \ldots, v_n\} \subseteq V$. For each i, we can write $v_i = \sum_{\text{finite}} c_{i,j} g_j$ ($c_{i,j} \in \mathbb{Q}$). Let $A_i = \{g_j \mid c_{i,j} \neq 0\}$. $\delta(S)$ is defined as follows:

$$
A = \bigcup_{i=1}^n A_i, \quad \delta(S) = \sum_{g \in A} g.
$$

According to Corollary 3, it suffices to show that δ satisfying

$$(\star) \qquad \delta(S) \in W \Leftrightarrow S \subseteq W$$

for each finite subset $S \subseteq V$ and for each $W \in \mathcal{V}_T$. Let W be an arbitrary element of $\mathcal{V}_T$. By Lemma 4, we can write $W = \langle g_{i_1}, \ldots, g_{i_n} \rangle$.
($\Rightarrow$) By applying the argument in the proof of Lemma 4 to $\delta(S) = \sum_{g \in A} g \in W$, we get that A is a subset of $\{g_{i_1}, \ldots, g_{i_n}\}$, hence A_i is. Since v_i is a linear combination of the elements of A_i, each v_i is in W.
($\Leftarrow$) $S \subseteq W$ implies that $A_i \subseteq \{g_{i_1}, \ldots, g_{i_n}\}$ for every i, so A is. Thus $\delta(S) \in W$.

A concrete inference algorithm of $\mathcal{V}_T^*$ is shown as follows:

Procedure 1: Learning $\mathcal{V}_T^*$;
Input: a positive presentation $\boldsymbol{v_1}, \boldsymbol{v_2}, \ldots, \boldsymbol{v_n}, \ldots$ for $V_1 \cup \ldots \cup V_m \in \mathcal{V}_T^*$;
Output: a sequence $V_1^{(1)} \cup \ldots \cup V_{m_1}^{(1)}, V_1^{(2)} \cup \ldots \cup V_{m_2}^{(2)}, \ldots$ of elements of $\mathcal{V}_T^*$;
begin
1. $S = \emptyset$;
2.   Put $n = 1$;
3.   **repeat**
4.   **if** there is no $V \in S$ such that $\boldsymbol{v_n} \in V$ **do**
5.   Set $A_n = \{ \boldsymbol{g_j} \mid \boldsymbol{g_j}$ occurs in $\boldsymbol{v_n}$ with nonzero coefficient $\}$;
6.   Remove all $V \in S$ such that $V \subset \langle A_n \rangle$ from S;
7.   Add $\langle A_n \rangle$ to S;
8.   **end do**;
9.   Output $\cup_{V \in S} V$;
10.   Add 1 to n;
11. **forever**;
end.

We consider an application of **Procedure 1** to the study of transaction databases in §6.1. We end this section with presenting an example of $\mathcal{V}$ and T.

Example 1. Let V be the subspace of the vector space consisting of Fourier series as follows:

$$ V = \left\{ \sum_{n=0}^{r} a_n \cos nx \;\middle|\; a_n \in \mathbb{Q}, \; r \in \mathbb{Z}_{\geq 0} \right\}. $$

The set $\{1, \cos x, \cos 2x, \ldots\}$ forms a basis of V. Let $T : V \ni f \mapsto \frac{d^2 f}{dx^2} \in V$. Then, $T(1) = 0, T(\cos x) = -\cos x, T(\cos 2x) = -4\cos 2x, \ldots$, so the class of unbounded unions of finite dimensional T-invariant subspaces is inferable from positive data. This situation can be generalized to that of Hilbert spaces. A vector space H over the field of real or complex numbers is called Hilbert space if an inner product is defined over H and H is complete with respect to the metric induced by the inner product. It is known that H has an orthonormal basis under a certain condition. This means that every element of H can be approximated by a linear combination of finite number of elements of the orthonormal basis within an arbitrary error. This example can be considered to be the situation treating an approximation cut off after the r'th term. A Hilbert space is one of the most typical and important examples of infinite dimensional vector spaces which appear in mathematics, and it is closely related to functional analysis and approximation theory. This might shed new light on the connection between learning theory and analysis.

5   Monomial Ideals of Polynomial Ring

We denote the set of polynomials of n variables with $\mathbb{Q}$-coefficients by $\mathbb{Q}[x_1, \ldots, x_n]$. A subset I of $\mathbb{Q}[x_1, \ldots, x_n]$ is called an *ideal* if it satisfies (i) $\forall f, g \in I \Rightarrow f \pm g \in I$

and (ii) $\forall f \in I$, $\forall h \in \mathbb{Q}[x_1, \ldots, x_n] \Rightarrow hf \in I$. We denote the set of all ideals by $\mathcal{I}$. For a subset $F \subseteq \mathbb{Q}[x_1, \ldots, x_n]$, the ideal generated by F, which is denoted by $\langle F \rangle$, is defined as follows:

$$\langle F \rangle = \left\{ \sum_{\text{finite}} h_i f_i \;\middle|\; h_i \in \mathbb{Q}[x_1, \ldots, x_n], f_i \in F \right\}.$$

Note that the correspondence $F \mapsto \langle F \rangle$ defines a closure operator on $\mathbb{Q}[x_1, \ldots, x_n]$. An ideal I is called finitely generated if there is a finite set F such that $\langle F \rangle = I$. The following theorem is well-known in algebra:

Theorem 7. (Hilbert's basis theorem, [3]) *Every ideal of* $\mathbb{Q}[x_1, \ldots, x_n]$ *is finitely generated.*

An interpretation of this statement from machine learning view point is that "$\mathcal{I}$ has a finite elasticity." Hence, $\mathcal{I}$ is a Noetherian closed set system with the closure operator $F \mapsto \langle F \rangle$. An ideal I is called *monomial ideal* if there exists a set of monomials F such that $I = \langle F \rangle$. Monomial ideals are characterized by the following fact.

Proposition 3. *Let I be an ideal of* $\mathbb{Q}[x_1, \ldots, x_n]$. *The followings are equivalent:*
(a) I is a monomial ideal,
(b) I is generated by the set of all monomials in I,
(c) for each $f \in I$, every monomial occurring in f is also in I,
(d) I is generated by a set of finitely many monomials.

Let $\mathcal{MI}$ be the class of all monomial ideals. Then,

Lemma 5. *1. $\mathcal{MI}$ is a closed set system. 2. Every $I \in \mathcal{MI}$ has a characteristic set.*

Proof. (1) See Remark 3. (2) According to Proposition 3 (d), there is a set of finite number of monomials G that generates I for each $I \in \mathcal{MI}$. This G forms a characteristic set of I.

Theorem 8. $\mathcal{MI}^*$ *is inferable from positive data.*

Proof. We denote the set of all monomials by $\mathcal{M}$. Let $S = \{s_1, \ldots, s_n\} \subseteq \mathbb{Q}[x_1, \ldots, x_n]$. For each i, we can write $s_i = \sum_{m \in \mathcal{M}} c_{i,m} m$, where all but finite $c_{i,m}$'s are equals to 0. Let $M_i = \{m \mid c_{i,m} \neq 0\}$. $\delta(S)$ is defined as follows:

$$M = \bigcup_{i=1}^{n} M_i, \quad \delta(S) = \sum_{m \in M} m.$$

According to Corollary 3, it suffices to show that δ satisfying $(\star)$ for each finite subset $S \subseteq \mathbb{Q}[x_1, \ldots, x_n]$ and for each $I \in \mathcal{MI}$. By Proposition 3 (d), I has a characteristic set that consists of finite number of monomials, say G.

($\Rightarrow$) Suppose $\delta(S) = \sum_{m \in M} m \in I$. Applying Proposition 3 (c) to $\delta(S)$, the set of all monomials occurring $\delta(S)$, that is, M is a subset of G. Hence M_i is. Since s_i is a linear combination of the elements of M_i, each s_i is in I.
($\Leftarrow$) $S \subseteq I$ implies that $M_i \subseteq G$ for every i, so M is. Thus $\delta(S) \in I$.

Example 2. Consider a monomial ideal $I = \langle x^3, xy, y^2 \rangle$ of the polynomial ring of two variables $\mathbb{Q}[x, y]$. I equals to the set

$$\left\{ \sum_{\text{finite}} c_m m \;\middle|\; c_m \in \mathbb{Q}, m \in \mathcal{M} \text{ is divisible by } x^3, xy \text{ or } y^2 \right\}.$$

Then, the set

$$\{\delta(\{x^3, xy, y^2\})\} = \{x^3 + xy + y^2\}$$

forms a characteristic set of I in $\mathbb{Q}[x, y]$.

In §6.2, we consider an application of above arguments to the study of transaction databases.

6 Application: Closed Set Systems and Transaction Databases

In this section, we apply our arguments of closed set systems considered in §§4 and 5 to the study of transaction databases.

6.1 Application 1: Vector Spaces and Transaction Databases

Let I be a countable set $\{p_1, p_2, \dots\}$ and we regard I as the set of items. A transaction database $\mathcal{D}$ over I is a sequence of finite subsets $X_1, X_2, \dots$ of I. For a set $X \subseteq I$, the support of X is defined by $\{i \in \mathbb{N} \mid X \subseteq X_i\}$ and denoted by $\mathcal{D}(X)$. A set $X \subseteq I$ is called closed if $\mathcal{D}(Y) \subset \mathcal{D}(X)$ for every $Y \supset X$. To avoid confusions, we call this DB-closed here. Note that every DB-closed set X is finite. In fact, if X is infinite, then $\mathcal{D}(X)$ is empty. So X can not be closed.

Let V be the set of all formal linear combinations of elements of I:

$$V = \left\{ \sum_{\text{finite}} c_k p_k \;\middle|\; c_k \in \mathbb{Q}, p_k \in I \right\}.$$

Then I forms a basis of V. I corresponds to $\mathcal{G}$ in §4. Take a linear transformation $T : V \to V$ that is the form $T(p_i) = a_i p_i$, where a_i's are distinct rational numbers (for example, a linear transformation T defined by $T(p_i) = i p_i$ is suitable.) By Lemma 4, we have

$$\mathcal{V}_T = \{\langle S \rangle \mid S \subset I, S \text{ is finite}\}.$$

Hence a finite subset of I, i.e. an itemset, can be regarded as a closed set. More precisely, it can be described as follows: Let $\mathcal{S}$ be the class of all itemsets. By

Lemma 4, for each $W \in \mathcal{V}_T$, there exists unique $S_W \in \mathcal{S}$ such that $W = \langle S_W \rangle$. Then the mapping

$$\Phi : \mathcal{V}_T \to \mathcal{S}, \quad W \mapsto S_W$$

is the inverse of

$$\Psi : \mathcal{S} \to \mathcal{V}_T, \quad S \mapsto \langle S \rangle,$$

so an itemset corresponds an element of $\mathcal{V}_T$, i.e. a closed set.

Now we apply our result in §4 to consider the inference of $\mathcal{V}_T^*$. Let $W_1 \cup \ldots \cup W_m \in \mathcal{V}_T^*$ be a target and $\boldsymbol{v_1}, \boldsymbol{v_2}, \ldots$ be a positive data of $W_1 \cup \ldots \cup W_m$. We assume that $W_1 \cup \ldots \cup W_m$ is not redundant: for each i, W_i is not included in $\cup_{j \neq i} W_j$. As we have seen in §4, $W_1 \cup \ldots \cup W_m$ is inferable from $\boldsymbol{v_1}, \boldsymbol{v_2}, \ldots$.

Here we define $X_i \subset I$ by

$$X_i = \{p_i \mid p_i \text{ appears in } \boldsymbol{v_i} \text{ with nonzero coefficient}\}.$$

Since every X_i is finite, the sequence $\{X_1, X_2, \ldots\}$ forms a transaction database, which we denote by $\mathcal{D}$. Put $C_i = \{p_i \mid p_i \in W_i\}$ $(= \Phi(W_i))$. From the correspondence as above, C_i be the subset of I that corresponds to W_i. Then it holds that:

Proposition 4. C_i *is DB-closed. Moreover, C_i is maximal with respect to set inclusion.*

Proof. Let $C_i \subset X \subset I$ be an arbitrary finite set. Since $W_1 \cup \ldots \cup W_m$ is not redundant, there is no W_k such that $\Psi(X) = \langle X \rangle \subseteq W_k$. This means that there is no j such that $X_j \supseteq X$. Hence $\mathcal{D}(X) = \emptyset$. On the other hand, $\mathcal{D}(C_i) \neq \emptyset$, since $\boldsymbol{v_1}, \boldsymbol{v_2}, \ldots$ is positive data and so there is a $\boldsymbol{v_k} = \sum_{p \in C_i} p$. Thus $\mathcal{D}(X) \subset \mathcal{D}(C_i)$, and then C_i is DB-closed. In addition, $\mathcal{D}(X) = \emptyset$ implies that $X \supset C_i$ can not be DB-closed. So C_i is maximal.

Furthermore, we have:

Proposition 5. *If X is a maximal DB-closed set of $\mathcal{D}$, then there exists unique i such that $C_i = X$.*

Proof. If $\mathcal{D}(X) = \emptyset$ then X is not DB-closed, so $\mathcal{D}(X) \neq \emptyset$. Let $j \in \mathcal{D}(X)$. There is a W_k such that $\boldsymbol{v_j} \in W_k$. By definition, it holds that $X \subseteq X_j \subseteq C_k$. The assumption and Proposition 4 imply $X = C_k$. If there is a $k' \neq k$ such that $X = C_k = C_{k'}$, then it means that $W_1 \cup \ldots \cup W_m$ is redundant, and this is a contradiction.

Thus "inference of $\mathcal{V}_T^*$" corresponds to "mining maximal closed sets of $\mathcal{D}$".

Remark 5. This example would be uninteresting, since a maximal DB-closed itemset is only a maximal itemset. But, if we choose the linear transformation T appropriately or make coefficients of $\boldsymbol{g_i}$'s have a meaning, it might be interesting.

Example 3. In practice, both the set of items I and the database $\mathcal{D}$ are finite. Here we consider such a case as an example. Let $I = \{p_1, p_2, p_3, p_4\}$ and $\mathcal{D}$ as follows:

	p_1	p_2	p_3	p_4
X_1	○	○	○	
X_2		○	○	
X_3				○
X_4		○	○	
X_5	○			○

Then DB-closed sets of $\mathcal{D}$ are

$$\{p_1\}, \{p_4\}, \{p_1, p_4\}, \{p_2, p_3\}, \{p_1, p_2, p_3\}$$

and maximal DB-closed sets are $\{p_1, p_4\}$ and $\{p_1, p_2, p_3\}$. Now let $v_i \in V$ ($i = 1, \ldots, 5$) be the element corresponding to X_i:

$$v_1 = p_1 + p_2 + p_3, v_2 = p_2 + p_3, v_3 = p_4, v_4 = p_2 + p_3, v_5 = p_1 + p_4.$$

Then **Procedure 1** outputs $\langle p_1, p_2, p_3 \rangle \cup \langle p_1, p_4 \rangle$ when $v_1, \ldots, v_5$ is taken as input. Since

$$\Phi(\langle p_1, p_2, p_3 \rangle) = \{p_1, p_2, p_3\}, \quad \Phi(\langle p_1, p_4 \rangle) = \{p_1, p_4\},$$

the maximal DB-closed sets of $\mathcal{D}$ can be given by **Procedure 1**.

6.2   Application 2: Monomial Ideals and Transaction Databases

We have consider an application of Theorem 8 to the study of transaction databases. Our advantage is that, by using Theorem 8, we can deal with transaction databases that contains data of quantities of items. Let $I = \{x_1, x_2, \ldots x_n\}$ be the set of items. A transaction considered in this section is a set of pairs of an item and its quantity $\{(x_{a_1}, p_1), (x_{a_2}, p_2), \ldots, (x_{a_k}, p_k)\}$. For convenience, we always assume that $a_1 < a_2 < \ldots < a_k$. A transaction database $\mathcal{D}$ is a sequence of transactions $\{T_1, T_2, \ldots\}$.

Definition 3. *1. Let $T_1 = \{(x_{a_1}, p_1), (x_{a_2}, p_2), \ldots, (x_{a_n}, p_n)\}$ and $T_2 = \{(x_{b_1}, q_1), (x_{b_2}, q_2), \ldots, (x_{b_m}, q_m)\}$ be itemsets. If there is an j_i such that $b_{j_i} = a_i$ and $p_i \leq q_{j_i}$ for each $i = 1, 2, \ldots, n$, then T_1 is said to be included in T_2, and denotes by $T_1 \preceq T_2$.*
2. Let T be an itemset and $\mathcal{D}$ be a transaction database. Define

$$\iota_{\mathcal{D}}(T) := \{T' : \text{transaction of } \mathcal{D} \mid T \preceq T'\}.$$

3. Let $\mathcal{D}' \subseteq \mathcal{D}$. Define

$$t_{\mathcal{D}}(\mathcal{D}') := \max_{\preceq}\{T : \text{itemset} \mid \forall T' \in \mathcal{D}', T \preceq T'\}.$$

For simplicity, we will write ι and t instead of $\iota_{\mathcal{D}}$ and $t_{\mathcal{D}}$, respectively. It is known that

Proposition 6. *Let $\mathcal{D}$ be a fixed transaction database and ι, t be the mappings above. Let $\mathcal{I}$ denote the set of all itemsets. The composition $C_{\mathcal{D}} = t \circ \iota : \mathcal{I} \to \mathcal{I}$ becomes a closure operator.*

An itemset can be interpreted as a monomial $x_{i_1}^{a_1} x_{i_2}^{a_2} \cdots \cdots x_{i_n}^{a_n}$. $\mathcal{M}$ denotes the set of all monomials of variable $x_1, x_2, \ldots$. Then $\mathcal{M}$ corresponds to the set of all itemsets $\mathcal{I}$ and a transaction database $\mathcal{D}$ can be regarded as a sequence of monomials $m_1, m_2, \ldots$ $(m_i \in \mathcal{M})$. Here we give an example of the correspondence.

	x_1	x_2	x_3	x_4
T_1	2			1
T_2		1	3	2
T_3	1			
T_4	1	2	1	

$\longleftrightarrow$

m_1	$x_1^2 x_4$
m_2	$x_2 x_3^3 x_4^2$
m_3	x_1
m_4	$x_1 x_2^2 x_3$

The mappings ι and t are interpreted as follows:

Proposition 7. *1. Let T_1 and T_2 be itemsets and let m_1 and m_2 be monomials corresponding to T_1 and T_2 respectively. Then $T_1 \preceq T_2 \Leftrightarrow m_1 | m_2$, where $m|n$ means that m divides n.*
2. Let $m \in \mathcal{M}$ be a monomial and $\mathcal{D} : m_1, m_2, \ldots$ be a transaction database. Then $\iota(m) = \{m_i \mid m | m_i\}$.
3. Let $S = \{n_1, n_2, \ldots, n_k\}$ be itemsets. $t(S)$ corresponds to the greatest common divisor of monomials of S, $GCD(S)$.

Now we fix a transaction database $\mathcal{D}$. As $\mathcal{D}$ is regarded to a sequence of monomials as above, the argument in §5 can be applied to $\mathcal{D}$. The algorithm becomes much simpler since the sequence presented as a positive data is a sequence of only monomials, instead of a sequence of polynomials.

Procedure 2-1:
Input: a sequence of monomials $m_1, m_2, \ldots$;
Output: a sequence of a set of monomials $S_1, S_2, \ldots$;
begin
1. $S = \emptyset$;
2.   Put $n = 1$;
3.   **repeat**
4.   **if** there is no $m' \in S$ such that $m' | m_n$ **do**
5.   Remove all $m' \in S$ such that $m_n | m'$ from S;
6.   Add m_n to S;
7.   **end do**;
8.   Set $S_n = S$ and output S;
9.   Add 1 to n;
10. **forever**;
end.

It is clear that

Proposition 8. *S_n in the output of **Procedure 2-1** is equals to the set $\{m_i \mid m_j \nmid m_i \ (i, j = 1, 2, \ldots, n)\}$.*

By translating Proposition 8 into the language of transaction databases, it follows that:

Proposition 9. S_n *is the set of all maximal closed itemset of* $\{m_1, m_2, \ldots, m_n\}$

Now we improve **Procedure 2-1**. We set a threshold k as follows.

Procedure 2-2:
Input: a sequence of monomials $m_1, m_2, \ldots$ and a threshold k;
Output: a sequence of a set of monomials $S_1, S_2, \ldots$;
begin
1. $S = \emptyset$, $M = \emptyset$;
2.   Put $n = 1$;
3.   **repeat**
4.   Set $m = m_n$.
5.   **if** there is an element of the form (c_m, m) in M, **then** $c_m = c_m + 1$;
6.   **else** set $c_m = 1$ and add (c_m, m) to M;
7.   **if** $c_m \geq k$ and there is no $m' \in S$ such that $m'|m$ **do**
8.   Remove all $m' \in S$ such that $m|m'$ from S;
9.   Add m to S;
10.   **end do**;
11.   Set $S_n = S$ and output S;
12.   Add 1 to n;
13.   **forever**;
end.

Note that we can take quantities as well as frequency into account in **Procedure 2-2**. This is an advantage of **Procedure 2-2**.

7   Conclusion

We have focused on the closed set systems based on the sets with some kind of algebraic structure. As a result, we showed that the scheme of learning from positive data can be applied to some objects in both algebra and analysis, such as polynomial rings or Hilbert spaces. Moreover, we have seen that such closed set systems can be applied to the study of transaction database in the last section. In particular, we showed that one can deal with a transaction database with the quantities of items, under the settings of §6.2. Our problem in the near future is experiment, i.e. to apply our method to various "actual" databases and see how it works.

Acknowledgement

The authors are grateful to professor Shin-ichi Minato for his constructive comments on the first version of this article. They also thank for the referees for their valuable comments.

References

1. Angluin, D.: Inductive Inference of Formal Languages from Positive Data. Information and Control 45, 117–135 (1980)
2. de Brecht, M., Kobayashi, M., Tokunaga, H., Yamamoto, A.: Inferability of Closed Set Systems From Positive Data. In: Washio, T., Satoh, K., Takeda, H., Inokuchi, A. (eds.) JSAI 2006. LNCS (LNAI), vol. 4384, pp. 265–275. Springer, Heidelberg (2007)
3. Cox, D., Little, J., O'Shea, D.: Ideals, Varieties, and Algorithms: An Introduction to Computational Algebraic Geometry and Commutative Algebra. Springer, Heidelberg (1992)
4. Gold, E.M.: Language Identification in the Limit. Information and Control 10, 447–474 (1967)
5. Kameda, Y., Tokunaga, H., Yamamoto, A.: Learning bounded unionsof Noetherian closed set systems via characteristic sets. In: Clark, A., Coste, F., Miclet, L. (eds.) ICGI 2008. LNCS (LNAI), vol. 5278, pp. 98–110. Springer, Heidelberg (2008)
6. Kapur, S., Bilardi, G.: On uniform learnability of language families. Information Processing Letters 44, 35–38 (1992)
7. Kobayashi, S.: Approximate Identification, Finite Elasticity and Lattice Structure of Hypothesis Space, Technical Report, CSIM 96-04, Dept. of Compt. Sci. and Inform. Math., Univ. of Electro-Communications (1996)
8. Lassez, J.L., Maher, M.J., Marriott, K.: Unification Revisited. In: Minker, J. (ed.) Foundations of Deductive Databases and Logic Programming, pp. 587–626. Morgan Kaufman, San Francisco (1988)
9. Motoki, T., Shinohara, T., Wright, K.: The Correct Definition of Finite Elasticity: Corrigendum to Identification of Unions. In: Proceedings of COLT 1991, p. 375, 587–626. Morgan-Kaufman, San Francisco (1988)
10. Shinohara, T., Arimura, H.: Inductive Inference of Unbounded Unions of Pattern Languages From Positive Data. Theoretical Computer Science 241, 191–209 (2000)
11. Takamatsu, I., Kobayashi, M., Tokunaga, H., Yamamoto, A.: Computing Characteristic Sets of Bounded Unions of Polynomial Ideals. In: Satoh, K., Inokuchi, A., Nagao, K., Kawamura, T. (eds.) JSAI 2007. LNCS (LNAI), vol. 4914, pp. 318–329. Springer, Heidelberg (2008)
12. Wright, K.: Identification of Unions of Languages Drawn from an Identifiable Class. In: Proc. of COLT 1989, pp. 328–388. Morgan-Kaufman, San Francisco (1989)

Mining Frequent k-Partite Episodes
from Event Sequences*

Takashi Katoh[1], Hiroki Arimura[1], and Kouichi Hirata[2]

[1] Graduate School of Information Science and Technology, Hokkaido University
Kita 14-jo Nishi 9-chome, Sapporo 060-0814, Japan
Tel.: +81-11-706-7678; Fax: +81-11-706-7890
{t-katou,arim}@ist.hokudai.ac.jp
[2] Department of Artificial Intelligence, Kyushu Institute of Technology
Kawazu 680-4, Iizuka 820-8502, Japan
Tel.: +81-948-29-7622; Fax: +81-948-29-7601
hirata@ai.kyutech.ac.jp

Abstract. In this paper, we introduce the class of *k-partite episodes*, which are time-series patterns of the form $\langle A_1, \ldots, A_k \rangle$ for sets A_i ($1 \leq i \leq k$) of events meaning that, in an input event sequence, every event of A_i is followed by every event of A_{i+1} for every $1 \leq i < k$. Then, we present a backtracking algorithm KPAR and its modification KPAR2 that find all of the frequent k-partite episodes from an input event sequence without duplication. By theoretical analysis, we show that these two algorithms run in polynomial delay and polynomial space in total input size.

1 Introduction

Episode Mining

It is one of the important tasks in data mining to discover frequent patterns from time-related data. For such a task, Mannila *et al.* [11] have introduced *episode mining* to discover frequent *episodes* in an event sequence. Here, an episode is formulated as a labeled acyclic digraphs in which labels correspond to events and arcs represent a temporal precedent-subsequent relation in an event sequence. Then, the episode is a richer representation of temporal relationship than a subsequence, which represents just a linearly ordered relation in sequential pattern mining (*cf.*, [3,13]). Since the frequency of the episode is formulated by a window that is a subsequence of an event sequence under a fixed time span, the episode mining is more appropriate than the sequential pattern mining when considering the time span.

For subclasses of episodes [5,8,7,9,11], a number of efficient algorithms have been developed so far. Mannila *et al.* [11] presented efficient mining algorithm for *parallel episodes* as a set of events and *serial episodes* a sequence of events. On the other hand, in order to capture the direct relationship between premises and consequences, Katoh *et al.* have introduced the episodes with the special events,

---

* This work is partially supported by Grand-in-Aid for JSPS Fellows (20·3406).

K. Nakakoji, Y. Murakami, and E. McCready (Eds.): JSAI-isAI, LNAI 6284, pp. 331–344, 2010.

a source as a premise and a sink as a consequence, as *sectorial episodes* [8], *diamond episodes* [7,9] and *elliptic episodes* [5]. Then, they have succeeded to find frequent their episodes concerned with the replacement of bacteria and the changes for drug resistance from bacterial culture data, which are valuable from the medical viewpoint [5,8,9].

Since their episodes have just a single source and a single sink, they can represent no relationship between plural premises and consequences. On the other hand, if we extend the diamond episodes with plural sources and sinks, we can deal with not only plural sources and sinks but also the precedent-subsequent relation between 3 sets of events as an episode with length 3. Furthermore, without restriction of the length of episodes, we can generalize such an episode by extending an event in a serial episode as a set of events. The generalized episode can represent the precedent-subsequent relation between sets of events that simultaneously occur.

Hence, in this paper, as a generalized form of episodes, we newly introduce *k-partite episodes* of the form $\langle A_1, \ldots, A_k \rangle$ $(k \geq 0)$, where A_i $(1 \leq i \leq k)$ are sets of events. The k-partite episode $\langle A_1, \ldots, A_k \rangle$ means that every event in A_i $(1 \leq i < k)$ is followed by every event in A_{i+1}. The name "k-partite" comes from the fact that we can represent a k-partite episode as a complete k-partite graph, by adding all of the transitive arcs and by ignoring the direction of arcs.

Main results

For the frequent k-partite episode mining problem, we present a backtracking algorithm KPAR and its modification KPAR2, which achieve polynomial delay and polynomial space complexity.

A key idea for the efficiency of these algorithms is an enumeration of the candidate k-partite episodes based on depth-first search. The algorithm KPAR searches for frequent k-partite episodes starting with the smallest empty episode, and then expands a candidate episode by attaching a new event one by one to the tail component. Once the algorithm reaches an infrequent episode, it stops to expand the current branch and backtracks for remaining branches in a search tree.

Let Σ be an alphabet of events, X a k-partite episode over Σ, S an event sequence of size N and of length n and w a window width. Then, *the matching problem of X in S against w* is to determine whether or not there exists a contiguous subsequence of S of length w that contains X satisfying the edge constraints. A straightforward generate-and-test method for the matching problem requires exponential time in the size M of a k-partite episode. By introducing the notion of the *leftmost tail occurrences*, we show that the matching problem is solvable in $O(|\Sigma|n)$ time, and we present an efficient scanning-based subprocedure COUNTBYSCAN that solves the matching problem of X in S in $O(w|\Sigma|n)$ time. Note here that $N \leq |\Sigma|n$. Then, we show that the algorithm KPAR runs in $O(|\Sigma|^2 wn)$ time per a k-partite episode and $O(|\Sigma|k)$ space.

In some real-world datasets, input sequences are often *sparse*, that is, the total number $N(=||S||)$ of the occurrences of events in S is much smaller than sn, where $s = |\Sigma|$ and $n = |S|$. In such a case, the performance of a scanning-based algorithm such as COUNTBYSCAN may degenerate, since it is necessary to traverse many windows overall in order to solve the matching problem. To cope with this problem, we give a practical speed-up technique for frequency counting. We show that the matching problem of a k-partite episode in $\mathcal{S}$ is solvable in $O(N)$ time, and we present a practical algorithm COUNTBYLIST that computes the frequency counts in $O(N^2 k)$ time by using the event lists. The modified algorithm KPAR2 with COUNTBYLIST runs in $O(|\Sigma|N^2 k)$ time per a k-partite episode and in $O(|\Sigma|k + N)$ space, where $O(N)$ coincides with the space for storing all event sequences.

As a corollary, we show that the frequent episode mining problem is solvable in polynomial delay and in polynomial space in the total input size.

Organization

This paper is organized as follows. In Section 2, we introduce episodes and other notions necessary to the later discussion. In Section 3, we introduce k-partite episodes and other notions and discuss their properties. In Section 4, we present the algorithms KPAR and KAPR2 to extract all of the frequent k-partite episodes. In Section 5, we give the experimental results for the algorithms given in Section 4, by applying to the randomly generated event sequences. In Section 6, we conclude this paper and discuss the future works.

2 Preliminaries

In this section, we introduce the frequent episode mining problem and the related notions necessary to later discussion. We denote the sets of all integers and all natural numbers by $\mathbf{Z}$ and $\mathbf{N}$, respectively. For a set S, we denote the cardinality of S by $|S|$. A *digraph* is a graph with directed edges (or, *arcs*). An *acyclic digraph* is a digraph without cycles.

2.1 An Input Event Sequence and Its Windows

Let $\Sigma = \{1, \ldots, m\}$ $(m \geq 1)$ be a finite alphabet with the total order $\leq$ over $\mathbf{N}$. Each element $e \in \Sigma$ is called an *event* [1]. An *input event sequence* (*input sequence*, for short) $\mathcal{S}$ on Σ is a finite sequence $\langle S_1, \ldots, S_n \rangle \in (2^\Sigma)^*$ of events $(n \geq 0)$, where $S_i \subseteq \Sigma$ is called the *i-th event set* for every $1 \leq i \leq n$. For any $i < 0$ or $i > n$, we set $S_i = \emptyset$. For an input event sequence $\mathcal{S}$, we denote the *length* n by $|\mathcal{S}|$ and define the *total size* $||\mathcal{S}||$ of $\mathcal{S}$ by $\sum_{i=1}^n |S_i|$. Clearly, $||\mathcal{S}|| = O(|\Sigma|n)$, but the converse does not hold in general, that is, $O(||\mathcal{S}||) \neq |\Sigma|n$. Without loss of generality, we can assume that every event in Σ appears at least once in $\mathcal{S}$.

2.2 Episodes

Mannila *et al.* [11] have formulated an episode as a partially ordered. On the other hand, We formulate an episode as a labeled acyclic digraph as follows. An

---

[1] Mannila *et al.* [11] originally referred to each element $e \in \Sigma$ itself as an *event type* and an occurrence of e as an *event*. However, we simply call both of them as *events*.

episode over Σ is a labeled acyclic digraph $X = (V, E, g)$, where V is a set of vertices, $E \subseteq V \times V$ is a set of arcs and $g : V \to \Sigma$ is a mapping associating each node with an event. It is not hard to see that two definitions of episodes by Mannila's partially ordered sets [11] and our labeled acyclic digraphs are essentially same each other.

Let $X = (V, E, g)$ be an episode. We define the *size* $||X||$ of X by $|V|$. For an arc set E on a vertex set V, let E^+ be the *transitive closure* of E such that $E^+ = \{ (u, v) \,|\, \text{there is a directed path from } u \text{ to } v \}$.

Definition 1 (embedding). For episodes $X_i = (V_i, E_i, g_i)$ $(i = 1, 2)$, X_1 *is embedded in* X_2, denoted by $X_1 \sqsubseteq X_2$, if there exists a mapping $f : V_1 \to V_2$ such that (i) *f preserves the labels of vertices*, i.e., for all $v \in V_1$, $g_1(v) = g_2(f(v))$, and (ii) *f preserves the precedence relation*, i.e., for all $u, v \in V$ with $u \neq v$, if $(u, v) \in E_1$ then $(f(u), f(v)) \in (E_2)^+$. The mapping f is called an *embedding* from X_1 to X_2.

Given an input sequence $\mathcal{S} = \langle S_1, \dots, S_n \rangle \in (2^\Sigma)^*$, an *window* in $\mathcal{S}$ is a contiguous subsequence $W = \langle S_i \cdots S_{i+w-1} \rangle \in (2^\Sigma)^*$ of $\mathcal{S}$ for some i, where $w \geq 0$ is the *width* of W.

Definition 2 (occurrence for an episode). An episode $X = (V, E, g)$ *occurs in* an window $W = \langle S_1 \cdots S_w \rangle \in (2^\Sigma)^*$, denoted by $X \sqsubseteq W$, if there exists a mapping $h : V \to \{1, \dots, w\}$ such that (i) h preserves the labels of vertices, i.e., for all $v \in V$, $g(v) \in S_{h(x)}$, and (ii) h preserves the precedence relation, i.e., for all $u, v \in V$ with $u \neq v$, if $(u, v) \in E$ then $h(u) < h(v)$. The mapping h is called an *embedding* of X into W.

A *window width* is a fixed positive integer $1 \leq w \leq n$. For any $-w + 1 \leq i \leq n$, we say that an episode X *occurs at* position i in an event sequence $\mathcal{S}$ if $X \sqsubseteq W_i$, where $W_i = \langle S_i, \dots, S_{i+w-1} \rangle$ is the i-th window of width w in $\mathcal{S}$. Then, we call i an *occurrence* or *label* of X in $\mathcal{S}$. In what follows, we denote the i-th window W_i by $\mathbf{W}_i^{\mathcal{S},w}$. Let $\mathbf{W}_{\mathcal{S},w} = \{ i \,|\, -w + 1 \leq i \leq n \}$ be the domain of the occurrences. For an episode X, we define the *occurrence window list* $\mathbf{W}_{\mathcal{S},w}(X)$ for X in $\mathcal{S}$ by the set $\{ -w + 1 \leq i \leq n \,|\, X \sqsubseteq W_i \}$ of the occurrences of X in $\mathcal{S}$.

2.3   Frequent Episode Mining Problem

Let $\mathcal{C}$ be a subclass of episodes, X an episode in $\mathcal{C}$, $\mathcal{S}$ an input sequence and w (≥ 1) a window width. Then, the *frequency* $freq_{\mathcal{S},w}(X)$ of X in $\mathcal{S}$ is defined by the number of w-windows, that is, $freq_{\mathcal{S},w}(X) = |\mathbf{W}_{\mathcal{S},w}(X)| = O(|\mathcal{S}|)$. A *minimum frequency threshold* is any positive integer $\sigma \geq 1$. Without loss of generality, we can assume that $\sigma \leq |\mathcal{S}|$. An episode X is *σ-frequent* in $\mathcal{S}$ if $freq_{\mathcal{S},w}(X) \geq \sigma$. We denote the set of all σ-frequent episodes occurring in $\mathcal{S}$ by $\mathcal{F}_{\mathcal{S},w,\sigma}$.

Definition 3. Frequent Episode Mining Problem for $\mathcal{C}$:
Let $\mathcal{C}$ be a subclass of episodes we consider. Given an input sequence $\mathcal{S} \in (2^\Sigma)^*$, a window width $w \geq 1$, and a minimum frequency threshold $\sigma \geq 1$, the task is to find all of the σ-frequent episodes X within class $\mathcal{C}$ that occur in $\mathcal{S}$ with a window width w without duplication.

Our goal is to design an efficient algorithm for the frequent episode mining problem for the class of k-partite episodes, which we will introduce in the next section, in the framework of enumeration algorithms [4]. Let N be the total input size and M the number of all solutions. An enumeration algorithm $\mathcal{A}$ is of *output-polynomial time*, if $\mathcal{A}$ finds all solutions $S \in \mathcal{S}$ in total polynomial time both in N and M. Also $\mathcal{A}$ is of *polynomial delay*, if the *delay*, which is the maximum computation time between two consecutive outputs, is bounded by a polynomial in N alone.

3 k-Partite Episodes

In this section, we introduce the class of k-partite episodes for any $k \geq 0$ and other notions and discuss their properties.

In this paper, we regard an event $e \in \Sigma$ as the episode $X = (\{v\}, \emptyset, g)$ such that $g(v) = e$. Similarly, we regard a set of event $\{e_1, \ldots, e_n\} \subseteq \Sigma$ ($n \geq 0$) as a *parallel episode*, that is, an episode $X = (\{v_1, \ldots, v_n\}, \emptyset, g)$ such that $g(v_i) = e_i$ for every $1 \leq i \leq n$ [11]. Then, we call an episode $X = (\emptyset, \emptyset, g)$ with an empty vertex set an *empty episode* and denote X by $\emptyset$.

Definition 4. For $k \geq 1$, a *k-serial episode* (or a *serial episode*) over Σ is an episode $X = (V, E, g)$ satisfying that $V = \{v_1, \ldots, v_k\}$, $E = \{(v_i, v_{i+1}) \mid 1 \leq i < k\}$ and $g(v_i) = a_i$ for every $1 \leq i \leq k$. We denote such a k-serial episode by a sequence $(a_1 \mapsto \cdots \mapsto a_k)$ of events $a_1, \ldots, a_k \in \Sigma$.

Definition 5. For $k \geq 1$, a *k-partite episode* (or a *partite episode*) over Σ is an episode $X = (V, E, g)$ satisfying the following conditions (i) – (iii):

(i) $V = V_1 \cup \cdots \cup V_k$, where $V_i \neq \emptyset$ and $V_i \cap V_j = \emptyset$ for every i and j $(1 \leq i < j \leq k)$.
(ii) X is *complete*, i.e., $E = (V_1 \times V_2) \cup \cdots \cup (V_{k-1} \times V_k)$ holds.
(iii) X is *partwise-linear*, i.e., for every $1 \leq i \leq k$, the set V_i contains no distinct vertices with the same labeling by g.

We denote such a k-partite episode by an k-tuple $X = \langle A_1, \ldots, A_k \rangle$, where $A_i \subseteq \Sigma$ is a set of events for every $1 \leq i \leq k$.

For example, we describe a 3-serial episode, a parallel episode, and a 3-partite episode on the alphabet $\Sigma = \{a, b, c\}$ in Fig. 1. In what follows, we denote the classes of k-serial, parallel, sectorial [8], diamond [7,9] and k-partite episodes over Σ by $\mathcal{SE}_k$, $\mathcal{PE}$, $\mathcal{SEC}$, $\mathcal{DE}$ and $\mathcal{PTE}_k$, respectively. For these subclasses of episodes, the following inclusion relation hold: (i) $\mathcal{SE}_1 \subseteq \mathcal{PE} \subseteq \mathcal{PTE}_1$, (ii) $\mathcal{SE}_2 \subseteq \mathcal{SEC} \subseteq \mathcal{PTE}_2$, (iii) $\mathcal{SE}_3 \subseteq \mathcal{DE} \subseteq \mathcal{PTE}_3$ and (iv) $\mathcal{SE}_k \subseteq \mathcal{PTE}_k$ ($k \geq 0$). In particular, for the 3-partite episode $X = \langle A_1, A_2, A_3 \rangle$, if both $|A_1| = 1$ and $|A_3| = 1$ then X is a diamond episode. Also, for the 2-partite episode $X = \langle A_1, A_2 \rangle$, if $|A_2| = 1$ then X is a sectorial episode.

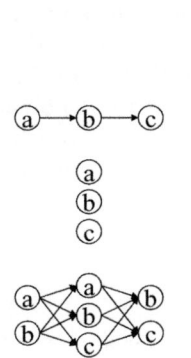

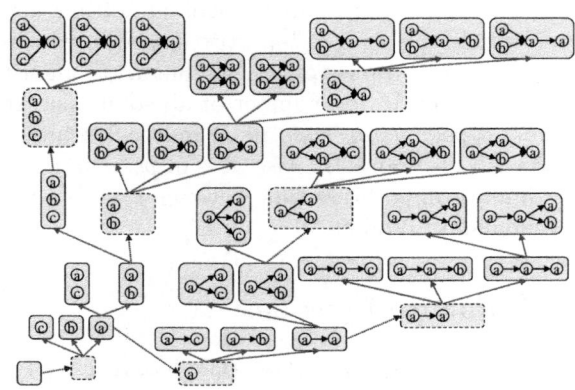

Fig. 1. The examples of a 3-serial episode (top), a parallel episode (center), and a 3-partite episode (bottom) on the alphabet $\Sigma = \{a, b, c\}$.

Fig. 2. The parent-child relationships on the alphabet $\Sigma = \{a, b, c\}$, where episodes in dashed boxes are pre-partite episodes.

4   Algorithm

4.1   Depth-First Enumeration of k-Partite Episodes

In this section, we present a polynomial-delay and polynomial-space algorithm KPAR for extracting all of the frequent k-partite episodes in an input event sequence. Throughout of this section, let $\mathcal{S} = (S_1, \ldots, S_n) \in (2^\Sigma)^*$ be an input event sequence, where $|\mathcal{S}| = n$ and $||\mathcal{S}|| = N$, $w \geq 1$ a window width and $\sigma \geq 1$ the minimum frequency threshold.

For a partite episode $X = \langle A_1, \ldots, A_{k-1} \rangle$ of the length $k-1 \geq 0$, we define the k-*pre-partite episode* of the length $k \geq 0$ as an episode $Y = \langle A_1, \ldots, A_{k-1}, \emptyset \rangle$. The main idea of our algorithm is to enumerate all of the frequent k-partite episodes by searching for the whole search space from general to specific by using depth-first search. For the search space, we introduce the parent-child relationship between k-partite episodes and k-pre-partite episodes, in order to output no k-pre-partite episodes.

Definition 6. The 0-partite episode $\perp = \langle \rangle$ is the *root*. For $1 \leq i \leq k$, the *parent* of the i-partite or i-pre-partite episode $X = \langle A_1, \ldots, A_i \rangle$ is defined by:

$$parent(\langle A_1, \ldots, A_i \rangle) = \begin{cases} \langle A_1, \ldots, A_{i-1} \rangle, & \text{if } A_i = \emptyset, \\ \langle A_1, \ldots, (A_i - \{\max A_i\}), & \text{otherwise.} \end{cases}$$

Also we define the set of all children of X by $Children(X) = \{Y \mid parent(Y) = X\}$. Then, we define the *family tree* for $\mathcal{PTE}_k$ by a rooted digraph $\mathcal{T}(\mathcal{PTE}_k) = (V, E, \perp)$ with the root $\perp$, where $V = \mathcal{PTE}_k \cup \{X \mid X$ is k-pre-partite episode $\}$ and $E = \{(X, Y) \mid X$ is the parent of $Y, Y \neq \perp\}$.

---

algorithm KPAR($\mathcal{S}, k, w, \Sigma, \sigma$)
input: input event sequence $\mathcal{S} = \langle S_1, \ldots S_n \rangle \in (2^\Sigma)^*$ of length $n \geq 0$,
maximum length of output partite episode $k \geq 0$, window width $w > 0$,
alphabet of events $\Sigma = \{1, \ldots, s\}$ ($s \geq 1$), the minimum frequency $1 \leq \sigma \leq n + k$;
output: the set of all σ-frequent k-partite episodes in $\mathcal{S}$ with window width k;
method:
1   **output** $\langle \rangle$;
2   KPARREC($\langle \emptyset \rangle, 1, \mathcal{S}, w, \Sigma, \sigma$);

procedure KPARREC($X, i, \mathcal{S}, k, w, \Sigma, \sigma$)
input: parent partite episode $X = \langle A_1, \ldots, A_i \rangle$,
and $\mathcal{S}$, k, w, Σ, and σ are same as in KPAR.
output: the set of all σ-frequent i-partite episodes
in $\mathcal{S}$ that are descendants of X and $i \leq k$;
method:
1   **if** ($i > k$) **then return**;
2    $f := $ COUNTBYSCAN($X, \mathcal{S}, w$);   // $f = |W_{\mathcal{S},w}(X)|$;
3   **if** ($f < \sigma$) **then return**;
4   **if** ($A_i \neq \emptyset$) **then**
5   **output** X;
6   KPARREC($\langle A_1, \ldots, A_i, \emptyset \rangle, i + 1, \mathcal{S}, w, \Sigma, \sigma$);
7   **end if**
8   **foreach** ($e \in \Sigma$ such that $e > \max(A_i)$) **do**
9   KPARREC($\langle A_1, \ldots, A_i \cup \{e\} \rangle, i, \mathcal{S}, w, \Sigma, \sigma$);

Fig. 3. The main algorithm KPAR and a recursive subprocedure KPARREC for mining frequent k-partite episodes in a sequence

Lemma 1. *The family tree $\mathcal{T}(\mathcal{PTE}_k) = (V, E, \perp)$ for $\mathcal{PTE}_k$ is a rooted tree with the root $\perp$.*

Proof. For any partite episode $X = \langle A_1, \ldots, A_k \rangle$ of length $k \geq 1$, it holds that $||parent(X)|| = ||X|| - 1$. For any pre-partite episode $X = \langle A_1, \ldots, A_{k-1}, \emptyset \rangle$ of length $k \geq 1$, it holds that $||parent(X)|| = ||X||$, and then, $parent(X) = \langle A_1, \ldots, A_{k-1} \rangle$ is a partite episode. Therefore, we can show that, for any vertex $X \in V$, X is reachable from $\perp$ by a path of the length at most $2||X||$. Moreover, we can show that $\mathcal{T}(\mathcal{PTE}_k)$ is acyclic. Hence, the statement holds. □

In Fig. 2, we describe the part of family tree $\mathcal{T}(\mathcal{PTE}_3)$ forms the spanning tree for all 3-partite episodes of $\mathcal{PTE}_3$ on the alphabet $\Sigma = \{a, b, c\}$.

In Fig. 3, we describe the algorithm KPAR and its subprocedure KPARREC for extracting frequent k-partite episodes from an input event sequence $\mathcal{S}$. The algorithm is a backtracking algorithm that traverses the spanning tree $\mathcal{T}(\mathcal{PTE}_k)$ based on depth-first search starting from the root $\perp$ using the parent-child relationships over $\mathcal{PTE}_k$.

4.2   Basic Algorithm with Frequency Counting by Scanning

Let $\mathcal{S} = (S_1, \ldots, S_n) \in (2^\Sigma)^*$ be an input event sequence of length n. For $1 \leq i \leq j \leq n$, we denote the subsequence $\mathcal{S}[i..j] = (S_i, \ldots, S_j)$ by $\mathcal{S}[i..j]$. Also

---

procedure CountByScan($X, \mathcal{S}, w$)
input: k-partite episode $X = \langle A_1, \ldots, A_k \rangle$, an input sequence $\mathcal{S} = \langle S_1, \ldots S_n \rangle$,
window width $w > 0$;
output: the frequency of X;
method:
1    $f := 0$;
2   **for** ($i := -w + 1, \ldots, n$) **do**
3   Test if $X \sqsubseteq \mathbf{W}_i^{\mathcal{S}, w} = \langle S_i, \ldots, S_{i+w-1} \rangle$ by the procedure in Lemma 2.
4   **if** the answer is "Yes" **then** $f := f + 1$;
5   **end for**
6   **return** f;

Fig. 4. An algorithm CountByScan for computing the frequency of a k-partite episode

let X be a k-partite episode over Σ and w a window width. Then, the *matching problem of X in $W = \mathcal{S}[i..j]$ against w* is to determine whether or not there exists a contiguous subsequence W' of W of length w such that $X \sqsubseteq W'$. Let $1 \leq i \leq n$ be any position. The *leftmost tail occurrence* of a k-partite episode $X = \langle A_1, \ldots, A_k \rangle$ w.r.t. the right boundary i is the smallest index $i \leq j \leq n$ such that X is contained in the prefix $\mathcal{S}[i..j]$, i.e., $X \sqsubseteq \mathcal{S}[i..j]$.

Lemma 2. *For a k-partite episode $X = \langle A_1, \ldots, A_k \rangle$ over Σ, an input sequence $\mathcal{S}$ of length n and a position $1 \leq r \leq n$, suppose that $P = (p_1, \ldots, p_k)$ is the list of positions such that, for every $1 \leq i \leq k$, p_i is the leftmost tail occurrence of i-partite episode $X_i = \langle A_1, \ldots, A_i \rangle$ w.r.t. the right boundary r. Also let $p_0 = r$. Then, we have the following statements:*

1. *P is increasing, i.e., it holds that $p_0 < p_1 < \cdots < p_k$.*
2. *For every $1 \leq i \leq k$, it holds that $p_i = \max_{e \in A_i} \min\{ j \mid p_{i-1} < j \leq n, e \in S_j \}$.*
3. *The list P can be computed in $O(|\Sigma|n)$ time.*

Thus, the matching problem for k-partite episodes against a window width w can be solved in $O(|\Sigma|w)$ time. The algorithm CountByScan in Fig. 4 computes the frequency count for a k-partite episode X based on the above lemma.

Lemma 3. *Let $\mathcal{S}$ be an input event sequence of length n and w a window width. Then, the algorithm CountByScan in Fig. 4 computes the frequency $freq_{\mathcal{S}, w}(X) = |\mathbf{W}_{\mathcal{S}, k}(X)|$ of a partite episode X in $O(|\Sigma|wn)$ time.*

To estimate the delay of the algorithm precisely, we adopt the compact representation of the current episode X by storing only the difference of X from its parent. Moreover, we use the alternating output technique of [14] to reduce the delay by factor of the depth of the search tree. Then:

Theorem 1. *Let S be an input event sequence of length n. Then, the algorithm* KPAR *in Fig. 3 with* COUNTBYSCAN *finds all of the σ-frequent k-partite episodes occurring in S without duplication in $O(|\Sigma|^2 wn)$ delay and in $O(|\Sigma|k)$ space.*

Proof. At each iteration of the algorithm KPARREC, the algorithm computes the frequency count f in $O(|\Sigma|wn)$ by Lemma 3 and executes other instructions than the invocation of KPARREC within the same cost. Since, each frequent episode has at most $O(|\Sigma|)$ infrequent children, the running time per a σ-frequent k-partite episode is $O(|\Sigma|^2 wn)$ time. □

4.3   Modified Algorithm with Frequency Counting on Event Lists

In Fig. 5, we describe a modified version of our algorithm KPAR2 and its sub-procedure KPARREC2 for extracting frequent k-partite episodes from input sequence S.

The key idea of KPAR2 is a subprocedure COUNTBYLIST that computes the frequency counts using so-called event lists. For an event $e \in \Sigma$, the *event list* of e in an input event sequence S is an increasing list $L[e] = (p_1, \ldots, p_m)$ of positions $1 \le p \le n$ such that $e \in S_i$. The *event list table* $\mathcal{L}$ in S is the set $\mathcal{L} = \{L[e]\}_{e \in \Sigma}$ of event lists for all events in Σ. Under the assumption that every event in Σ appears at least once in S, the total number of elements in $\mathcal{L}$ equals to N. A position $1 \le i \le n$ is said to be *appropriate* in S if S_i is not an empty set. Without loss of generality, we can assume that the list $\mathcal{A} = (i_1, \ldots, i_h)$ $(h \ge 0)$ of the appropriate positions in S is given.

Lemma 4. *The event list table $\mathcal{L}$ in S can be computed in $O(N)$ time.*

Proof. We encode a event list $L[e] = (p_1, \ldots, p_h)$ $(h \ge 0)$ by a triple $Head = 0$, $POS[0..h-1]$ and $NEXT[0..h-1]$ such that, for each pointer $\pi \in [0..h-1]$, $POS[\pi]$ is the position pointed by π and $\pi' = NEXT[\pi]$ is the next pointer of π. Then, the algorithm in Fig. 6 computes $\mathcal{L}$ in S. The time complexity of the algorithm is obviously $O(N)$ time. □

---

algorithm KPAR2(S, k, w, Σ, σ)
input: an input event sequence $S = \langle S_1, \ldots S_n \rangle \in (2^\Sigma)^*$ of length $n \ge 0$,
an integer $k \ge 0$, a window width $w > 0$, an alphabet Σ,
a minimum frequency $1 \le \sigma \le n + k$;
output: the set of all σ-frequent k-partite episodes in S with window width w;
method:
1   Compute the event list table $\mathcal{L} = \{L[e]\}_{e \in \Sigma}$ in S;
2   **output** $\langle \rangle$;
3   KPARREC2($\langle \emptyset \rangle, 1, S, \mathcal{L}, w, \Sigma, \sigma$);
// KPARREC2 is same as KPARREC except that this calls COUNTBYLIST
// with $\mathcal{L}$ instead of COUNTBYSCAN;

Fig. 5. The modified algorithm KPAR2 for mining frequent k-partite episodes in a sequence

---

method:
1   **foreach** $e \in \Sigma$ **do** $last_e := -1; free_e := 0; Head_e := 0;$ **end foreach**
2   **for** $(i := i_1, \ldots, i_h)$ **do**
     // $(i_1, \ldots, i_h) = \mathcal{A}$ is the list of the appropriate positions in $\mathcal{S}$;
3   **foreach** $e \in S_i$ **do**
4            $p_e := free_e; free_e := free_e + 1; POS_e[p_e] := i;$
5   **if** $last_e \neq -1$ **then** $NEXT_e[last_e] := p_e;$
6            $last_e := p_e;$
7   **end foreach**
8   **foreach** $e \in S_i$ **do** $NEXT_e[last_e] := last_e;$
9   **return** $\mathcal{L} = (Head_e, POS_e, NEXT_e)_{e \in \Sigma};$

Fig. 6. The algorithm for computing the event list table $\mathcal{L}$ in $\mathcal{S}$

---

procedure LEFTMOSTTAIL(A: a set of events, $\mathcal{S}$: an input sequence of total size N):
method:
1    $i := 0;$
2   **foreach** $e \in A$ **do** $\pi_e := 0; LM[e] := 0;$ **end foreach**
3   **while** $i < n$ **do**
4   Let i be the next appropriate position such that $i > i_{last}$ and $S_i \neq \emptyset;$
5   **foreach** $e \in S_i \cap A$ **do**
6            $LM[e] := POS_e[\pi_e]; \ell_{\max} := \max(\ell_{\max}, LM[e]); \pi_e = NEXT_e[\pi_e];$
7   **end foreach**
8   **output** $(h, \ell_{\max})$ as the pair of the position $h = i + 1$
             and the leftmost tail position w.r.t. $h;$
9        $i_{last} := i;$
10   **end while**

Fig. 7. A streaming algorithm LEFTMOSTTAIL that computes all leftmost tail positions of a set A of events by scanning an input sequence from left to right

By regarding a set $A \subseteq \Sigma$ of events as a parallel episode, we can define the leftmost tail occurrence q of A w.r.t. a position p in $\mathcal{S}$ as before. Then, we have that $p = \max_{e \in A} \min\{ j \mid p \leq j \leq n, e \in S_j \}$.

Lemma 5. *Suppose that $S_0 = \Sigma$. Then, the algorithm* LEFTMOSTTAIL *in Fig. 7 computes the set of all distinct pairs (h, ℓ_h) of a position $1 \leq h \leq n$ and the corresponding leftmost tail position ℓ_h w.r.t. h in the increasing order of h in $O(N)$ time.*

In Fig. 9, we describe the algorithm COUNTBYLIST for computing frequency of episodes with an incremental version INCLEFTMOSTTAIL of LEFTMOSTTAIL in Lemma 5. The algorithm COUNTBYLIST is an algorithm that computes the size of occurrence window list $\mathbf{W}_{\mathcal{S},k}(X)$ for a k-partite episode X by sweeping from the heads of the event lists $\mathcal{L}$ to the tails of $\mathcal{L}$.

---

procedure INCLEFTMOSTTAIL(A: a set of events, ℓ: the left boundary,
$\qquad\qquad\qquad\qquad (LM[\cdot], \pi[\cdot], i_{last}, \ell_{max})$: internal state):
method:
1 $i := i_{last}$;
2 **if** $i \geq \ell$ **then return** $(\ell_{max}, (LM[\cdot], \pi[\cdot], i_{last}, \ell_{max}))$;
3 **while** $i < n$ **do**
4 $\qquad$ Let i be the smallest appropriate position such that $i > i_{last}$ and $S_i \neq \emptyset$;
5 $\qquad$ **foreach** $e \in S_i \cap A$ **do**
6 $\qquad\qquad LM[e] := POS_e[\pi[e]]; \ \ell_{max} := \max(\ell_{max}, LM[e]); \ \pi[e] = NEXT_e[\pi[e]]$;
7 $\qquad$ **end foreach**
8 $\qquad i_{last} := i$;
9 $\qquad$ **if** $i \geq \ell$ **then return** $(\ell_{max}, (LM[\cdot], \pi[\cdot], i_{last}, \ell_{max}))$;
10 **end while**
11 **return** $(n + 1, (LM[\cdot], \pi[\cdot], i_{last}, n + 1))$; $\quad$ //*failed!*

Fig. 8. An incremental version of LEFTMOSTTAIL that computes the leftmost tail position of a set A of events w.r.t. a given position ℓ

---

procedure COUNTBYLIST($X = \langle A_1, \ldots, A_K \rangle, w, \mathcal{S}, \mathcal{L}$)
input: k-partite episode X, window width $w > 0$,
an input sequence $\mathcal{S} = \langle S_1, \ldots S_n \rangle$, and occurrence lists $\mathcal{L}$ for events in $\mathcal{S}$;
output: the number f of windows in which X occurs;
method:
1 **for** $k := 1, \ldots, K$ **do**
2 $\qquad$ **foreach** $e \in A_i$ **do** $LM[k][e] := 0; \pi_e := 0; i_{last} := 0;$ **end foreach**
3 $\qquad State[k] := (LM[k][\cdot], \pi[\cdot], i_{last}, 0)$;
4 **end for**
5 $i := 0$;
6 **while** $i < n$ **do**
7 $\qquad$ Let i be the next appropriate position
$\qquad\qquad$ such that $i > i_{last}$ and $S_i \neq \emptyset$; $\ell_{max}[0] := i$;
8 $\qquad$ **for** $k := 1, \ldots, K$ **do**
$\qquad\qquad$ // *Note: The vector $State[k]$ represents the internal state for k.*
9 $\qquad\qquad (\ell_{max}[k], State[k]) := $ INCLEFTMOSTTAIL($A_k, \ell_{max}[k - 1], State[k]$);
10 $\qquad \ell_{max} := \ell_{max}[K]$;
$\qquad\qquad$ // *Note: ℓ_{max} is the leftmost tail occurrence for episode X w.r.t. position i;*
11 $\qquad$ **if** $\ell_{max}[k] > \ell_{last}$ **then**
12 $\qquad\qquad cont := i - \max(0, \ell_{max}[k] - w + 1); \ f := f + count$;
13 $\qquad$ **end if**
14 $\qquad \ell_{last} := i + w$;
15 $\qquad i_{last} := i$;
16 **end while**
17 **return** f;

Fig. 9. An algorithm COUNTBYLIST for computing the number of windows in which a k-partite episode occurs

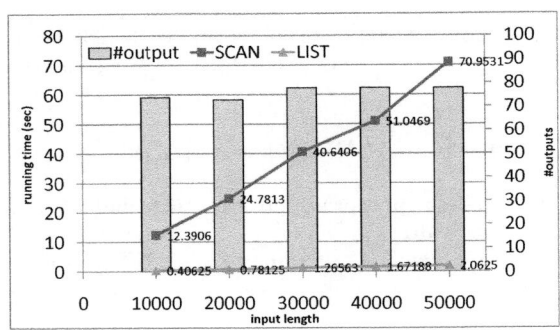

Fig. 10. Running time for the input length n, where $s = 4$, $p = 0.1$, $k = 4$, $w = 4$, and $\sigma = 0.01n$

Lemma 6. *Let $\mathcal{S}$ be an input sequence of length n and w a window width. Then, the algorithm* COUNTBYLIST *in Fig. 9 computes the frequency $freq_{\mathcal{S},w}(X)$ of a k-partite episode X in $O(N^2 K) = O(N^2 w)$ time, where $N = ||\mathcal{S}||$ is the total size of an input event sequence $\mathcal{S}$.*

Proof. In the algorithm COUNTBYLIST, the outer while-loop from line 6 to line 16 is executed $O(N)$ times and the inner for-loop from line 8 to line 9 is executed $O(K)$ times and the call to INCLEFTMOSTTAIL takes $O(N)$ time in the worst case. Therefore, the running time of COUNTBYLIST is $O(N^2 K)$ time.   □

By using alternating output technique [14], we show the following theorem on the complexity of the modified algorithm KPAR2.

Proposition 1. *Let $\mathcal{S}$ be an input event sequence of length n over Σ. Then, we can implement the algorithm* KPAR2 *to find all of the σ-frequent k-partite episodes in $\mathcal{S}$ without duplication in $O(|\Sigma| w N^2)$ delay and $O(|\Sigma| k + N)$ space, where $N = ||\mathcal{S}||$ is the total size of input event sequence $\mathcal{S}$.*

5   Experimental Results

In this section, we give the experimental results for the following combinations of the algorithms given in Section 4, by applying to the randomly generated event sequences.

Data: As randomly generated data, we adopt an event sequence $\mathcal{S} = (S_1, \ldots, S_n)$ over an alphabet $\Sigma = \{1, \ldots, s\}$ from four parameters (n, s, p), by generating each event set S_i $(i = 1, \ldots, n)$ under the probability $P(e \in S_i) = p$ for each $e \in \Sigma$.

Method: We implemented the following two algorithms given in Section 4:

SCAN : the algorithm KPAR with COUNTBYSCAN in Fig. 3
LIST : the algorithm KPAR2 with COUNTBYLIST in Fig. 5

All experiments were run in a PC (AMD Mobile Athlon64 Processor 3000+, 1.81GHz, 2.00GB memory, Window XP, Visual C++) with window width $w \geq 1$, maximum length of output partite episode $k \geq 0$ and minimum frequency threshold $\sigma \geq 1$.

Experiments. Fig. 10 shows the running time of the algorithms SCAN and LIST for the randomly generated event sequences from the parameter ($10000 \leq n \leq 50000, s = 4, p = 0.1$), where $k = 4$, $w = 4$ and $\sigma = 0.01n$. Then, time complexity of these algorithms seem to be linear in the input size and thus expected to scales well on large datasets. Furthermore, SCAN is 35 times as faster as LIST.

6   Conclusion

This paper studied the problem of frequent k-partite episode mining, and presented polynomial-delay and polynomial-space algorithms KPAR and KPAR2 that find all frequent k-partite episodes in an input sequence. Then, we have implemented our two algorithms KPAR and KPAR2 and given empirical results to compare the time and space efficiencies of the algorithms.

It is a future work to give more detailed analysis of the algorithms with different parameters. Also, it is a future work to apply the proposed algorithm to bacterial culture data [5,9]. Although Lemma 6 says that the time complexity of COUNTBYLIST is $O(N^2K)$ time, this can be improved to $O(Nw)$ time by more detailed analysis on the amortized time of the repeated execution of the subprecedure INCLEFTMOSTTAIL. This is another future work.

References

1. Agrawal, R., Srikant, R.: Fast algorithms for mining association rules in large databases. In: Proc. 20th VLDB, pp. 487–499 (1994)
2. Arimura, H.: Efficient algorithms for mining frequent and closed patterns from semi-structured data. In: Washio, T., Suzuki, E., Ting, K.M., Inokuchi, A. (eds.) PAKDD 2008. LNCS (LNAI), vol. 5012, pp. 2–13. Springer, Heidelberg (2008)
3. Arimura, H., Uno, T.: A polynomial space and polynomial delay algorithm for enumeration of maximal motifs in a sequence. In: Deng, X., Du, D.-Z. (eds.) ISAAC 2005. LNCS, vol. 3827, pp. 724–737. Springer, Heidelberg (2005)
4. Avis, D., Fukuda, K.: Reverse search for enumeration. Discrete Applied Mathematics 65, 21–46 (1996)
5. Katoh, T., Hirata, K.: Mining frequent elliptic episodes from event sequences. In: Proc. 5th LLLL, pp. 46–52 (2007)
6. Katoh, T., Hirata, K.: A simple characterization on serially constructible episodes. In: Washio, T., Suzuki, E., Ting, K.M., Inokuchi, A. (eds.) PAKDD 2008. LNCS (LNAI), vol. 5012, pp. 600–607. Springer, Heidelberg (2008)
7. Katoh, T., Arimura, H., Hirata, K.: A polynomial-delay polynomial-space algorithm for extracting frequent diamond episodes from event sequences. In: Theeramunkong, T., Kijsirikul, B., Cercone, N., Ho, T.-B. (eds.) PAKDD 2009. LNCS (LNAI), vol. 5476, pp. 172–183. Springer, Heidelberg (2009)

8. Katoh, T., Hirata, K., Harao, M.: Mining sectorial episodes from event sequences. In: Todorovski, L., Lavrač, N., Jantke, K.P. (eds.) DS 2006. LNCS (LNAI), vol. 4265, pp. 137–145. Springer, Heidelberg (2006)

9. Katoh, T., Hirata, K., Harao, M.: Mining frequent diamond episodes from event sequences. In: Torra, V., Narukawa, Y., Yoshida, Y. (eds.) MDAI 2007. LNCS (LNAI), vol. 4617, pp. 477–488. Springer, Heidelberg (2007)

10. Katoh, T., Hirata, K., Harao, M., Yokoyama, S., Matsuoka, K.: Extraction of sectorial episodes representing changes for drug resistant and replacements of bacteria. In: Proc. CME 2007, pp. 304–309 (2007)

11. Mannila, H., Toivonen, H., Verkamo, A.I.: Discovery of frequent episodes in event sequences. Data Mining and Knowledge Discovery 1, 259–289 (1997)

12. Pei, J., Wang, H., Liu, J., Wang, K., Wang, J., Yu, P.S.: Discovering frequent closed partial orders from strings. IEEE TKDE 18, 1467–1481 (2006)

13. Pei, J., Han, J., Mortazavi-Asi, B., Wang, J., Pinto, H., Chen, Q., Dayal, U., Hsu, M.-C.: Mining sequential patterns by pattern-growth: The PrefixSpan approach. IEEE TKDE 16, 1–17 (2004)

14. Uno, T.: Two general methods to reduce delay and change of enumeration algorithms, NII Technical Report, NII-2003-004E (April 2003)

15. Srikant, R., Agrawal, R.: Mining sequential patterns: Generalizations and performance improvements. In: Apers, P.M.G., Bouzeghoub, M., Gardarin, G. (eds.) EDBT 1996. LNCS, vol. 1057, pp. 3–17. Springer, Heidelberg (1996)

16. Uno, T., Asai, T., Uchida, Y., Arimura, H.: An efficient algorithm for enumerating closed patterns in transaction databases. In: Suzuki, E., Arikawa, S. (eds.) DS 2004. LNCS (LNAI), vol. 3245, pp. 16–31. Springer, Heidelberg (2004)

17. Zaki, M.J.: Scalable Algorithms for Association Mining. IEEE TKDE 12, 372–390 (2000)

18. Zaki, M.J., Hsiao, C.-J.: CHARM: An efficient algorithm for closed itemset mining. In: Proc. 2nd SDM, pp. 457–478. SIAM, Philadelphia (2002)

Learning from Positive Data Based on the MINL Strategy with Refinement Operators

Seishi Ouchi* and Akihiro Yamamoto

Graduate School of Informatics, Kyoto University
Yoshida Honmachi, Sakyo-ku, Kyoto, Japan 606-8501
akihiro@i.kyoto-u.ac.jp

Abstract. In the present paper we clarify the combination of the MINL (MINimal Langugae) strategy and refinement operators in the model of identification in the limit from positive data, by giving a learning procedure in a general form adopting both of the two. The MINL strategy is to choose minimal concepts consistent with given examples as guesses, and has been adopted in many previous works in the model. The minimality of concepts is defined w.r.t. the set-inclusion relation, and so the strategy is semantic-based. Refinement operators have developed in the field of learning logic programs to construct logic programs as hypotheses consistent with logical formulae given as examples. The operators are defined based on inference rules in first-order logic and so are syntactical. With the proposed procedure we give such a new class of tree pattern languages that every finite unions of the languages is identifiable from positive data without assuming the upperbound of the number of unions. Moreover, we revise the algorithm so that we can show that the class is polynomial time identifiable.

1 Introduction

The goal of this research is to give a new view to learning formal langugages in the model of identification in the limit from positive data [1,5], by adopting the MINL (MINimal Langugae) strategy and refinement operators in one learning procedure. The MINL strategy, which means to choose minimal concepts consistent with given examples as guesses in a learning process, was adopted in many learning proedures in the model (see e.g. [1,16]). Refinement operators [8,14] were developed for learning logic programs from logical formulae given as examples. The MINL strategy is semantical since the minimality of concepts is defined w.r.t. the set-inclusion relation, while refinement operators are syntactically defined with inference rules in first-order logic. In the field of symbolic logic, it is usual to investigate the relations between syntacts and semantics. Both of the MINL strategy and refinement operators are concerned with ordering in generating guesses and they should be much related.

Shinohara et al. [15] combined the two for designing learning procedure for specific classes of pattern languages, but in the present paper we give a procedure

---

* Current Affiliation: NTT Communications Corporation.

K. Nakakoji, Y. Murakami, and E. McCready (Eds.): JSAI-isAI, LNAI 6284, pp. 345–357, 2010.
© Springer-Verlag Berlin Heidelberg 2010

in a general form to reveal the relation between the two more clearly. Moreover, we apply the combination to give such a new class of tree pattern languages that every finite unions of the concepts in the class is identifiable from positive data without assuming the upperbound of the number of unions. Generally speaking, learning bounded unions of concepts has been investigated by using the finite elasticity property of a concept class [11,17], and leaning unbounded unions have treated with Higman's Thoerem [16]. In both types of the classes, it is hard to construct precise learning algorithms from theory. The concept class that we present in the paper has not the finite elasticity property, but unbounded unions of the concepts in it are identifiable. Moreover, we present a polynomial update time identification algorithm by revising redundancies of the generic one.

2 Preliminaries

The symbol $\mathbb{N}$ denotes the set of natural numbers $\{0, 1, 2, \ldots\}$, and $\mathbb{N}^+$ the set of positive natural numbers $\{1, 2, 3, \ldots\}$. The set of rational numbers is denoted by $\mathbb{Q}$. For a set A, $card(A)$ denotes the cardinality of A. For sets A and B, we write $A \subseteq B$ if A is a subset of B and $A \subset B$ if A is a proper subset of B. For an infinite sequence $\sigma = \langle s_1, s_2, \ldots \rangle$, $\sigma[n]$ denotes the sequence $\langle s_1, \ldots, s_n \rangle$, and $content(\sigma)$ denotes the set $\{s_1, s_2, \ldots\}$.

Let X be a recursively enumerable set of *objects*. A *concept* is a subset of X. A *concept class* (or *identification problem*) is a triplet $(\mathcal{C}, \mathcal{H}, L(\cdot))$, where $\mathcal{C}$ is a set of concepts over X, $\mathcal{H}$ is a recursively enumerable set of *hypotheses*, $L(\cdot)$ is a mapping from $\mathcal{H}$ to $\mathcal{C}$, which satisfies the following conditions:

 – $\mathcal{C} = \{L(h) \mid h \in \mathcal{H}\}$.
 – There is a recursive function $f : X \times \mathcal{H} \to \{0, 1\}$ such that $f(w, h) = 1$ iff $w \in L(h)$. Such $\mathcal{C}$ is called an *indexed family of recursive concepts*.

If $L(h) = C$ for a hypothesis h and a concept C, we call h a *description* of C. We often do not distinguish a concept class and a set of concepts, and use the symbol $\mathcal{C}$ for them.

Let $(\mathcal{C}, \mathcal{H}, L(\cdot))$ be a concept class. A *learning machine* M is a procedure which receives a member of a given $C \in \mathcal{C}$ as a *positive example*, and outputs a hypothesis in $\mathcal{H}$ calculated from already received positive examples, at evey time of iteration.

$M(\sigma[i])$ denotes the ith guess of a learning machine M which receives a sequence of positive examples σ. If $L(M(\sigma[n])) \supseteq content(\sigma[n])$ for any σ and $n \in \mathbb{N}^+$, we say that M is *consistent*. If $L(M(\sigma[n])) \supseteq content(\sigma[n+1])$ implies $M(\sigma[n]) = M(\sigma[n+1])$ for any σ, we call M is *conservative*.

We call σ a *positive presentation* of a concept $C \in \mathcal{C}$ if $content(\sigma) = C$. If there exists a time $N \in \mathbb{N}^+$ such that $M(\sigma[i]) = M(\sigma[N])$ for every $i \geq N$, we say M *converges* to $M(\sigma[N])$ for σ. If M converges to an description of $C \in \mathcal{C}$ for any positive presentation of C, we say M *identifies in the limit* a concept C *from positive data*. M *identifies in the limit* a concept class $\mathcal{C}$ *from positive data* iff M identifies in the limit any $C \in \mathcal{C}$ from positive data. A concept class

C is *identifiable in the limit from positive data* iff there exists a learning machine that identifies in the limit C from positive data. A concept class C is *polynomial time identifiable* in the limit from positive data iff there exists a consistent and conservative learning machine M that identifies in the limit C from positive data and the time of M producing a guess h after receiving a positive example $e_1, \ldots, e_n$ is at most a polynomial of $|h|$ and $|e_1| + \cdots + |e_n|$, where $|e|$ denotes the size of e.

A *finite tell-tale set* of a concept $C \in C$ is a finite subset T of C such that $T \subseteq C' \in C$ implies $C' \not\subset C$. The concept class, C is identifiable in the limit from positive data iff there exists an effective procedure which enumerates a finite tell-tale set of $L(h)$ from $h \in \mathcal{H}$ [1].

A *characteristic set* T of a concept C in C is a finite subset of C such that $T \subseteq C'$ implies $C \subseteq C'$ for every $C' \in C$ [2,7]. The concept class C has the *characteristic set property* if every $C \in C$ has a characteristic set in C. Kobayashi [7] proved that if every language $C \in C$ has a characteristic set then the class C is identifiable in the limit from positive data. Characteristic sets contributes not only proving the learnability of a concept class C but also designing learning algorithms.

For a finite set E of objects, a hypothesis h is *minimal* w.r.t. E if $E \subseteq C' \subset L(h)$ for no other concept $C' \in C$.

Definition 1. We say a machine M for learning concepts in C adopts the *MINL strategy* if it outputs a minimal hypothesis h, if exists, for every set E of any subset of any concept in C.

Sakakibara et al. [13] pointed out that, if the concept class has the characteristic set property, there is a learning machine which adopts the MINL strategy and identifies in the limit C from positive data.

Proposition 1 ([13]). *If C has the characteristic set property, any learning machine for C adopting the MINL strategy identifies every concept in C in the limit.*

3   Combining the MINL Strategy and Refinement Operators

3.1   Main Results

Refinement operators were first introduced by Shapiro [14] for his theory and system of learning logic programs, and precisely investigated by Laird [8]. Based on the works we give a revised definition of refinement so that it fits our purpose of combining it with the MINL strategy.

Definition 2. Let $(C, \mathcal{H}, L(\cdot))$ be a concept class. A mapping $\rho \colon \mathcal{H} \to 2^{\mathcal{H}}$ is called a *refinement operator* on the class if it satisfies the following four:

[R-1] For every $h \in \mathcal{H}$, $\rho(h)$ is recursively enumerable.
[R-2] $g \in \rho(h) \Rightarrow L(g) \subseteq L(h)$.
[R-3] There exists no sequence $h_1, \ldots, h_n$ of hypotheses such that $h_1 = h_n$ and $h_{i+1} \in \rho(h_i)(1 \leq i \leq n - 1)$.

For a hypothesis h, every element of $\rho(h)$ is called a *refinement* of h.

The condition [R-3] is weaker than the corresponding one in [14], which requests that there is no infinite sequence $h_1, h_2, \ldots$ such that $h_i \in \rho(h_{i+1})(i \in \mathbb{N}^+)$. The definition in [8] does not have any condition on such existence of infinite sequences.

For a hypothesis h and $k \in \mathbb{N}$ the set $\rho^k(h)$ is inductively defined as

$$\rho^0(h) = \{h\}, \text{ and}$$
$$\rho^{k+1}(h) = \{h' \in \mathcal{H} \mid \text{there exists } h'' \in \rho^k(h) \text{ such that } h' \in \rho(h'')\} \text{ for } k \geq 0.$$

We also define $\rho^+(h) = \bigcup_{k \in \mathbb{N}^+} \rho^k(h)$ and $\rho^*(h) = \bigcup_{k \in \mathbb{N}} \rho^k(h)$. For a set of hypotheses H, we write $\rho(H) = \{g \in \mathcal{H} \mid \text{there exists } h \in H \text{ such that } g \in \rho(h)\}$.

For our convenience we relate a refinement operator to a directed graph. With a refinement operator ρ for $(\mathcal{C}, \mathcal{H}, L(\cdot))$, we define a directed graph by letting $\mathcal{H}$ be the set of nodes and every pair of $h \in \mathcal{H}$ and $g \in \rho(h)$ be its edge. The graph is called a *refinement graph* induced by ρ. A *refinement path* is a sequence $h_1, h_2, \ldots$ such that $h_{i+1} \in \rho(h_i)(i \geq 1)$. We also define two classes of refinements as in [8].

Definition 3. A refinement operator ρ on $(\mathcal{C}, \mathcal{H}, L(\cdot))$ is

- *the locally finite* if $\rho(h)$ is finite for $h \in \mathcal{H}$ and there is a algorithm which outputs the list of elements in $\rho(h)$, and
- *semantically complete* if, for every h and every $C_i \subset L(h)$, there exists a finite sequence $h_1, \ldots, h_n$ such that $h_1 = h$, $L(h_n) = C_i$, and $h_{i+1} \in \rho(h_i)$ $(1 \leq i \leq n-1)$.

We give the main theorem of the paper.

Theorem 1. *Let ρ be a refinement operator on a concept class $(\mathcal{C}, \mathcal{H}, L(\cdot))$ which satisfies the following four conditions.*

[A-1] *ρ is locally finite.*
[A-2] *ρ is semantically complete.*
[A-3] *A finite set $T \subseteq \mathcal{H}$ is given such that $\{\rho^*(h) \mid h \in T\} = \mathcal{H}$.*
[A-4] *There is no infinite sequence $h_1, h_2, \ldots$ such that $\rho(h_i) = h_{i+1}$, and $L(h_i) = L(h_{i+1})$ for $i \geq 1$.*

Then the concept class is identifiable in the limit from positive data. Moreover, there is a conservative learning machine for the identification.

The theorem is proved by showing the conservative procedure illustrated in Fig. 1.

The next lemma is directly proved from the assumptions in Theorem 1.

Lemma 1. *Let C be any concept in $\mathcal{C}$. If all of the assumptions [A-1]–[A-4] hold, there is a hypothesis $h \in \mathcal{H}$ satisfying the following three:*

[M-1] *$L(h) = C$,*
[M-2] *there exists $k \in \mathbb{N}$ such that $h \in \bigcup_{g \in T} \rho^k(g)$, and*
[M-3] *for all $h' \in \rho(h)$ it holds that $C \not\subseteq L(h')$.*

Procedure Learn-with-Refinement-and-MINL($\mathcal{H}, \rho$)
Require A positive presentation $\sigma = e_1, e_2, \ldots$ of C
Ensure An enumeration of $\mathcal{H}$
    1: $S := \emptyset$, $L(h_0) = \emptyset$
    2: **for** $i = 1$ to ∞
    3:   $S := S \cup \{e_i\}$
    4:   **if** $e_i \in L(h_{i-1})$
    5:      $h_i := h_{i-1}$
    6:   **else if** $\text{MINL}(T, S, i) =$ "no hypothesis"
    7:      $h_i := h_{i-1}$
    8:   **else**
    9:      $h_i := \text{MINL}(T, S, i)$
   10:    output h_i as a guess

Algorithm $\text{MINL}(T, S, n)$
Require A finite set T in Theorem 1, any finite subset S of any concept C, and a natural number n.
Ensure a hypothesis in $\mathcal{H}$ or "no hypothesis
    1: $H_0 := T$
    2: **for** $j = 0$ to n
    3:   **if** $\exists h \in H_j.[S \subseteq L(h) \wedge \forall g \in \rho(h). S \not\subseteq L(g)]$
    4:      Return the hypothesis which satisfies the condition above
           and is firstly enumerated
    5:   $H_{j+1} := \{g \in \mathcal{H} \mid g \in \rho(H_j) \wedge S \subseteq L(g)\}$
    6: return "no hypothesis"

Fig. 1. Learning with the MINL strategy and a refinement operator

Proof of Theorem 1. At first we can assume a total ordering on $\mathcal{H}$, which gives the enumeration of all hypotheses in it. The ordering might be irrelevant to the refinement operator. From the condition [A-1] and the definition of Algorithm $\text{MINL}(T, S, n)$, the set H_i in the algorithm is finite. We give a numbering for each hypothesis in H_i in the following way:

$$T = \{h_1, \ldots, h_{k_0}\},$$
$$\rho^1(T) = \{h_{k_0+1}, \ldots, h_{k_1}\},$$
$$\vdots$$
$$\rho^j(T) = \{h_{k_{j-1}+1}, \ldots, h_{k_j}\}, \ldots,$$

where hypotheses in each $\rho^j(T)$ are indexed according to the assumed total order.

Let C be any concept in $\mathcal{C}$ and h_N be the hypothesis which satisfies all of [M-1]–[M-3] in Lemma 1 and appears first in the numbering. Let S be any finite subset of positive examples of any intended concept C.

At first we consider every hypothesis h_i for $i < N$. We consider the following three cases:

- In the case that $C \subset L(h_i)$, the condition [A-2] assures a refinement path from h_i to a hypothesis h representing C, and every hypothesis g on the path includes any subset of C. Thus there exists $g \in \rho(h_i)$ such that $S \subseteq L(g)$.
- If $C = L(h_i)$ and $S \nsubseteq L(g)$ for all $g \in \rho(h_i)$, then it holds that $S \nsubseteq L(g) \Rightarrow C \nsubseteq L(g)$, but this contradicts the definition of h_N.
- If $C \nsubseteq L(h_i)$, there exists an example e such that $e \in C$ and $e \notin L(h_i)$, and therefore it holds that $S \nsubseteq L(h)$ after e appears in the positive presentation σ.

Concluding the analysis of the three cases, we know that for every hypothesis h_i with $i < N$ there exists t_i such that

$$t \geq t_i \Rightarrow [S \nsubseteq L(h_i) \vee \exists g \in \rho(h_i).\, S \subseteq L(g)].$$

Next we consider the refinements of h_N. From the condition [M-3] it holds that, for every $g \in \rho(h_N)$, there exists a positive example e such that $e \in C$ and $e \notin L(g)$ and it holds that $S \nsubseteq L(g)$ after e appears in the positive presentation σ. Thus there exists t_N such that

$$t \geq t_N \Rightarrow [S \subseteq L(h_N) \wedge \forall g \in \rho(h_N).\, S \nsubseteq L(g)].$$

This means that at any time $t \geq \max\{t_1, \ldots, t_N\}$ there exists j such that $h_N \in H_j$ in Algorithm $\mathrm{MINL}(T, S, n)$ and it returns h_N.

Now we show that the outputs of the procedure converges to some hypothesis representing C. We have shown that the procedure never outputs h such that $C \subset L(h)$. Let h be the hypothesis at a time $\max\{t_1, \ldots, t_N\}$. If $C = L(h)$, the procedure never changes its guess. If $C \nsubseteq L(h)$, there exists a positive example e such that $e \in C$ and $e \notin L(h_i)$ and so the procedure outputs h_N when e is provided. $\qquad\square$

For preserving consistency of our learning procedure, we need existence of a minimal language for any set of examples, but this assumption makes the algorithm MINL simpler.

Corollary 1. *Assume* [A-1]–[A-4], *and additionally assume*

[A-5] *There exists a minimal hypothesis for any set of examples.*

Then we can substitute n with ∞ in lines 6 and 9 of Learn-with-Refinement-and-MINL in Fig. 1. In the case the procedure identifies C conservatively and consistently in the limit from positive data.

The conjunction of assumptions [A-1]–[A-4] does not imply the characteristic-set property. Let us define concepts of subsets of the set of rationales $\mathbb{Q}$ as follows:

$$C_0 = \{x \in \mathbb{Q} \mid 0 \leq x \leq 1\},$$

$$C_{2n} = \{x \in \mathbb{Q} \mid 0 \leq x \leq 1 + \frac{1}{n}\},$$

$$C_{2n-1} = \{x \in \mathbb{Q} \mid 0 \leq x \leq 1 + \frac{1}{n}\} - \{\frac{1}{n}\}.$$

We use a natural number $i \in \mathbb{N}^+$ as a hypothesis representing C_i. For any finite subset S of C_0, there exists m such that $1/m \in \{1/n \mid n \in \mathbb{N}^+\} - S$, and $C_0 \not\subseteq C_m$. This leaves the characteristic set property from $\mathcal{C}$. Next we define a refinement operator ρ as

$$\rho(0) = \emptyset, \ \rho(2n) = \{0, 2n-1, 2n+2\}, \ \text{and} \ \rho(2n-1) = \emptyset \ \text{for} \ n \geq 1.$$

This operator satisfies all of [A-1]–[A-4], and therefore $\mathcal{C}$ is identifiable in the limit from positive data.

4   Learning Unbounded Unions of Concepts

As an application of Theorem 1 we treat unbounded unions of concepts.

Definition 4. For a concept space $(\mathcal{C}, \mathcal{H}, L(\cdot))$ where $\mathcal{C} = \{C_1, C_2, \ldots\}$, we define the concept space as

$$\mathcal{C}^* = \{\bigcup_{i \in A} C_i \mid A \subset \mathbb{N}^+ \wedge card(A) < \infty\},$$

$$\mathcal{H}^* = \{\bigcup_{i \in A} \{H_i\} \mid A \subset \mathbb{N}^+ \wedge card(A) < \infty\}, \ \text{and}$$

$$L(\bigcup_{i \in A} \{H_i\}) = \bigcup_{i \in A} L(H_i)$$

We call the concept space *the unbounded union* of $\mathcal{C}$.

4.1   Tree Pattern Languages

We follow the previous works [3,6] for terminology and concepts on tree pattern languages. Let Σ be a finite set of *function symbols*, and let V be a countable set of *variables*. All function symbols are connected with a non-negative integer (that is called *arity*) by a mapping $arity(\cdot) : \Sigma \to \mathbb{N}$. Σ must contain at least one function symbol whose arity is 0. We call such symbols *constant symbols*. The triplet $(\Sigma, V, arity(\cdot))$ is called a *signature*. *Tree patterns* on the signature is defined as follows:

(1) $c \in \Sigma$ and $x \in V$ are tree patterns.
(2) When $f \in \Sigma$ is a n-ary function symbol and $t_1, \ldots, t_n$ are tree patterns, $f(t_1, \ldots, t_n)$ is a tree pattern.

$|p|$ denotes the number of symbols in a tree pattern p. $|p|_{>0}$ denotes the number of function symbols whose arity is greater than 0 in a tree pattern p. $var(p)$ denotes the set of variables appearing in a tree pattern p. Tree patterns without a variable called *ground tree patterns*. $\mathcal{TP}$ denotes the set of all tree patterns.

A *substitution* is a finite set $\{x_1/q_1, \ldots, x_n/q_n\}$. x_i are distinct variables, and q_i are tree patterns different from x_i. If p is a tree pattern and $\theta = \{x_1/q_1, \ldots, x_n/q_n\}$ is a substitution, $p\theta$ is a tree pattern obtained by replacing all occurrences of $x_1, \ldots, x_n$ in p by $q_1, \ldots, q_n$. Let p, q be tree patterns. If there exists a substitution θ such that $p = q\theta$, We write $p \preccurlyeq q$ (or $q \succcurlyeq p$). If $p \preccurlyeq q$ and $p \succcurlyeq q$, we write $p \equiv q$. Such tree patterns are equivalent up to renaming of variables. We assume this equivalence when we treat a set of tree patterns as hypothesis space of a identification problem.

Let p be a tree pattern. A *tree pattern language* expressed by p is a set of tree patterns $L(p) = \{t \mid t \preccurlyeq p, t \text{ is ground tree pattern}\}$. A member of $L(p)$ is called a ground tree pattern generated by p.

A *principal symbol* of a tree pattern p is defined as follows:

(1) If p is variable or constant symbol, p's principal symbol is p.
(2) If $p = f(q_1, \ldots, q_n)$ where f is a function symbol with arity $n > 0$ and $q_1, \ldots, q_n$ are tree patterns, p's principal symbol is f.

Definition 5. A sub tree pattern of a tree pattern p is inductively defined as follows:

1. p itself is a sub tree pattern of p.
2. If $q = f(r_1, \ldots, r_n)$ is a sub tree pattern of p where f is a function symbol with arity $n > 0$ and $r_1, \ldots, r_n$ are tree patterns, then $r_1, \ldots, r_n$ are sub tree patterns of p.

Definition 6. An occurrence is a sequence of natural numbers $I = \langle i_1, \ldots, i_n \rangle$. For a tree pattern p, we define an sub tree pattern $p(I)$ of the occurrences I as follows.

(1) If $I = \langle \rangle$, then $p(I) = p$.
(2) If $I = \langle n_1, \ldots, n_m, i \rangle$ and $p(\langle n_1, \ldots, n_m \rangle) = f(q_1, \ldots, q_i, \ldots, q_n)$, then $p(I) = q_i$.

A occurrence of p is an occurrence I such that there is a sub tree pattern $p(I)$. The depth of p, denoted by $depth(p)$, is the maximal length of occurrences of p. When the length of an occurrence I of p is n, we say the principal symbol of p occurs at depth n.

Proposition 2. *For tree patterns p and q, holding $p = q$ is equivalent to holding that the principal symbols of $p(I)$ coincides with that of $q(I)$ for every occurrence I.*

Proposition 3. *For tree patterns p and q, holding $p \succcurlyeq q$ is equivalent to holding that, for every occurrence I .*

(a) *if $p(I)$ is a variable x, $q(I) = q(J)$ for every J such that $p(J) = x$, and*
(b) *if the principal of $p(I)$ is a function symbol or constant symbol, then it coincides with that of $q(I)$.*

It is well-known that $L(\mathcal{TP})$ is identifiable from positive data because it has the finite thickness property, and that the so-called "anti-unification" algorithm firstly given by Plotkin [12] provides a polynomial time learning procedure. Arimura et al. [4] provided a polynomial time procedure which identifies $L(\mathcal{TP})^k$ under some restriction on the signature.

We introduce some classes of tree patterns.

Definition 7. A tree pattern p is *regular* if each variable occurs no more than once in it. The tree pattern is *constant free* if p has no constants in it.

The sets of regular tree patterns, constant-free patterns, and constant-free regular patterns are respectively denoted by $\mathcal{RTP}$, $\mathcal{TP}_{cf}$, and $\mathcal{RTP}_{cf}$.

Theorem 2. *Neither $L(\mathcal{TP})^*$ nor $L(\mathcal{RTP})^*$ are identifiable from positive data.*

Proof. Let $\Sigma = \{f(\cdot), a\}$ and consider $L(\{f(x)\})$. For any positive example S, there is a hypothesis consistent with S and represented in the form $L(\{f(a), f(f(a)), f(f(f(a))), \ldots, f(f \cdots f(a) \cdots)\})$. Such concepts are properly included by $L(\{f(x)\})$ and therefore $L(\{f(x)\})$ has no finite tell-tale. $\square$

Theorem 3. *There is a signature with which $L(\mathcal{TP}_{cf})^*$ is not identifiable from positive data.*

The theorem is proved by the fact that $L(\{f(x,y)\})$ has no finite tell-tale in the signature $\Sigma = \{f(\cdot, \cdot), a\}$.

The next is a fundamental lemma for tree pattern languages.

Lemma 2 (Lassez, et al.[10]). *For $p, q \in \mathcal{TP}$, $L(p) \subseteq L(q) \Leftrightarrow p \preccurlyeq q$.*

We can easily show the compactness for $L(\mathcal{RTP}_{cf})$.

Lemma 3. *Let $p, q_1, \ldots, q_n \in \mathcal{RTP}_{cf}$. If $L(p) \subseteq L(q_1) \cup \cdots \cup L(q_n)$, then there exists i $(1 \le i \le n)$ such that $p \preccurlyeq q_i$.*

Proof. Assume that $L(p) \subseteq L(q_1) \cup \cdots \cup L(q_n)$ and $p \not\preccurlyeq q_i$ for all $i(1 \le i \le n)$. We can see that $p, q_1, \ldots, q_n \in \mathcal{RTP}_{cf}$. From Proposition 3(b), there is an occurrence I_i of q_i for $p \not\preccurlyeq q_i$ satisfying either of the followings:

1. The principal function symbol of $q_i(I_i)$ is a function symbol f with arity $n > 0$, and $p(I_i)$ is a variable. (Note that $q_i \in \mathcal{RTP}_{cf}$ implies the arity $n > 0$).
2. The principal function symbol of $q_i(I_i)$ is a function symbol f, and that of $p(I_i)$ is a function symbol or constant symbol different from f.

Let t be a ground tree which is obtained by substituting a constant symbol for every variable in p. From the analysis above any ground tree $q_i\theta$ has a principal symbol at I_i different with at occurrence that of t. This means that $t \notin L(q_i)$ for $1 \le i \le n$, and contradicts the assumption. $\square$

By Lemmas 2, and 3, we obtain the next lemma.

Lemma 4. *Let $P, Q \in \mathcal{RTP}_{cf}^*$ and $\sqsubseteq$ be the Hoare ordering induced from $\preccurlyeq$. Then $L(P) \subseteq L(Q) \Leftrightarrow P \sqsubseteq Q$.*

From lemma we design a refinement operator ρ on $\mathcal{RTP}_{cf}^*$ such that $\rho^* = \sqsubseteq$.

Definition 8. For $p \in \mathcal{RTP}_{cf}$, we define $\hat{\rho}(p)$ is the set of all tree patterns q such that $q = p\{x/f_i(x_1, \ldots x_n)\}$ where x is a variable appearing in p, $arity(f_i) = n \neq 0$, and $x_1, \ldots, x_n$ are variables not occurring in p. We regard $\hat{\rho}$ as a mapping $\hat{\rho} : \mathcal{RTP}_{cf} \to 2^{\mathcal{RTP}_{cf}}$.

Proposition 4. *The mapping $\hat{\rho}$ in Definition 8 is a refinement operator on $\mathcal{RTP}_{cf}$, and is locally finite and semantically complete.*

Definition 9. We define a mapping $\rho : \mathcal{RTP}^*_{cf} \to 2^{\mathcal{RTP}^*_{cf}}$ by letting $\rho(P)$ for $P \in \mathcal{RTP}^*_{cf}$ be the set of all Q which satisfies at least one of the following two:

(1) Q is the set obtained by removing redundancy from $P \cup \hat{\rho}(p) - \{p\}$ for some $p \in P$.
(2) $Q = P - \{p\}$ for some $p \in P$.

Lemma 5. *The mapping ρ in Definition 9 is a refinement operator.*

Theorem 4. *$L(\mathcal{RTP}_{cf})^*$ is identifiable in the limit from positive data.*

Proof. It is sufficient to show that the refinement operator ρ in Definition 9 satisfies all of the assumptions [A-1]–[A-4] in Theorem 1. The previous lemmas shows that ρ satisfies [A-1] and [A-2], and the proof of Lemma 5 assures [A-4]. We can see that ρ satisfies [A-3] by putting $T = \{\{x\}\}$.

We can prove that $L(\mathcal{RTP}_{cf})^*$ has infinite elasticity. This means we cannot obtain the above Theorem in the method by Shinohara et al. [16], which is based on Higman's Theorem.

4.2   Polynomial Time Learning

By using the previous theorem, we improve the MINL algorithm in Fig. 1. The obtained algorithm is illustrated in Fig. 2 by replacing bredth-first search with depth-first, with the refinement operator defined in Definition 9. On Lines 5 and 6 the improved algorithm the refinement is applied to $H_1 \cup H_2 \cup H_3$. The improved algorithm and this makes $L(\mathcal{RTP}_{cf})^*$ to be polynomial time identifiable.

Lemma 6. *In the while loop of Algorithm MINL-RTP-CF* in Fig. 2, if $i = n$ then H_1 consists of tree patterns whose total number of function symbols is n.*

Lemma 7. *Let $P \in \mathcal{RTP}^*_{cf}$. For every $p \in P$, it holds that*

$$L(P) - L(P - \{p\} \cup \hat{\rho}(p)) = \{t \mid t \text{ is a ground tree}, t \preccurlyeq p, |t| = |p|\}.$$

Proof. From the definition of $\hat{\rho}$, it holds that

$$L(P) - L(P - \{p\} \cup \hat{\rho}(p)) \supseteq \{t \mid t \text{ is a ground tree}, t \preccurlyeq p, |t| = |p|\}.$$

Let $t \in L(P) - L(P - \{p\} \cup \hat{\rho}(p))$. Then $t \in L(p)$ and $t \notin L(\hat{\rho}(p))$. If $|t| > |p|$, $\hat{\rho}(p)$ contains any constant free regular tree pattern q such that $q \preccurlyeq p$ and $|q| = |p| + 1$ and so $t \in L(\hat{\rho}(p))$. This is a contradiction, it holds that $|t| = |p|$, and therefore we obtain that

$$L(P) - L(P - \{p\} \cup \hat{\rho}(p)) \subseteq \{t \mid t \text{ is a ground tree}, t \preccurlyeq p, |t| = |p|\}. \qquad \square$$

Algorithm MINL-RTP-CF*(T, S)
Require A positive example S for $L(\mathcal{RTP}_{cf})^*$
Ensure a hypothesis for which $L(h)$ is minimal w.r.t. S
  1: $H_1 := \{x\}$, $H_2 := \emptyset$, $H_3 := \emptyset$
  2: **for** $i = 0$ to ∞
  3:   **while** $H_1 \neq \emptyset$
  4:     Choose one $h \in H_1$
  5:     **if** $S \subseteq L(H_1 - \{h\} \cup H_2 \cup H_3)$
  6:       $H_1 := H_1 - \{h\}$
  7:     **else if** $[\hat{\rho}$ is in Theorem 8$]S \subseteq L(H_1 - \{h\} \cup \hat{\rho}(h) \cup H_2 \cup H_3)$
  8:       $H_1 := H_1 - \{h\}$, $H_2 := H_2 \cup \hat{\rho}(h)$
  9:     **else**
10:       $H_1 := H_1 - \{h\}$, $H_3 := H_3 \cup \{h\}$
11:   **if** $H_2 = \emptyset$
12:     return H_3
13:   **else**
14:     $H_1 := $ (the set obtained by removing redundancy from H_2), $H_2 := \emptyset$

Fig. 2. Improved MINL algorithm for $L(\mathcal{RTP}_{cf})^*$

Proposition 5. *For any finite set S of positive example, Algorithm MINL-RTP-CF* terminates and returns a hypothesis P which represents a minimal concept for S.*

Proof. Let n be the maximal number of non-constant function symbols occurring in any ground tree S.

In order to show that the algorithm terminates, we consider the while loop of the case $i = n$. Let $h \in H_1$. From the definition of $\hat{\rho}$, the number of function symbols occurring in tree patterns in $\hat{\rho}(h)$ is $n + 1$. Then no tree patterns in $\hat{\rho}(h)$ any ground tree in S. Therefore the results of the conditions on Lines 5 and 7 are same. Since no elements is added to H_2, $H_2 = \emptyset$ when the while loop is finished, and so the algorithm terminates when $i \leq n$.

To show that the output P represents a minimal concept, we let $Q = \rho(P)$. From the definition of the refinement, $P - Q = \{p\}$. This p is added to H_3 during the execution. Note that $S \subseteq L(H_1 \cup H_2 \cup H_3)$ is the loop invariant. At the moment when p is added to H_3 on Line 8, both $S \subseteq L(H_1 \cup H_2 \cup H_3)$ and $S \not\subseteq L(H_1 - \{p\} \cup \hat{\rho}(p) \cup H_2 \cup H_3)$ should hold. Then from Lemma 7, there exists $t \in S$ such that $t \preceq p$ and $|t| = |p|$. Since $t \notin L(Q)$, $S \not\subseteq L(Q)$. Because we can take any refinement operator which is semantically complete, $L(P)$ is minimal w.r.t. S. $\qquad\square$

Proposition 6. *For a set S of positive examples as its inputs, Algorithm MINL-RTP-CF* returns, outputs a minimal concepts in time complexity $O(s^3 n^4 \alpha^2 \beta^2)$, where $s = card(S)$, n is the maximum arity of functions appearing in S, α is the number of functions in Σ, and β is the maximum arity of functions in Σ.*

Proof. At first, we consider the cost of each iteration in the case $i = n$. When the if condition makes false, $card(H_1) < s$.

Note that $card(\hat{\rho}(p)) \leq n\alpha\beta$ because at most $n\beta$ variables appears in p. Thus, when $i = n - 1$, at most $s(n - 1)\alpha\beta$ tree patterns are added to H_2, and at most $sn\alpha\beta$ tree patterns are added when $i = n - 1$. In similar analysis, we get $card(H_3) \leq s$, and therefore, when $i = n$, $card(H_1 \cup H_2 \cup H_3) \leq 2sn\alpha\beta$. Since the subsumption check for a tree pattern and a ground tree is executed in $O(n)$, the both of the conditions in Line 5 and Line 7 are evaluated in $s \cdot O(n) \cdot O(sn\alpha\beta) = O(s^2 n^2 \alpha\beta)$.

The computation of $\hat{\rho}(p)$ is realized by replacing each variable with a function symbol of α kinds, and so is executed in $O(n\alpha\beta)$ because $|p| < n + n\beta$ for every $p \in H_1$. So, every iteration of the while loop is in $O(s^2 n^2 \alpha\beta) + O(n\alpha\beta) + O(s^2 n^2 \alpha\beta) = O(s^2 n^2 \alpha\beta)$.

Since $card(H_1)$ iteration is executed, the total cost for $i = n$ is $O(s^3 n^3 \alpha^2 \beta^2)$. From Proposition 5, the total cost of the algorithm is $O(s^3 n^4 \alpha^2 \beta^2)$. □

In Procedure 2 there is no restriction of how to choose h from H_1 on Line 4. We need not care how to choose h because of the next theorem.

Theorem 5. *There is a unique minimal concept for every set S of examples up to renaming of variables.*

The theorem follows from the next lemma.

Lemma 8. *Let P be the hypothesis in $\mathcal{RTP}_{cf}$ for a minimal concepts for a set S of examples. Let S' be the set of constant-free regular tree patterns obtained by replacing every occurrence of constants with distinct variables. Then for all $p \in P$ there exists $s \in S'$ such that $p \equiv s$.*

5   Concluding Remarks

We have shown a learning procedure by combining the MINL strategy and using refinement operators in the model of identification in the limit from positive data. By revising our learning procedure according to [9], we can show that, under the four assumptions [A-1]–[A-4], the concept class $\mathcal{C}$ can be learned only with super-set queries.

We showed that the procedure makes the class $L(\mathcal{RTP}_{cf})^*$ identifiable from positive data. We also revise the procedure so that we can show that the class is polynomial time identifiable. We can easily show that $L(\mathcal{RTP}_{cf})^*$ has the characteristic set property, but our result gives a new method for showing the identification of unbounded unions.

Comaring with the previous researches [3,4,6] on learning unions of tree patterns, our result differs from them on the points that we removed the restriction of setting upperbound of numbers of unions, and that the framework of learning is the identification in the limit from positive data [1,5] without queries or mistake bound.

References

1. Angluin, D.: Inductive inference of formal languages from positive data. Information and Control 45, 117–135 (1980)
2. Angluin, D.: Inference of reversible languages. Journal of the ACM 29, 741–765 (1982)
3. Arimura, H., Ishizaka, H., Shinohara, T.: Learning unions of tree patterns using queries. In: Zeugmann, T., Shinohara, T., Jantke, K.P. (eds.) ALT 1995. LNCS, vol. 997, pp. 66–79. Springer, Heidelberg (1995)
4. Arimura, H., Shinohara, T., Otsuki, S.: A polynomial time algorithm for finding finite unions of tree pattern languages. In: Brewka, G., Jantke, K.P., Schmitt, P.H. (eds.) NIL 1991. LNCS (LNAI), vol. 659, pp. 118–131. Springer, Heidelberg (1993)
5. Gold, E.M.: Language identification in the limit. Information and Control 10, 447–474 (1967)
6. Goldman, S.A., Kwek, S.: On learning unions of pattern languages and tree patterns in the mistake bound model. Theoretical Computer Science 288, 237–254 (2000)
7. Kobayashi, S.: Approximate identification, finite elasticity and lattice structure of hypothesis space. In: Technical Report, CSIM 96-04, Department of Computer Science and Information Mathematics. University of Electro-Communications (1996)
8. Laird, P.D.: Learning from Good and Bad Data. Kluwer Academic Publishers, Dordrecht (1988)
9. Lange, S., Zilles, S.: On the learnability of erasing pattern languages in the query model. In: Gavaldá, R., Jantke, K.P., Takimoto, E. (eds.) ALT 2003. LNCS (LNAI), vol. 2842, pp. 129–143. Springer, Heidelberg (2003)
10. Lassez, J.L., Maher, M.J., Marriott, K.: Unification revisited. In: Minker, J. (ed.) Foundations of Deductive Databases and Logic Programming, pp. 587–625. Morgan Kaufmann, San Francisco (1988)
11. Motoki, T., Shinohara, T., Wright, K.: The correct definition of finite elasticity. In: Proceedings of the fourth annual workshop on Computational learning theory (COLT 1991), p. 375 (1991)
12. Plotkin, G.D.: A note on inductive generalization. Machine Intelligence 5, 153–163 (1970)
13. Sakakibara, Y., Kobayashi, S., Yokomori, T.: Computational Learning Theory, Baifukan (2001) (in Japanese)
14. Shapiro, E.Y.: Inductive inference of theories from facts. In: Research Report YALEU/DCS/RR-192. Department of Computer Science, Yale University (1980)
15. Shinohara, T., Arikawa, S.: Pattern inference. In: Lange, S., Jantke, K.P. (eds.) GOSLER 1994. LNCS, vol. 961, pp. 259–291. Springer, Heidelberg (1995)
16. Shinohara, T., Arimura, H.: Inductive inference of unbounded unions of pattern languages from positive data. Theoretical Computer Science 241, 191–209 (2000)
17. Wright, K.: Identification of unions of languages drawn from an identifiable class. In: Proceedings of the Second Annual Workshop on Computational Learning Theory (COLT 1989), pp. 328–333 (1989)

Computing Minimal Models by Positively Minimal Disjuncts

Ken Satoh

National Institute of Informatics and Sokendai
ksatoh@nii.ac.jp

Abstract. In this paper, we consider a method of computing minimal models in propositional logic. We firstly show that *positively minimal* disjuncts in DNF (Disjunctive Normal Form) of the original axiom corresponds with minimal models. A disjunct D is positively minimal if there is no disjunct which contains less positive literal than D. We show that using superset query and membership query which were used in some learning algorithms in computational learning theory, we can compute all the minimal models.

We then give a restriction and an extension of the method. The restriction is to consider a class of positive (sometimes called monotone) formula where minimization corresponds with diagnosis and other important problems in computer science. Then, we can replace superset query with sampling to give an approximation method. The algorithm itself has been already proposed by [Valiant84], but we show that the algorithm can be used to approximate a set of minimal models as well.

On the other hand, the extension is to consider circumscription with varied propositions. We show that we can compute equivalent formula of circumscription using a similar technique to the above.

1   Introduction

In this paper, we consider a method of computing minimal models in propositional logic. The method is based on techniques from computational learning theory [Angluin88, Bshouty95].

There are many proposals computing minimal models [Ben-Eliyahu96, Ben-Eliyahu00, Satoh00] and theoretical analysis of problems related with computing minimal models (see [Cadoli93] for survey), but none of them relates computing minimal models with computational learning theory.

Let A be a formula (of any form) and n be the number of propositions. Then, we show an algorithm for computing all the minimal models using superset query and membership query. The number of superset queries is bound by a polynomial of $|DNF(A)|$ where $|DNF(A)|$ is the number of conjunctions of a minimal DNF of A, and the number of membership queries is bound in polynomial of n and $DNF(A)$.

Then, we discuss a restriction and an extension of the methods. The restriction is to consider computing minimal models of positive formula which is related with

K. Nakakoji, Y. Murakami, and E. McCready (Eds.): JSAI-isAI, LNAI 6284, pp. 358–371, 2010.

important problems such as diagnosis [Reiter87], data mining [Gunopulos97] and other fields in computer science [Eiter95, Ibaraki91]. We show an approximation algorithm using sampling and membership query and analyze the method in PAC (probably approximately correct) learning framework [Valiant84]. Let $\epsilon < 1$, $\delta < 1$ be an arbitrary positive number. We can efficiently find an approximate a set of minimal models such that the probability that difference ratio between the approximate set and the true set of minimal models is more than ϵ is at most δ. The number of sampling is bound in polynomial of $\frac{1}{\epsilon}$, $\frac{1}{\delta}$, $|DNF(A)|$ and the necessary number of membership queries is bound in polynomial of n, $|DNF(A)|$.

The extension is to consider minimal models in propositional circumscription [McCarthy86] where varied propositions are allowed. We give a method of computing a formula which is logically equivalent to the result of circumscription.

2 Preliminaries

We represent an interpretation I as a set of true propositions in I; $p \in I$ iff $I \models p$. Let A be a formula. A model M is called *minimal model M* w.r.t. A if M satisfies the following condition.

- $M \models A$
- There is no model M' s.t. $M' \models A$ and $M' \subset M$ where $\subset$ is a strict subset relation.

We denote a set of all the models of A as $models(A)$ and a set of all the minimal models w.r.t. A as $min(A)$.

A DNF formula is a disjunction of conjunctions of literals and a CNF formula is a conjunction of clauses. Note that there are many DNF representations of the same formula and many CNF representations of the same formula. We denote the DNF size of a formula F as $|DNF(F)|$ meaning the minimum possible number of conjunctions in any logically equivalent DNF representation to F. We call DNF representation of F with the size of $|DNF(F)|$ *minimal DNF* of F. Similarly, we denote the CNF size of a formula F as $|CNF(F)|$ meaning the minimum possible number of clauses in any CNF representation of F and call CNF representation of F with the size of $|CNF(F)|$ *minimal CNF* of F.

3 Correspondence between Minimal Models and Minimal DNF Formula

Let D be a conjunction of literals. We denote a set of propositions each of which appears in D as a positive literal as $pos(D)$, and a set of propositions each of which appears in D as a negative literal as $neg(D)$.

We define positively minimal disjunct D of a DNF formula as a disjunct s.t. there is no disjunct D' in the DNF formula s.t. $pos(D') \subset pos(D)$ where $\subset$ is a strict subset relation.

Theorem 1. *I is a minimal model w.r.t. A if and only if there exists a positively minimal disjunct D in a DNF of A s.t. $I = pos(D)$.*

Proof. $\Rightarrow$) Suppose that I is a minimal model w.r.t. A. Then, I must satisfy some disjunct D in a DNF of A. Suppose that D is not positively minimal. Then, there is a disjunct D' in the DNF of A s.t. $pos(D') \subset pos(D)$. Then, there exists an interpretation I' such that $I' = pos(D')$. This contradicts with the assumption that I is a minimal model.

$\Leftarrow$) Suppose that $I = pos(D)$ for a positively minimal disjunct D in a DNF of A. Suppose that I is not a minimal model. Then, there exists I' such that $I' \models A$ and $I' \subset I$. Then, there must be a disjunct D' in the DNF of A such that $I' \models D'$. Then, $pos(D') = I' \subset pos(D)$ and this contradicts with the assumption that D is positively minimal. $\square$

If we can give a DNF of A, we can compute minimal models of A. However, DNF representation can be very large. We can reduce the size of DNF representation and cost of selecting positively minimal disjunct by considering a minimal DNF representation of A since the above theorem holds even if DNF is minimal.

Corollary 2. *I is a minimal model w.r.t. A if and only if there exists a positively minimal disjunct D in a minimal DNF of A s.t. $I = pos(D)$.*

Example 1. Let $\langle p, q, r, s \rangle$ be a set of propositions and A be $(p \wedge q \wedge r) \vee (p \wedge \neg q \wedge \neg r) \vee (\neg p \wedge q \wedge \neg r)$. This is a minimal DNF. There are two positively minimal disjuncts, $(p \wedge \neg q \wedge \neg r)$ and $(\neg p \wedge q \wedge \neg r)$ and disjuncts give two minimal models $\{p\}$ and $\{q\}$.

4   Computing Minimal Models by Superset Query and Membership Query

Finding minimal DNF formula is quite a difficult task. Also, what we have to find is not a minimal DNF formula itself but positively minimal disjuncts in a minimal DNF formula. In this section, we show an algorithm to find models which satisfy a positively minimal disjunct using *superset query* and *membership query*.

Let A, F_1, F_2 be formulas and M be an interpretation. We define a *superset query* $SQ(F_1, F_2)$ and *membership query* $MQ(M, A)$ as follows.

$SQ(F_1, F_2)$ returns "yes" if $models(F_1) \supseteq models(F_2)$
otherwise it returns a model M s.t. $M \in models(F_2) \backslash models(F_1)$.

$MQ(M, A)$ returns "yes" if $M \in models(A)$ otherwise it returns "no".

$SQ(F_1, F_2)$ corresponds with checking whether $F_2 \wedge \neg F_1$ is unsatisfiable. If $F_2 \wedge \neg F_1$ is satisfiable, $SQ(F_1, F_2)$ returns a model of $F_2 \wedge \neg F_1$. $MQ(M, A)$ is equivalent to checking $M \models A$. Therefore, we write $M \overset{?}{\models} A$ instead of $MQ(M, A)$ from now on.

```
ComputeMinimal(A)
A: a formula
begin
    S := ∅;
    while SQ(Φ_S, A) returns a model M
    do the following
    begin
        M' := pmin(A, M);
        S := S ∪ {M'};
    end
    S_out := MinSet(S);
    output S_out
end
```

```
pmin(A, M)
A: a formula
M: a model of A
begin

    1. Select p ∈ M s.t. M↓_p ⊨ A by using M↓_p ⊨? A.
    2. If there exists no p ∈ M s.t. M↓_p ⊨ A, return M.
    3. else M := M↓_p and go to 1.

end
```

Fig. 1. Computing Minimal Models

Let I be an interpretation and p a proposition s.t. $I \models p$. Then, we define $I{\downarrow}_p$ as $I \backslash \{p\}$.

Let S be a set of interpretations $\{I_1, ..., I_k\}$. We define Φ_S as a DNF formula $D_1 \vee ... \vee D_k$ such that D_i contains positive literals only and $pos(D_i) = I_i$ $(1 \le i \le k)$.

Let S be a set of interpretations. We define $MinSet(S)$ as $\{M \in S |$There is no $M' \in S$ s.t. $M' \subset M\}$.

Figure 1 shows an algorithm which computes a set of all minimal models. The algorithm is inspired by algorithms proposed in [Angluin88, Bshouty95].

The following lemma is related with Claim 4.8 [Khardon and Roth96]. Let $min^*(A) = \{M | M \models A$ and for every $p \in M$, $M{\downarrow}_p \not\models A\}$.

Lemma 3. *The number of calls of* **pmin** *is at most* $|DNF(A)|$.

Proof. The number of calls of **pmin** is at most $|min^*(A)|$. Therefore, it is sufficient to show that $|min^*(A)| \le |DNF(A)|$. Let $M \in min^*(A)$ and assume that M satisfies a disjunct D in a minimal DNF of A. Then $M \subseteq pos(D)$. Let $M_{min}(D) = pos(D)$ and suppose that $M_{min}(D) \subset M$. Then, there exists proposition such that $p \in M_{min}(D) \backslash M$. Then, $M_{min}(D) \subseteq M{\downarrow}_p \subset M$. Since $M_{min}(D) \models D$, so is $M{\downarrow}_p$. This contradicts the assumption that $M \in min^*(A)$. Therefore, $M = pos(D)$. This means that for every $M \in min^*(A)$, there must be a disjunct D in a minimal DNF of A s.t. $M = pos(D)$. Thus, $|min^*(A)| \leq |DNF(A)|$. $\qquad\square$

Note that $min^*(A)$ can be regarded as a set of "locally minimal models" such that there are no models which is 1-bit smaller than a model in $min^*(A)$.

Lemma 4. *The number of calls of SQ is at most $|DNF(A)|$.*

Proof. Suppose that $SQ(\Phi_S, A)$ returns M. We show that there exists $M' \in min^*(A)\backslash S$ s.t. $M' \subseteq M$. Suppose that there are no such M'. This means that for every $M' \in min^*(A)$, $M' \not\subseteq M$ or already $M' \in S$. However, since $M \models A$, there exists a minimal model M_{min} of A (which is also in $min^*(A)$) s.t. $M_{min} \subseteq M$. Therefore, M' must be in S. Then, $M \in \Phi_S$ and this contradicts the initial assumption that $SQ(\Phi_S, A)$ returns M.

Every time the above M finds, we call **pmin**. Let M_{pmin} be an output from **pmin**. Then, M_{pmin} is in $min^*(A)$. Since we add M_{pmin} to S after calling **pmin** in **ComputeMinimal**, the number of unadded element in $min^*(A)$ is reduced at least by 1. Since $|min^*(A)| \leq |DNF(A)|$ is shown in the proof of Lemma 3, the above situation happens at most $|DNF(A)|$ times. $\qquad\square$

Lemma 5. *The number of checking $M\downarrow_p \overset{?}{\models} A$ in a call of **pmin** is at most n^2 where n is a number of propositions.*

Proof. In every iteration in **pmin**, we call membership queries at most n times to find p s.t. $M\downarrow_p \models A$. After each iteration, numbers of true proposition in the model is reduced by 1 and therefore the number of iteration is at most n. $\qquad\square$

Note that the final S after the iteration in **ComputeMinimal** is not guaranteed to be a set of all minimal models since Φ_S could include a non-minimal model as the following execution trace is shown. However, we can obtain all the minimal models by minimality checking in the end of **ComputeMinimal** by calling *MinSet*.

Example 2. We show an execution trace of **ComputeMinimal**. Let $\langle p, q, r, s \rangle$ be a set of propositions and A be the same formula in Example 1.

1. Since $\Phi_\emptyset$ is **false**, we call $SQ(\textbf{false}, A)$. Suppose that $SQ(\textbf{false}, A)$ returns $\{p, q, r, s\}$(Fig.2)[1].
   Then, we call **pmin**$(A, \{p, q, r, s\})$ and $\{p, q, r\}$ is returned(Fig.3)[2]. We add $\{p, q, r\}$ to S and S becomes $\{\{p, q, r\}\}$ and $\Phi_{\{\{p,q,r\}\}}$ becomes $(p \wedge q \wedge r)$.
2. Then, we call $SQ((p \wedge q \wedge r), A)$ and suppose that $\{p, s\}$ is returned(Fig.4). We call **pmin**$(A, \{p, s\})$ and $\{p\}$ is returned(Fig.5). We add $\{p\}$ to S and S becomes $\{\{p, q, r\}, \{p\}\}$ and $\Phi_{\{\{p,q,r\},\{p\}\}}$ becomes $(p \wedge q \wedge r) \vee p$.
3. Then, we call $SQ((p \wedge q \wedge r) \vee p, A)$ and suppose that $\{q\}$ is returned(Fig.6). We call **pmin**$(A, \{q\})$ and $\{q\}$ is returned(Fig.7). and $\Phi_{\{\{p,q,r\},\{p\},\{q\}\}}$ becomes $(p \wedge q \wedge r) \vee p \vee q$.

---

[1] In Fig.2, we represent an interpretation as binary values of p, q, r and s in stead of set representation, and represent a model of A as double circled node.

[2] In Fig.3, a node with italic characters expresses a model Φ_S where $S = \{\{p, q, r\}\}$.

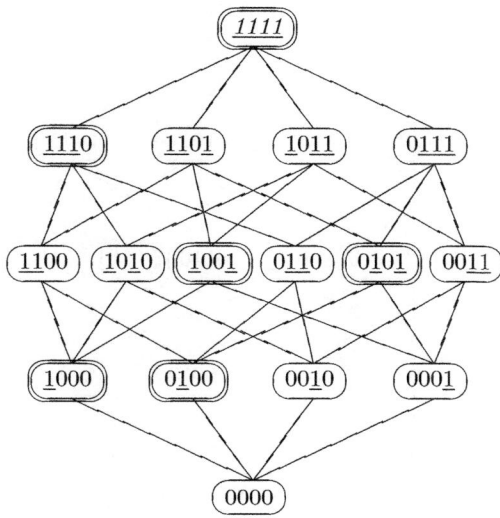

Fig. 2. After $SQ(\textbf{false}, A)$ returns $\{p, q, r, s\}$

4. Finally, we call $SQ((p \wedge q \wedge r) \vee p \vee q, A)$ and no model is returned and we exit the **while** loop.
5. We call $MinSet(\{p, q, r\}, \{p\}, \{q\})$ and obtain $\{\{p\}, \{q\}\}$ as S_{out}.

So, we now have the following lemma.

Lemma 6. *Let S_{out} be an output of the above algorithm.*

$$S_{out} = min(A).$$

Proof. Let S be the final set of models after the iteration in **ComputeMinimal**. It is sufficient to show that $min(A) \subset S$ since $S_{out} = MinSet(S)$. Suppose that there exists $M_{min} \in min(A)$ but $M_{min} \notin S$. Since there is no M s.t. $M \in A$ but $M \notin \Phi_S$, there must be a model M' such that $M' \subset M_{min}$. This contradicts the assumption that M_{min} is a minimal model w.r.t. A. □

From the above lemmas, we have the following theorem.

Theorem 7

  - *The algorithm halts after at most $|DNF(A)|$ superset queries and $n^2 \cdot |DNF(A)|$ membership queries where n is a number of propositions.*
  - *The output S_{out} is a set of all the minimal models of A*

Note that the bound of $|DNF(A)|$ is not always good. For example, the n-variable "even parity" function has the single minimal model which is an empty set, but the algorithm might traverse 2^{n-1} locally minimal ones.

Bshouty et al. [Bshouty94] and Balcázar and Guijarro [Balcazar02] proposed a method of learning DNF formula with proper subset and superset queries. But

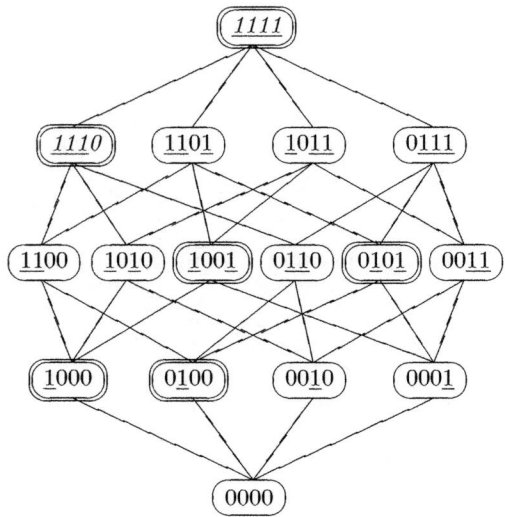

Fig. 3. After $\mathbf{pmin}(A, \{p, q, r, s\})$ returns $\{p, q, r\}$

it is difficult to compare our method with them since our purpose in this paper is not to learn exact DNF formula, but to learn positively minimal disjuncts in a minimal DNF formula.

We also note that there is a research using superset queries for "preference elicitation" in combinatorial auctions [Blum03].

5   Approximating Minimal Models for Positive Formula by Sampling and Membership Query

In this section, we consider a restricted class of *positive* formula.

Let F be a formula. F is *positive* if for every interpretation I and J, if $I \models F$ and $I \subseteq J$, $J \models F$. Note that a minimal DNF and a minimal CNF for a positive formula is unique.

To compute minimal models of positive formula is related with computing minimal hitting sets (or in other words, minimal transversals [Eiter95]) and related with various fields such as diagnosis [Reiter87], data mining [Gunopulos97], hypergraph theory [Eiter95] and distributed systems (theory of coteries [Ibaraki91]). Therefore, research on this restricted class is very important.

If A is a positive formula, a superset query can be replaced by sampling for approximating minimal models because of the following reason. For a positive formula, superset queries can be replaced by equivalence queries in the above algorithm of Fig. 1 since a counter example found by equivalence query is always a counter example by superset query for a positive formula. Then equivalence queries can be approximately simulated by sampling.

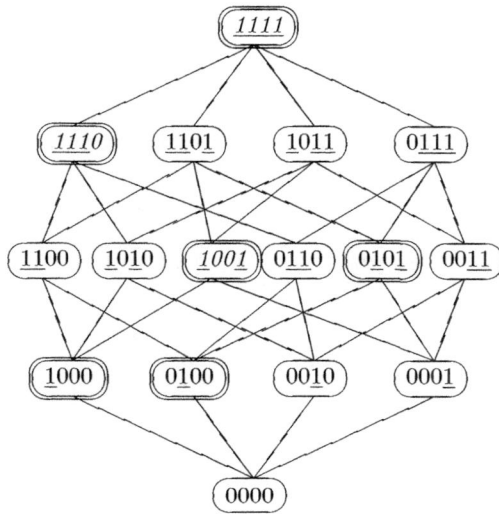

Fig. 4. After $SQ(p \wedge q \wedge r, A)$ returns $\{p, s\}$

For this purpose, we assume that there is a probability distribution **P** over interpretations. We would like to have a set of minimal models such that the probability that the difference of minimal models is more than we expect is low.

Figure 8 shows such an algorithm. Note that compared with **ComputeMinimal**, in **ApproximateMinimal**, **pmin** always returns a minimal model since a considered formula is positive and so final check of minimality of models in the algorithm is not necessary. This is essentially equivalent to the algorithm proposed by [Valiant84]. So, the algorithm itself is not new, but we show that the algorithm can be used to approximate set of minimal models as well.

Intuitively, in the algorithm we try to find counter examples by sampling and we count the number of sampled data and if the number of sampling is more than a certain number which guarantees law probability of existence of counter examples for the current hypothesis, we are done. If we find a counter example corresponding with a model returned by $SQ(\Phi_S, A)$ in the previous algorithm, we try to find a minimal model by **pmin**.

Let $f \Delta h$ be a difference set between f and h (that is, $(\overline{f} \cap h) \cup (f \cap \overline{h})$).

Lemma 8. *Let A be a positive formula. The probability that the above algorithm outputs a set of models S such that $\mathbf{P}(models(A)\Delta models(\Phi_S)) \geq \epsilon$ is at most δ.*

Proof. Let f, h be finite subsets of a finite set X and **P** be a probability distribution over X. Suppose that $\mathbf{P}(f \Delta h) \geq \epsilon$ and let $m = \dfrac{1}{\epsilon} \ln \dfrac{1}{\delta}$.

If we choose randomly m samples from X according to **P**, then the probability that no sample belongs to $f \Delta h$ is at most $(1 - \epsilon)^m$.

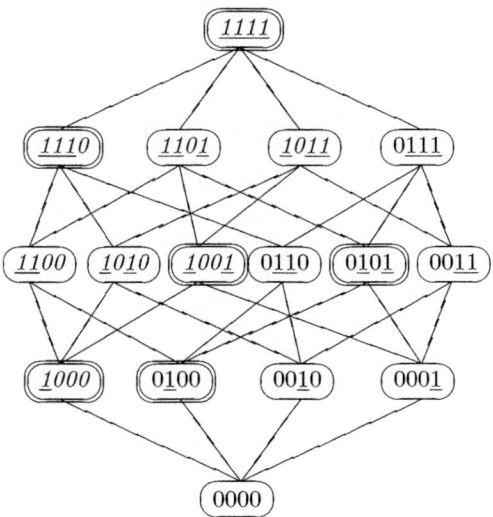

Fig. 5. After **pmin**$(A, \{p, s\})$ returns $\{p\}$

Then, since $\ln(1 - \epsilon) < -\epsilon$,

$m \ln(1 - \epsilon) < -m\epsilon$

$(1 - \epsilon)^m < 2^{-m\epsilon} = < 2^{\ln \delta} = \delta$

Let us regard f as $models(A)$ and h as $models(\Phi_S)$. If we conclude that $models(\Phi_S)$ subsumes $models(A)$ by checking that m samples does not belong to $models(A)\Delta models(\Phi_S)$, then the probability that the algorithm produces S such that $\mathcal{P}(models(A)\Delta models(\Phi_S)) \geq \epsilon$ is at most δ. □

Lemma 9. *The probability that the above algorithm outputs a set of models S such that $\mathbf{P}(min(A)\Delta S) \geq \epsilon$ is at most δ.*

Proof. Suppose that $M \in (min(A)\Delta S)$. Since $S \subseteq min(A)$ always holds, $M \in min(A)$ and $M \notin S$. This means $M \models A$ and $M \not\models \Phi_S$. Therefore, $M \in (models(A)\Delta models(\Phi_S))$. This means that $(min(A)\Delta S) \subseteq (models(A)\Delta models(\Phi_S))$. Then, if $\mathbf{P}(min(A)\Delta S) \geq \epsilon$, $\mathbf{P}(models(A)\Delta \Phi_S) \geq \epsilon$. Therefore, by Lemma 8, the probability that $\mathbf{P}(min(A)\Delta S) \geq \epsilon$ is at most δ. □

Theorem 10. *The above algorithm stops after taking at most $(\frac{1}{\epsilon} \ln \frac{1}{\delta}) \cdot |DNF(A)|$ samples according to $\mathbf{P}$ and asking at most $n^2 \cdot |DNF(A)|$ membership queries and produces S with the probability at most δ such that $\mathcal{P}(min(A)\Delta S) \geq \epsilon$.*

Proof. By Lemma 9, we only need to get at most $\frac{1}{\epsilon} \ln \frac{1}{\delta}$ examples according to $\mathbf{P}$ to check whether a counter example exists or not, in order to satisfy the accuracy condition. Since the number of counter examples is $|DNF(A)|$ by Lemma 3, we only need to get at most $(\frac{1}{\epsilon} \ln \frac{1}{\delta}) \cdot |DNF(A)|$ examples as a total. The number of membership queries is the same as Theorem 7. □

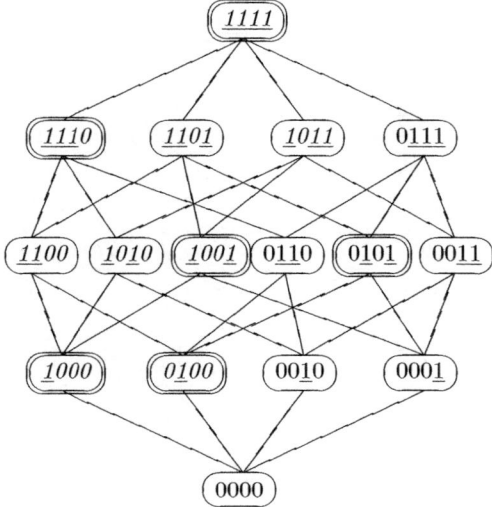

Fig. 6. After $SQ((p \wedge q \wedge r) \vee p, A)$ returns $\{q\}$

Note that even if the number of minimal models of A is very small, if the number of models of A is large enough to find out most of models of A by sampling, we can find a approximate set of minimal models using **pmin**. Therefore, we can obtain a good approximation of minimal models. If the number of models of A is too small to find out any model by sampling, falsity is a good approximation of A and therefore, falsity is also a good approximation of minimal models of A.

6   Computing Circumscription

In this section, we give a method of computing equivalent formula with propositional circumscription [McCarthy86] by extending the above algorithm.

Let P and Q be tuples of propositions, $\langle P_1, P_2, ..., P_n \rangle$ and $\langle Q_1, Q_2, ..., Q_n \rangle$. We define $P \leq Q$ as $\bigwedge_{i=1}^{n} P_i \supset Q_i$. We define $P < Q$ as $P \leq Q$ and $Q \not\leq P$.

Let A be a formula. We divide a set of propositions used in A into disjoint two tuples of propositions P, Z which are called *minimized propositions, varied propositions*. We only consider here minimized proposition and varied proposition since we can translate circumscription with fixed proposition can be translated into circumscription without fixed proposition according to [deKleer89].

Circumscription of P for A with Z varied is defined as follows.

$$Circum(A; P; Z) = A(P, Z) \wedge \neg \exists p \exists z (A(p, z) \wedge p < P).$$

For a model theory of circumscription, we define an order of interpretations to minimize P with Z varied is defined as follows. Let I be an interpretation and P be a tuple of propositions. We define $I[P]$ as $\{p \in P | I \models p\}$ or, equivalently, $I \cap P$. Let I_1 and I_2 be interpretations.

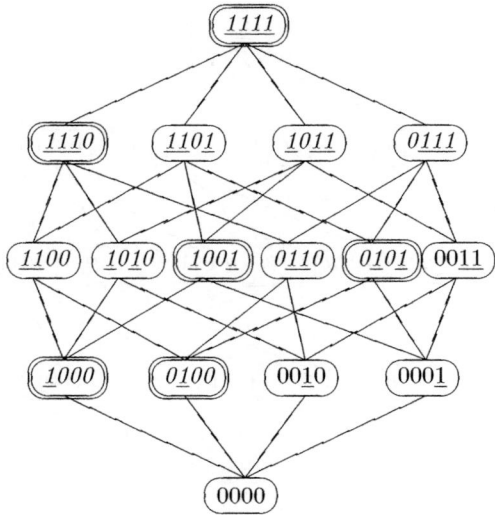

Fig. 7. After **pmin**$(A, \{q\})$ returns $\{q\}$

ApproximateMinimal(A)
A: a positive formula
begin
   $S := \emptyset$;
   $m := 0$;

1. Produce M according to the probability distribution **P**
2. If $M \models A$ and $M \not\models \Phi_S$ then
       $M' := \mathbf{pmin}(A, M)$;
       $S := S \cup \{M'\}$;
3. $m := m + 1$;
4. If $m >= \dfrac{1}{\epsilon} \ln \dfrac{1}{\delta}$ then **output** S
    else **go to** 1.

end

Fig. 8. Approximating Minimal Models

We define $I_1 \leq^{P;Z} I_2$ as $I_1[P] \subseteq I_2[P]$. We define $I_1 <^{P;Z} I_2$ as $I_1 \leq^{P;Z} I_2$ and $I_2 \not\leq^{P;Z} I_1$. A minimal model M of $A(P, Z)$ w.r.t. P with Z varied is defined as follows.

1. M is a model of $A(P, Z)$.
2. There is no model M' of $A(P, Z)$ such that $M' <^{P;Z} M$.

Note that if Z is empty, then the above definition coincides with the previous definition of minimal model.

```
ComputeCircum(A, P)
A: a formula P: a set of minimized propositions
begin
    S := ∅;
    while SQ(Φ_S, A) returns a model M
    do the following
    begin
        M' := pmin(A, M);
        S := S ∪ {M'};
    end
    S_out := MinSet_P(S);
    output ⋁_{M∈S_out} reflect(A, M)
end
```

Fig. 9. Computing Circumscription

According to [Lifschitz85], I is a minimal model of $A(P, Z)$ w.r.t. P with Z varied if and only if I is a model of $Circum(A; P; Z)$.

The following corresponds with Theorem 1 for circumscription. Here, positively P-minimal disjunct D in a DNF formula is define as a disjunct s.t. there is no disjunct D' in the DNF formula s.t. $(pos(D') \cap P) \subset (pos(D) \cap P)$.

Theorem 11. *Let P be a set of minimized propositions. I is a minimal model of A w.r.t. P with Z varied if and only if there exists a positively P-minimal disjunct D in a DNF of A s.t. I satisfies the following conditions.*

- *$pos(D) \subseteq I$*
- *$(neg(D) \cap I) = \emptyset$*
- *$(I \backslash pos(D)) \subseteq Z$*

Proof. $\Rightarrow$) Suppose that I is a minimal model of A w.r.t. P with Z varied. Then, I must satisfy a disjunct D in a DNF of A. Then, $pos(D) \subseteq I$ and $(neg(D) \cap I) = \emptyset$.

Suppose that D is not positively P-minimal. Then, there is a disjunct D' in the DNF of A s.t. $(pos(D') \cap P) \subset (pos(D) \cap P)$. Then, there exists an interpretation I' such that $I' = (pos(D') \cap P)$. This contradicts with the assumption that I is a minimal model.

Suppose that there exists p such that $p \in I$ and $p \notin pos(D)$ and $p \notin Z$. Then, $p \in P$. This means that $I\downarrow_p$ can be a model of D and it contradicts minimality of I. Therefore, $I \backslash pos(D) \subseteq Z$.

$\Leftarrow$) Suppose that I satisfies the above three conditions but I is not a minimal model. From the conditions, $I \models D$ and $I[P] = (pos(D) \cap P)$. Then, there exists I' such that $I' \models A$ and $I'[P] \subset I[P]$. Then, there must be a disjunct D' in the DNF of A such that $I' \models D'$ and $I'[P] = (pos(D') \cap P)$. Then, $(pos(D') \cap P) \subset (pos(D) \cap P)$ and this contradicts with the assumption that D is positively P-minimal.   □

Now, we compute minimal models of circumscription using superset query and membership query. We use the following lemma.

Lemma 12. *Suppose that J is a minimal model of A w.r.t. P with Z varied. Then, there exists a minimal model I of A w.r.t. $P \cup Z$ s.t. $J[P] = I[P]$.*

Proof. Since J is a minimal model of A w.r.t. P with Z varied, there is no model J' of A s.t. $J'[P] \subset J[P]$. We can assume that there is a model J' of A s.t. $J'[P] = J[P]$ and $J'[Z] \subseteq J[Z]$ and there is no model J'' of A s.t. $J''[P] = J[P]$ and $J''[Z] \subset J'[Z]$. Then, J' is a minimal model of A w.r.t. $P \cup Z$.   □

Computing minimal models of A w.r.t. $P \cup Z$ gives partial interpretation of minimal models of A w.r.t. P with Z varied. Therefore, we first compute all the combination of interpretation of P in minimal models of A w.r.t. P with Z varied from models of $Circum(A; P \cup Z)$ and then reflect interpretation of P into A and form a disjunction of these formulas. Figure 9 shows an algorithm which performs such a computation.

Let S be a set of interpretations and P be a set of minimized propositions. We define $MinSet_P(S)$ as $\{M \in S | \text{There is no } M' \in S \text{ s.t. } M'[P] \subset M[P]\}$. Let A be a formula and M be a model of A. $reflect(A, M)$ is defined as

$$A \wedge \bigwedge_{p \in M} p \wedge \bigwedge_{p \in P \setminus M} \neg p.$$

Theorem 13

- *The algorithm halts after at most $|DNF(A)|$ superset queries and $n^2 \cdot |DNF(A)|$ membership queries where n is a number of propositions.*
- *Output $\bigvee_{M \in S_{out}} reflect(A, M)$ is logically equivalent to $Circum(A; P; Z)$.*

Proof. The number of queries can be estimated in the same way as the proof of Theorem 7. Since $MinSet_P(S)$ in the algorithm gives all the models each of whose interpretation of P is minimal, a set of these interpretations of P coincides with a set of interpretations of P in all the models of $Circum(A; P; Z)$ by Lemma 12. Then, $reflect(A, M)$ complements interpretations of Z corresponding with the interpretation of $M[P]$ in $Circum(A; P; Z)$.   □

7   Conclusion

The contributions of this paper are as follows.

- We propose a method of computing minimal models using membership query and superset query and analyze its computational complexity.
- We propose a method of approximating minimal models for a positive formula using membership query and sampling and analyze its computational complexity.
- We extend the first method to compute minimal models in circumscription.

The future works include computer experiments to analyze an average behavior of the algorithm and theoretical comparison with other methods.

Acknowledgments. I thank anonymous reviewers for their valuable comments on the paper.

References

[Angluin88] Angluin, D.: Queries and Concept Learning. Machine Learning 2, 319–342 (1988)

[Balcazar02] Balcázar, J.L., Castro, J.: A New Abstract Combinatorial Dimension for Exact Learning via Queries. Journal of Computer and System Sciences 64, 2–21 (2002)

[Ben-Eliyahu96] Ben-Eliyahu, R., Dechter, R.: On Computing Minimal Models. Annals of Mathematics and Artificial Intelligence 18, 3–27 (1996)

[Ben-Eliyahu00] Ben-Eliyahu-Zohary, R.: A Demand-Driven Algorithm for Generating Minimal Models. In: Proc. of AAAI-2000, pp. 267–272 (2000)

[Blum03] Blum, A., Jackson, J.C., Sandholm, T., Zinkevich, M.: Preference Elicitation and Query Learning. In: COLT-2003, pp. 13–25 (2003)

[Bshouty94] Bshouty, N.H., Cleve, R., Gavaldà, Kanna, S., Tamon, C.: Oracles and Queries that are Sufficient for Exact Learning. Journal of Computer and System Sciences 52, 421–433 (1994)

[Bshouty95] Bshouty, N.H.: Exact Learning Boolean Functions via the Monotone Theory. Information and Computation 123, 146–153 (1995)

[Cadoli93] Cadoli, M., Schaerf, M.: A Survey on Complexity Results for Non-monotonic Logics. Journal of Logic Programming 17(2-4), 127–160 (1993)

[deKleer89] de Kleer, J., Konolige, K.: Eliminating the Fixed Predicates from a Circumscription. Artificial Intelligence 39, 391–398 (1989)

[Eiter95] Eiter, T., Gottlob, G.: Identifying the Minimal Transversals of a Hypergraph and Related Problems. SIAM Journal on Computing 24(6), 1278–1304 (1995)

[Gunopulos97] Gunopulos, D., Khardon, R., Mannila, H., Toivonen, H.: Data Mining, Hypergraph Transversals, and Machine Learning. In: Proc. of PODS-1997, pp. 209–216 (1997)

[Ibaraki91] Ibaraki, T., Kameda, T.: A Boolean Theory of Coteries. In: Proc. of 3rd IEEE Symposium on Parallel and Distributed Processing, pp. 150–157 (1991)

[Khardon and Roth96] Khardon, R., Roth, D.: Reasoning with Models. Artificial Intelligence 87, 187–213 (1996)

[Lifschitz85] Lifschitz, V.: Computing Circumscription. In: Proc. of IJCAI 1985, pp. 121–127 (1985)

[McCarthy86] McCarthy, J.: Applications of Circumscription to Formalizing Common-Sense Knowledge. Artificial Intelligence 28, 89–116 (1986)

[Reiter87] Reiter, R.: A Theory of Diagnosis from First Principles. Artificial Intelligence 38, 49–73 (1987)

[Satoh00] Satoh, K., Okamoto, H.: Computing Circumscriptive Databases by Integer Programming: Revisited. In: Proc. of AAAI 2000, pp. 429–435 (2000)

[Valiant84] Valiant, L.G.: A Theory of the Learnable. CACM 27, 1134–1142 (1984)

Author Index